GUIDE TO VINTAGE TRADE STIMULATORS & COUNTER GAMES

With Values

Richard M. Bueschel

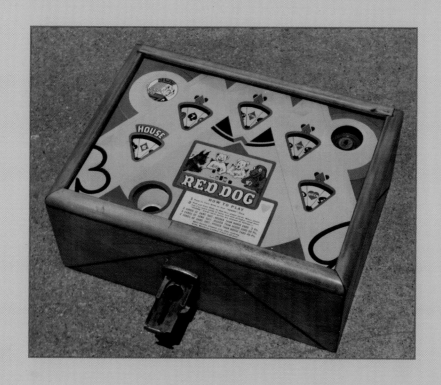

4880 Lower Valley Road, Atglen, PA 19310 USA

Dedicated to

Clarence Kolk of Stony Lake, Shelby, Michigan. He got a kick out of everything I did, including this book as it was being developed in the summer of 1996. At a youthful seventy-eight, we lost him too soon.

Bueschel, Richard M.
　　Collector's guide to vintage trade stimulators & counter games : with price guide / Richard M. Bueschel
　　　p.　　cm.
　　Includes index.
　　ISBN 0-7643-0119-5 (hc)
　　1. Coin operated machines--Collectors and collecting. I. Title.
TJ1557.B845 1997
629.8'323--dc21　　　　　　　　　　97-21756
　　　　　　　　　　　　　　　　　CIP

Copyright © 1997 by Richard M. Bueschel

All rights reserved. No part of this work may be reproduced or used in any form or by any means—graphic, electronic, or mechanical, including photocopying or information storage and retrieval systems—without written permission from the copyright holder.

Book Design by: Laurie A. Smucker

ISBN: 0-7643-0119-5
Printed in China
1 2 3 4

Published by Schiffer Publishing Ltd.
4880 Lower Valley Road
Atglen, PA 19310
Phone: (610) 593-1777; Fax: (610) 593-2002
E-mail: schifferbk@aol.com
Please write for a free catalog.
This book may be purchased from the publisher.
Please include $3.95 for shipping.
Try your bookstore first.

We are interested in hearing from authors with book ideas on related subjects.

CONTENTS

Acknowledgments 4
Introduction 7
The Universe of Trade Stimulators and Counter Games 8
History 10
Race Games 22
Roulette 31
Wheel 38
Pointer 53
Card Machines 60
Dial 81
Dice 86
Coin Drop 105
Target 115
Single Reel 122
Baby Bell 128
Cigarette 138
Specialty Reel 149
Novelty and Skill 161
Building a Collection 184
Resources & Restoration 195
Collectible Trade Stimulators and Counter Games
    (dated by manufacturers) 210
Collectible Trade Stimulators and Counter Games
    (by name alphabetically) 238
Price Guide 254
Index 259

ACKNOWLEDGMENTS

When my first coin machine collector book was published by Schiffer Publishing Ltd. in the summer of 1995 it seemed to me to be a nice survey of what it was to be a coin machine collector, and covered all of the types of machines people could collect. The *Collector's Guide To Vintage COIN MACHINES* is filled with advice, and is packed with color photographs of slot machines, jukeboxes, pinball games, arcade machines, trade stimulators and counter games, vending machines, and coin-operated scales. It also goes into some detail about building a collection, and includes the resources collectors have at their disposal to repair, maintain, rebuild, and restore their machines, plus a price guide. That says it. What else is there to say?

The world doesn't work that way. When you think you haven't done enough, there is always the possibility you overkilled and didn't leave much more to say. But when you think you have probably said it all, there's generally a lot more to be said. Peter Schiffer saw that. *Collector's Guide To Vintage COIN MACHINES* was barely out at the time of the ABA Convention in Chicago in June 1995, yet based on advance sales Peter knew what he wanted next. More! If the first book covers coin machines in general, what about books dedicated to the seven basic forms of coin machines. How about a series of individual books about collectible slot machines, jukeboxes, pinball games, arcade machines, trade stimulators and counter games, vending machines, and coin-operated scales. How about the whole universe of coin machines? He asked. I said okay. It's been a flurry of activity ever since.

No easy task, this. But the reach of *Collector's Guide To Vintage COIN MACHINES* more than confirms the need for these individual books. It became a worldwide seller, and even led to a monthly column in the Japanese publication *Yugi-Nippon* by yours truly called "The World Of Coin Machines." An almost exclusively American hobby (not completely, as there are a few British, French, German, and other European collectors, but that is about it) leaped the oceans and grabbed the attention of the world, not to mention an expanding North American base. That led to questions by readers, and interest in particular forms of coin operated machines. My task is to expand that scope and deliver the books that will go into the various forms of machines in greater detail, and get more specific about the finding and collecting of these machines, as well as their needs in maintenance, repair, and restoration. Plus prices.

I picked Trade Stimulators and Counter Games first because they are clever, generally small, and are a good entry level coin-op collectible as many can still be found at attractive prices. You can get a vintage counter game at some of the large collectible coin machine shows for around $200, and it looks like that cost pattern will continue, unless the hobby is deluged with new collectors. That is the danger of a book like this. We might make coin-ops sound so attractive that hundreds of new collectors suddenly join the fold. But that's okay, too, as that would lead to more machine finding and identification. To aid in that never ending search I have included a list of the trade stimulator and counter game machines that were made in the last century or so, of which at least one third have never been found. Maybe you are the person that will make the find of a rare and unknown discovery. You should also realize that while some counter games from the 1950s and later still sell for low prices, considering they are coin machines, some of the trade stimulators and counter games of the past are far, far more valuable, and are among some of the most valuable vintage coin machines in existence. It is hoped that this volume will get you interested in the genre and help you make historic finds. Finding machines is not only a lot of fun; they can also be valuable.

In my first Schiffer book *Collector's Guide To Vintage COIN MACHINES* I started out by thanking the current coin machine industry for putting up with me as I labored and researched their machines. I can't do that in this volume for the simple fact that trade stimulators and counter games haven't been made for years. Oh, perhaps you can count the non-payout electronic poker bartop machines of the present era as counter games, and they probably are. But the classic mechanical games of the past have all been produced. The job now is to find them and restore them to working condition. That means that something must be learned about these ancient devices, and in that I had the help of numerous librarians across the country that got interested in the subject, and took the time to dig for critical information. Authors can't constantly be flying around the country pouring over tons of material in repositories situated coast to coast to get to the heart of a story. It takes too much time and money, and books like these do not deliver the return. They are very often a labor of love. They need the input of others willing to provide that gratis labor. To their credit, many did.

The special people interviewed and the librarians that contributed significant data for this book include the late Grant Shay of Chicago, the former advertising manager for the Mills Novelty Company; the late and the great collector and historian Frederick Fried, New York City; gambling paraphernalia collector and researcher Robert Rosenberger, of Cincinnati; Davis Erhardt, head, Long Island Division, Queen's Borough Public Library in New York; Mrs. Elmer S. Forman, reference librarian, The Cincinnati Historical Society, Cincinnati, OH; Yeatman Anderson II, rare books curator, The Public Library of Cincinnati and Hamilton County, in Cincinnati; and the Allen County Historical Society, of Lima, OH.

Next I must thank the collectors of these games for their unlimited access and unbounded help in putting this book together. No leading collectors stinted when asked for advice, information and pictures of their games. When I say "leading" I don't mean people of great financial means with enormous collections. Trade stimulators and counter games aren't that way. "Leading" is a matter of enthusiasm, with most of the successful collectors earning more in the way of results for their knowledge and time than the money spent on their collections. And when it comes to enthusiasm these collectors certainly have it.

I particularly want to thank Tom Gustwiller, of Ottawa, and Bill Howard of Akron, OH. Both are trade stimulator and counter game collectors and both share in the production and marketing of Tom's 1995 book *For Amusement Only*. It would have been easy and logical for them to avoid sharing their collections and knowledge in order to curtail the coverage of a competitive book. But when asked they didn't react that way, and immediately committed to sharing their thoughts and photographs to make this coverage as broad as possible. They have added immeasurably to this book, and I highly recommend their book for additional information. You will find it listed in the Resources & Restoration chapter.

Thanks also to Ken and Jackie Durham of Washington, D.C., for their insight, photos, and proofreading of the game and name lists to catch errors and possible additions. Ken publishes a newsletter featuring trade stimulators and counter games, and is the leading bookseller to the hobby. You will also find him listed in the resources section. He can also be found on the internet at http://GameRoomAntiques.com.

Further thanks to major collectors Jack Freund of Slots of Fun, Springfield, WI; Alex Warschaw, San Antonio, TX; and Bill Whelan of Slot Dynasty, Daly City, CA for sharing their knowledge and pictures of the machines in their collections. Bill Whelan set out to find and photograph all of the rare and one-of-a-kind machines he could find in the western states, including his own collection, with the result that dozens of games are to be seen here that have never before appeared in print. That includes many from the outstanding Joe Welch collection in San Bruno, CA. Special thanks to Brian Witherall of Witherall's Americana Auctions, 3620 West Island Court, Elk Grove, CA 95758, for sharing the fine photography of the rare and valuable trade stimulators that have crossed their auction block over the past few years.

For graphics appearing throughout this book I thank Rick Akers, Oklahoma City, OK; Kent Anderson, Excelsior, MN; George Angeloso, Clay, NY; Charlie Ball, North Little Rock AR; Barton Battaile, Lexington, KY; Paul Beichler, Home Arcade, Lisle, IL; Jack Biehl, Mt. Clements, MI; Jim C., Garden Grove, CA; Uwe H. Breker, Auction Team Köln, Germany; G. S. Brierley, Clwyd., U. K.; Richard A. Carlson, Portland, OR; John Carini, South Milwaukee, WI; Loretta Collins, Pleasantville, NY; Joe Constantino, Vineland, NJ; Rod Cornelius, Auckland, New Zealand; the late Elmer Cummings, Sioux Falls, SD; Mike Curran; John Currie, Reno, NV; Bob Davis, Oregon, OH; Bill Daugharty, Saginaw, MI; Charles Deibel, Jeanette, PA; Dave DeRosa, Schenectady, NY; Lee Dunbar, Dunbar's Gallery, Milford, MA; Ken Durham, Washington, DC; Mme. Caroline Emmanuelle, Grisy-Suisnes, France; Harold Eunch, Perrysburg, OH; Richard Fague, San Francisco, CA; Marshall Fey, Reno, NV; John Fignar, Saratoga Springs, NY; John "Jukebox Johnny" Finkler, Wales, WI; Jeff Frahm, St. Louis, MO; Jack Freund, Springfield, WI; the late Frederick Fried, New York, NY; Richard J. Gilbert, Oconomowoc, WI; Jon Glenn, Forsyth, MO; Bernie Gold, Great Neck, NY; Jim and Betty Good, Cincinnati, OH; John Gooding, Huron, OH; Tony Goodstone, Los Angeles, CA; J. P. Goosman, Pickering, Ontario, Canada; Mike Gorski, Westlake, OH; Duane "Bud" Gott, Eagan, MN; Murray Gottlieb, Forest Hills, NY; Dennis Green, Akron, OH; Steve Gronowski, Barrington Hills, IL; Mike Gumula, Graham, NC; Tom Gustwiller, Ottawa, OH; Pete Hansen, DesPlaines, IL; John Harper, Santa Ana, CA; Harrah's Museum, Reno, NV; Stan Harris, Philadelphia, PA; Terry Harte, Alta Loma, CA; Mark and Judy Hartman, Kirkwood, MO; Eric Hatchell, Clifton, VA; Roger Hilden, Minneapolis, MN; Rick Hoffman, York, PA; Bill Howard, Akron, OH; Barbara and Bill Huber, Greenville, OH; Thomas Jacobson, Chicago, IL; Joseph Jakubuwski, Punta Gorda, FL; Norman D. Johnson, Bowling Green, OH; Bill Johnson, Rosamonde, CA; Jack Kelly, St. Joseph, MI; Dave Kenney, Rockford, IL; John Kieckhefer, Prescott, AZ; Bob Klepner, Victoria, Australia; Rodger Knutson, Seattle, WA; Gary Kothera, Middlefield, OH; Bob Kretchko Antiques, Pleasantville, NY; Bill Krugman, Calumet City, IL; Jim Krughoff, Darien, IL; Rick Lee, Lincoln, MA; Bob Legan, Euclid, OH; Carl Lepiane, Los Gatos, CA; Mr. and Mrs. David E. Leverentz, Fort Wayne, IN; Jay Lowe, Lancaster, PA; Larry Lubliner, Highland Park, IL (and now Farmington Hills, MI); Rory Lucas, Perth, Australia; Milt Margolis, Springfield, VA; Bob McGrath, Indian Rocks Beach, FL; Thomas McHameds, Fremont, CA; Walter Mintel, Jr., Madison, NY; Tim Morsher, Kelly Island, OH; Stan Muraski, Rockford, IL; Bob Nelson, Wichita, KS; Bill Nesnay, Matawan, NJ; the late Jerry Noel, Hamburg, WI; Tom O'Connor, Delmar, NY; Paul Olson, Sartell, MN; Hal O'Rourke, Lanexa, VA; David Pace, Pace Auctions, DesPlaines, IL; Bob Packer, San Jose, CA; Allan Pall, River Forest IL; Kent Patterson, Edwardsville, IL; Rich Penn, Waterloo, IA; Bill Petrochuk, Akron, OH; Clark Phelps, Midvale, UT; Michael Pinkowsky, Stratford, CT; the late John Ragan, St. Maries, ID; Steve A. Reno; Herb Rhyner, Roselle Park, NJ; Bruce Robinson, Ligounier, PA; Marty Roenigk, East Hampton, CT; Dorothy Roll, Memphis, TN; Maurice E. Rowe, Greenfield Museum, Greenfield, OH; Ken Rubin, Brooklyn Heights, NY; Alan Sax, Long Grove IL; Anthony Schneller, Metairie, LA; Kenneth Schnoll, Paramus, NJ; James Seilz, Jasper, IN; Steve Sonter, Coos Bay, OR; Ed Smith, Pecatonica, IL; Jerry Soling, Pound Ridge, NY; Stardust Slot Museum, Las Vegas, NV; Frank Stratman, Okeechobee, FL; Ronald Stroup, Houston, TX; Gary Sturtridge, Tonganoxie, KS; Robert and Debbie Swerdlow,

Baldwin, NY; Alan and Francine Swire, West Henrietta, NY; Gary Taplin, Stamford, CT; Blaine Thomas, Manhattan, KS; Bill Triola, St. Johns, MI; Dennis Valdez, Hercules, CA; Robert Vicic, Euclid, OH; Ira Warren, Northridge CA; Alex Warschaw, San Antonio, TX; Jim Waters, Madison, WI; Joe Weaver, Phoenix, AZ; Joe Welch, San Bruno, CA; Randy Wellman; Jack West, Cincinnati, OH; Bill and Carole Whelan, Daly City, CA; John Wilcox, Oroville, CA; Carlton Wilkins, Bloomfield, CT; Brian Witherall, Witherall's Americana Auctions, Elk Grove, CA; John Zalesny, Shell Beach, CA and Frank Zygmunt, Darien, IL.

Finally, pricing, a particularly thankless job as there are always critics of the process. Owners and sellers want their game values high; collectors and buyers want the values low. Collectors and dealers courageous enough to stick their necks out for the "One Shot" price guide concept at the back of this book are Ken Durham, Tom Gustwiller, Jack Freund, Alex Warschaw, Bill Whelan and yours truly, Dick Bueschel. I also take the final responsibility of compiling and averaging the price lists, so if you have a gripe against the pricing I'm your guy.

 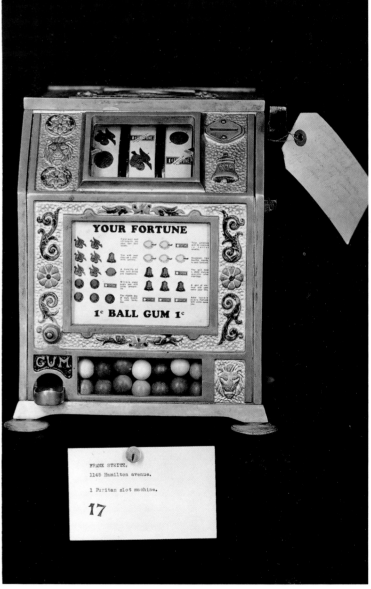

One of the great fascinations in trade stimulator and counter game collecting has to do with minor details. These are St. Louis, Missouri, police "mug shots" of Midwest Novelty Company LION PURITAN BABY VENDOR machines confiscated in a series of raids on September 22, 1933. Machine No. 16 was picked up at 1012 Hamilton Avenue, while No. 17 was found in the next block at 1145 Hamilton Avenue. While they look much alike, they are distinctly different. No. 16 with plain fruit symbols has a closed gumball cup at lower left. No. 17 has fortune sayings superimposed over the fruit symbols with an open gumball cup with a dispensing push rod at lower left. Different production runs and upgraded foundry match plates account for the variations.

INTRODUCTION

This is a book about collecting a specific type of vintage coin operated machine first called the trade stimulator, and later called the counter game. Early on, when this book was being put together, suggestions were made to change the class name of these delicate and often intricate chance machines to non-payout or miniature slot machines, or perhaps vintage gambling and penny arcade games, to make it easier for the general public to understand. But antiquers and collectors aren't the general public. We're different, and we want to know the real story. It's more fun when we know exactly what you're dealing with. So the original class names remain, mostly because that's what people called them at the time and it's what dedicated collectors call them today.

It would be an easy segue to say that trade stimulators and counter games were the same games under different names. But they really aren't. While both categories of these games are chance machines that award the player minimal prizes in one form or another for high scores or skillful accomplishments without making actual payouts of cash or products, they differ in when and where they were placed, and what they offered the accomplished or purely lucky player. To really appreciate the various trade stimulators and counter games that you may encounter, you need to know something about why they were made and how they were used. Which also makes this a history book.

But this isn't the "In 1492 Columbus sailed the ocean blue" kind of history you used to get in school. It's living history of the experiences of American marketing and what it took to make a buck in a highly competitive world as evidenced by surviving artifacts. Trade stimulators and counter games are games, and while they appeared to be placed where they were in order to give the players an enjoyable run for their money, it was the money that really counted to the bartender, restaurant proprietor, grocer, cigar seller, pharmacist and other retailers of goods and services that operated them. In the high powered and high priced and relatively high income producing world we live in, it is somewhat difficult to envision a world where coins counted, and livings were made by scratching for every penny.

These small coin machines, generally placed on a bar or countertop, were designed to get the pennies (and on up to nickels, dimes, and quarters) that would have stayed in people's pockets unless they attracted the attention and desire of the customer to enjoy the play and maybe get something for nothing (ignoring the fact that a coin had to be played). They were also operated to give a storekeeper the advantage over a competitive store that didn't have them, and did so by providing discounts on merchandise. They may not look that way, but the bottom line is that's what they did.

In this attempt to help stores pay the rent, hundreds (probably thousands!) of clever ideas and formats were tried in over a century of usage in order to grab those coins. It's their ingeniousness and the creative ways these machines work, displaying their advertising and providing a show for the player, that makes them the most desirable and rewarding collectible vintage coin-ops to their many devoted collectors. If this book has a mission, it is twofold: to show you a comprehensive display of the collectible trade stimulators and counter games, and to try and make you a collector of these remarkable machines. We are unabashedly trying to hook you with their features and color, and by providing answers you may have about finding, maintaining, and playing with your own games. Greater collector interest means greater game finding and sharing, and those of us who are already devotees of these remarkable machines would like that. You will too.

Richard M. Bueschel
414 N. Prospect Manor Avenue
Mt. Prospect, IL 60056-2046
Tel. 1-847/253-0791
FAX: 1-847-253-7919
email: buschlhist@aol.com

May 19, 1997

Author Richard M. Bueschel of Mt. Prospect, Illinois. This picture was taken some years back, when Dick still had a shock of brown hair and color in his moustache. It is used to show the enthusiasm that trade stimulator and counter game collectors have for their field of endeavor. Dick is holding an original Clawson AUTOMATIC DICE machine of 1890 that was found in the mid-1970s by fellow collector Bob Strauss in a barn loft in Joplin, Missouri. *Photo courtesy Bob Strauss.*

THE UNIVERSE OF TRADE STIMULATORS AND COUNTER GAMES

The way to pick up change is to place a coin grabbing device in an area where people congregate. In the 1880s the people who congregated were men, and the place they congregated was the saloon. There were a lot of reasons for this, but the simple one is that there were no entertainment centers for the laboring classes (there wasn't much of a middle class in those days) and the saloons offered companionship, cheap drinks (sometimes free food), and, of all wonders, a john and a telephone. The new technological advances of electricity for lighting and doing all sorts of other clever things arrived in parallel with the telephone and indoor plumbing, and to most of the populace at the time the plumbing seemed to be a better boon to humanity than either of the other two.

Only the well to do, and the new public building and hotel construction of the day, had the telephone and telegraph on their premises. Or indoor plumbing for that matter. While a person on the street could drop in to a city hall or public library to check out the public toilets, once they got away from the center of town they were pretty much on their own. So if nature called, or they wanted to make a telephone call to a public or private person of significance, the saloon was the answer. Ladies weren't on the street much in those days (they would soon arrive in droves as trains and streetcars made urban travel and "downtown" shopping possible, but that wasn't for a few years) so the market was men. The saloons catered to their patrons by providing bartop games of strength and skill, with losers buying drinks. By the mid-1880s some small backshop makers created coin operated race, dice, and card games that offered "Free Drinks," while cigar makers added "Cigar Machines" to their lines that offered "Free Cigars." The larger of these early trade stimulator makers included Clement C. Clawson in Newark, New Jersey; The Amusement Machine Company in New York City; Milton O. Griswold in Rock Island, Illinois; a cigar maker named Charles T. Maley in Cincinnati, Ohio; the Decatur Fairest Wheel Company in Decatur, Illinois; an in-town competitor in the Drobisch Brothers who had a woodworking shop in Decatur; Leo Canda in Cincinnati; Charlie Fey in San Francisco, the Waddell Wooden Ware Works, a cigar and store counter maker, in Greenfield, Ohio; and the Brunhoff Manufacturing Company of Cincinnati. Over a century later, when the performance of this unique industry was finally tabulated, it turned out that these pioneer producers were so prolific at the time that they were among the top thirty trade stimulator and counter game producers since the industry began.

By the mid-1890s trade stimulators had made the jump from saloons to stores to stimulate cigar and merchandise sales. Their popularity also attracted the attention of the larger slot, vending and arcade coin machine producers, with the big names in the business soon becoming the major producers of the machine class. The Mills Novelty Company of Chicago, Illinois, became the largest producer, starting in the late 1890s, still making counter games as late as 1946 in the post-WWII period. The Caille Brothers Company of Detroit, Michigan, the prime Mills competitor prior to WWI, made the games from 1901 to 1912, becoming the fourth largest producer over the years. Other makers that made the top thirty list by the end of WWI were the Paupa & Hochriem Company, of Chicago, Illinois; The Cowper and Watling Manufacturing Companies, as well as H. C. Evans and the Industry Novelty Company (which later became O. D. Jennings & Company) in Chicago; and a machine rebuilder and revamper called Silver King, in Indianapolis, Indiana.

Things changed with the coming of Prohibition, and the disappearance of the saloon. Checkout counters in stores and restaurants, and cigar counters, became the primary locations, with a whole new sea change of machines becoming available, now being called counter games. The 1920s added the Exhibit Supply Company, Pace Manufacturing Company, Ad-Lee, Sanders Manufacturing, and A. B. T., all of Chicago; and the Field Paper Products Company of Peoria, Illinois, as significant producers.

But the big days for counter games were yet to come. It was the Great Depression, followed by the election of Franklin Delano Roosevelt and the Repeal of Prohibition in 1933, that propelled the industry to new heights. The opening of the taverns provided the impetus for production, with counter games often legal in areas where slot machines were not. Groetchen Tool started making the games in 1930, followed by Buckley Manufacturing in 1931, and Daval, Bally and Pierce Tool in 1932, with all of these firms located in Chicago. Another top thirty starter in 1932 was the Great States Manufacturing Company in Kansas City. When a performance tally was finally made, and in spite of the fact that the newcomers only produced the games for a dozen or so years or less, Groetchen ranked second, Daval third, Buckley sixth and Bally seventh among the top thirty producers in over a century.

Most of the counter games still being found are Groetchen, Daval, or Buckley, a testimony to their accelerated production by these firms in the pre- and early post-WWII years. Every one of these firms is now out of business, as are all the trade stimulator and counter game makers included in the list later in this volume. If you don't count the electronic poker machines now being made, this is a business that has come and gone, so it has a beginning and an end. That's the perfect profile for a vintage collectible, because every machine found from this day forward is a new piece of history.

The Chas. T. Maley Novelty Company of Cincinnati, Ohio, was a pioneer in direct response trade stimulator advertising, promoting his machines with humor. The small copy below the thieves robbing the saloon in this March 1894 mailer says: "Ike Killer - Say Jim, wous der use ter bodder wid der combination? Dere's no money dere. Jim Gun - All right! You bust de slotters, and ef we kin carry any more'n you git, it'll be time eauf to blast de box."

HISTORY

Advertising woodcuts sold trade stimulators in the 1890s. This is the dice throwing Clawson Slot Machine Company AUTOMATIC FORTUNE TELLER of 1890, made in a variety of models that told fortunes or offered "Free Drinks" and "Free Cigars."

Our public amusements, and often our arts, come from the bottom. More often than not we have to credit the soulful low end of acceptable society for coming up with those things that entertain, amuse, charm, and move us, and stay with us to define our national culture. After the Civil War, and into the World War I period, it was most often the American saloon that gave us the games and songs that set the pattern for our entertainments. Given the name "The Poor Man's Club" at the time (the rich had their own private clubs and didn't mess with penny, nickel, dime, and quarter amusements, preferring to wager the big bucks at the track or at their own gaming tables), the saloons were filled with the Great Unwashed, the ethnics and the drinkers, as well as the dapper dandies that were growing up American. There were both married and singles bars, and the only singles that were there were men.

To keep to this rambunctious crowd entertained, and keep the money coming in for the drinks and snacks (as the multifunctional saloon was also the original Fast Food outlet of its day), the bartender needed bartop or free standing patron amusements that didn't take his time or demanded that he make judgment calls. So most saloons had all sorts of wagering gadgets on the bartops, or standing on cast iron pedestals, from dice pads to hand grips to prove who was the stronger, even an open gambling wheel here and there. All worked by the customers or, in the cases of some of the dice games and gambling wheels, after paying the barkeep a penny or nickel for a plunge, depending on the economic status of the neighborhood. This system soon commanded too much of the barkeep's attention, if only in squabbles over who was right or wrong, or what constituted a winning show of the wheel. The stage was set for the wonder of the age: the coin machine.

The first barroom games to benefit from coin control were the strength testers, grips, pulls, and other athletic devices of the 1880s that provided a ready platform for patron competitions, with the loser buying the drinks. Other forms of coin-ops that were immediately popular were breath mint, gum, and peanut venders to take the sting out of a whiskey breath, or even perfume squirts that overpowered the smell of beer with a scent that seemed to come straight from one of the Pretty Waiter Girls, who were as often as not available for added upstairs or back room entertainment. Thus the beer drinking, cigar smoking, peanut eating, sometimes girl chasing (although most of the bar patrons were really afraid of women, otherwise they wouldn't have been there) and whiskey swilling saloon customer was a whirl of smells and stomach churning ingredients when he went home (if he did; that public john could keep him standing on the sawdust saloon floor for days, and sometimes did) in spite of the temptation pro-

Clawson Slot Machine Company AUTOMATIC DICE SHAKER of 1890 sits at the far right end of the bar in the El Dorado Saloon, Sacramento, California, around 1901. Coin machine at the far left is a Mills Novelty QUARTOSCOPE, a 1901 peep show most likely equipped with "Girlie" stereo cards.

vided by the fact the saloon kept on going as few had legal opening and closing hours imposed on them. It was a different world.

And an opportunistic one. With a room full of drinking men willing to challenge each other over their skills and strengths, and a long bar that could hold change grabbers, it wasn't long before the makers of the barroom novelties saw the opportunity for chance taking devices that usually kept the money and didn't give anything out, yet could pull in a lot of extra change for the location. The "boys" were already putting down cash for wagers, inserting coins in pull testers and vending machines, and taking chances on small race games and dice tossers. The field was wide open for anything that could take in the money and entertain the patrons at the same time. And the magic ingredient that would make it work was greed, the chance at getting something for nothing, and the opportunity to hang around the bar longer than one's money would allow. The basic concept was simple: "Free Drinks and Free Cigars."

Even before the breakthrough idea of a "free" giveaway (at very high odds) coin-operated machines had inched their way into the nation's bars, and a veritable cottage industry of saloon bartop game and amusement makers had evolved. The devices were cataloged and sold through the major saloon equipment supply houses of New York, Cincinnati, Chicago, St. Louis, Kansas City, and elsewhere. Just as often they were sold directly from the manufacturer through advertising in *The National Police Gazette* and saloon oriented publications such as *Mixed Drinks*, *Fair Play*, and *Champion of Fair Play* among others, or through drummers that carried samples in their carpet bags across the country by rail. Typically, the producers were small foundries or machine shops that were represented by entrepreneurs who acted as agents for the myriad makers of bartop games, with no major machine producers among them.

Some of the better known early producers were M. Siersdorfer & Company, a small foundry in Louisville (using a Cincinnati address); Lindall E. Cowper, working in his father's gravel business and dabbling in art glass, later taking on saloon novelties (to become the Cowper Manufacturing Company) in Chicago; and a small machine shop in Brooklyn run by a flamboyant and creative Irishman named Patrick Kennedy, Jr. (who was a machine inventor and salesman) and a second generation German-American machinist named Charles J. Diss (who was the shop man that made things work), named Kennedy & Diss Machinists. We're talking about the "Brown Period" here, when men were men and machines were in stained wooden cabinets unless they were all flashed up in plated, marbleized, or painted cast iron. This is the "Romantic Age" of the trade stimulator, dating from the mid-1880s until 1920, and the coming of Prohibition.

Brooklyn was the first hotbed of small manufacturing and saloon novelty production for the simple fact that there were more saloons there per capita than anywhere else in the nation (with the possible exception of San Francisco; Virginia City, Nevada; Dodge City, Kansas and Hurley, Wisconsin), assuring a ready and readily supplied market. It was the Kennedy & Diss Machinists firm that really got

The Clawson Slot Machine Company AUTOMATIC DICE SHAKER was still on the bar a year or so later, after the El Dorado Saloon added the Caille-Schiemer LOG CABIN proto-pinball, circa 1902. Same bartender and telephone.

things going, advertising themselves in 1888 as providing "dies and presses, rubber, metal and button molds, tin dies, models and special machinery." Their sudden spike in fortune in 1889 through 1890 was attributable to one thing only; the coming of coin machines. Kennedy was a pioneer in the art.

With a cluster of other coin machine makers already around him in Brooklyn at the time, Kennedy took a form that had existed since the 1860s and added excitement by making a larger, more random circular bartop horse race game operated under coin control called the AUTOMATIC RACE TRACK, the firm's first indigenous coin-controlled product. It put Kennedy and Diss solidly in the corner of capability in the new field of coin machines. They were soon approached by a young and already seasoned inventor of New York City named William W. Rosenfield, then only twenty-three years old. Rosenfield had the knack of attracting investors to his projects, and he was about to accomplish the *coup* of his lifetime. Promoting the enormous profitability inherent in the new coin machines, Rosenfield got the attention of a group of investors. The idea was to create the first multi-product nationally recognized coin-controlled chance machine manufactory with sales through local agents in major cities across the country.

But first they needed a product. So Bill Rosenfield took his idea for a small bartop device that spun reels with card symbols to Kennedy & Diss Machinists in Brooklyn to gain their expertise in machine development and engineering to lead to a commercialized product. The three men proceeded with the development of what they called a "Rotating Toy" to get around the patent restrictions applied to gambling machines. By April 1890 they were far enough along to consider legal protection, drawing up the papers on April 21 and filing for a patent on April 23. It was to become the first multi-reel coin-operated chance machine and the most significant and widely copied form of coin-controlled device in the history of the industry. Rosenfield formed his own Amusement Machine Company in New York City to popularize the coin machine as a gambling device, with the firm's "Rotating Toy" (soon to become known as the "Card Machine") leading off as their most popular product, spurred on by its offer of "Free Drinks and Free Cigars" as the prize. It was an instant American icon.

While the saloons were gaining income and activity with their coin machines, America's shopkeepers were also facing a sales crisis. The economic depression of 1893, the severest the country had seen since the 1870s in the aftermath of the Civil War, cut sales to the bone. With recovery came added competition and a tougher marketing environment. Where in years past most towns had a local storekeeper and a general store, by the mid-1890s there were three or four or more stores in the same town. In the cities the general stores that had carried everything gave way to grocery and hardware and soft goods stores lined up, one after another, on the same street in neighborhood shopping centers, competing against the larger department stores situated in the center of town and the mail

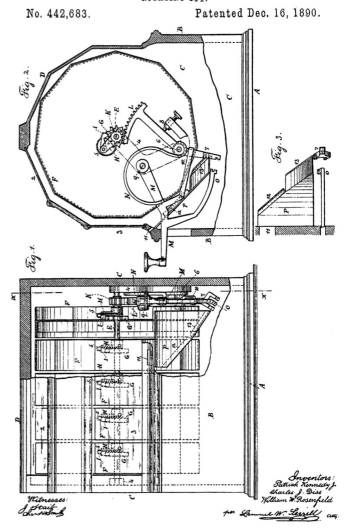

The invention that led to an industry. The Bill Rosenfield "Rotating Toy" of 1890 created the multiple reel card machine that led to the proliferation of trade stimulators and, a decade or so later, the automatic payout three-reel slot machine.

order houses that sold literally everything out of a catalog. For the first time customers had choices, and could pick an outlet based on its lines and services.

Along came the trade stimulator to the rescue. Or at least to provide a temporary patch. The trade stimulator served a valuable commercial function, for if customers could be entertained while getting full value for their money on "no blanks" machines that assured them at least as much in trade as they played, and sometimes more, store traffic would theoretically grow. It often did. The machines became common and were soon a part of North American daily life. The variety was so endless, and some of the machines so unique, they were talked about, mentioned in the daily press and, when their stores shut down as they often did (particularly around the WWI period) they were sometimes saved. It is this inventory of cached machines that still show up one at a time in small rural auctions and at estate sales that have constituted a source of supply for latter day collectors for years.

The first commercial trade stimulators were created as an adjunct to product marketing and were often given away to the storekeeper as a promotional perk to sell other products, specifically gum and cigars. Smoking in public, regarded as crass in Europe, had become a uniquely American habit, and it was the cigar that ruled supreme. Presidents, starting with U. S. Grant, smoked cigars in public, so why couldn't the average American? It led to an enormous industry spread out all over the country with small cigar manufacturers making brands that had local followings, and in some cases, national. The image of a man with a cigar in his mouth was that of masculinity and affluence. Cigar sales became as significant as bread and butter, and offered a line leader that brought spenders into stores. One result was the "cigar store," an American phenomena that still survives in major cities selling magazines, cigarettes, lottery tickets, and increasingly these days, cigars, which are now staging a comeback.

The late 1890s and early 1900s cigar stores, adding to the thousands of cigar counters in hotels, bars, restaurants, and railroad stations, demanded a whole new level of distribution and created competition at both the wholesale and retail level. Cigar makers, tobacco distribution houses, and store counter and fixture suppliers all found themselves backing into coin machine production and distribution to keep even with the other guy. A substantial number of cigar makers made machines that they gave away as free premiums tied to specific case orders, with some of them ending up as coin machine makers after dropping their cigar lines. Charles T. Maley of Cincinnati is an example. Starting out as a cigar wholesaler in the late 1880s, by 1892 he was giving away a dice machine and a coin drop called the NICKEL TICKLER to spur his cigar sales. He soon found himself in the coin machine business and ended up forming the Charles T. Maley Novelty Company, one of the largest dealers in cigar machines, saloon dicers, and payout slot machines in the country by the end of 1893.

Maley was matched by the Waddell Wooden Ware Works (often called "The Four Ws"), a store sales cabinet maker located in Greenfield, Ohio. Waddell introduced a machine called THE BICYCLE WHEEL in 1896 that came in both cigar and "bonus discount" versions in a variety of models. They followed these in July 1897 with one of the most elegant trade machines of the period, a boxed miniature bicycle machine called THE BICYCLE that spun a wheel with numbers on its periphery that indicated the awards won by the customer. It was added to their line of cash registers and store fixtures, and sold well for years.

The most successful trade stimulator of all time was created by James G. Huffman, a cigar maker in Decatur, Illinois. In April 1894 he introduced a machine called THE FAIREST WHEEL (loosely named after the first Ferris Wheel at the World's Columbian Exposition in Chicago in 1893) that spun a bicycle type wheel on a hub once a nickel was dropped in a coin chute at the top. Numbers on the perim-

Second product of the Amusement Machine Company in New York City, the BABY CARD MACHINE of 1891 was produced in great numbers, and led to hundreds of copycat models, large and small.

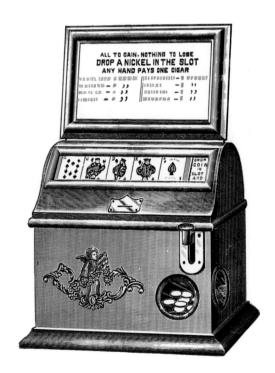

The Leo Canda PERFECTION CARD machine of 1898 was the most copied trade stimulator, made by over a dozen producers. Mills Novelty Company of Chicago called it the LITTLE PERFECTION, and sold a later "Flat Top" version of the machine into the 1930s.

eter of the wheel allowed the player to get from one up to five cigars for his money depending on where the wheel stopped. The idea was to give away THE FAIREST WHEEL as a trade premium for his Huffman & Son cigar business. When his machine achieved national prominence and generated sales orders on its own, Huffman formed the Decatur Fairest Wheel Works and got out of the cigar business. It made him a wealthy man, with a long string of FAIREST WHEEL models being made and sold for the next dozen years. The machine became ubiquitous on American cigar counters until well into the 1920s.

Other makers followed suit, with cigar and gum machines called THE WIZARD CLOCK, VICTOR, THE KELLEY, and others following for years afterward. The reason these machines were so successful was their economics. Storekeepers often sold their cigars for a nickel apiece, six for a quarter. Gum sold for a penny, or six for a nickel. The machines took the same sales position and made even more profit for the storekeeper. With the wheels, and other cigar machines, the cigars sold for a nickel each, with the winning awards still not approaching the expense of that sixth free cigar given with five sold for a quarter. The retailer came out ahead.

"Trade Stimulators are the result of a definite demand among merchants and businessmen for a device that would stimulate sales and, at the same time, entertain their patrons. The element of chance which enters into the successful operation of the machines by the patrons acts as a source of entertainment and leads on the play where it would otherwise lag. Every play, where the machine pays in trade checks or cigars, means the sale of a commodity, for which the checks are good, at the retail price. Thus it is easily seen why the machines have proved immensely popular among merchants. If you run any kind of a commercial enterprise which permits the introduction of trade stimulators you will find, as other businessmen have found, that your business will be doubled and tripled after their introduction."

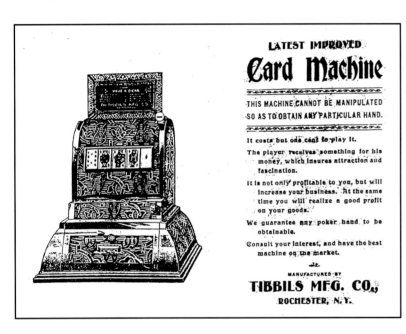

Card machines were often made of heavy cast iron to discourage lifting and manipulation. The Tibbils CARD MACHINE of 1891 was the biggest and heaviest.

The "Giant" Card Machine.

The Leo Canda GIANT card machine of 1894 lifted its display high so other saloon patrons could follow the game, and line up to play. Advertising panels in the middle promote whiskey and cigars.

The burgeoning market for trade stimulators and other forms of coin machines soon attracted larger producers, with major makers entering the market. Primarily concentrating on payout slot machines, arcade machines, and vending, the major producers relegated their card, cigar, and gum machines to a secondary position in their offerings, calling all of them trade stimulators. They are most clearly defined in the Mills Novelty Company pocket size 1907 yellow catalog called "Mills Trade Stimulators," where they are described as "machines that create business." The catalog states that:

This optimistic view was also held by the second-largest machine maker of the period, The Caille Brothers Company of Detroit. Their 1909 catalog says much the same thing:

"These machines appeal to the known spirit of speculation, so strong in every man. They interest and gratify that element in him at small cost to himself, and they give him a fair show and a good run for his money against slight odds. The card or trade machine is simply an innocent form of amusement for a penny or nickel with a chance to win merchandise. The merchant who is anxious to increase his business can do so by means of machines. They are made expressly for the purpose of stimulating trade in a manner equally satisfying to merchant and customer."

For the next twenty years or so trade stimulators were an important part of the lines of the coin machine producers, and were also produced in short runs by hundreds of smaller shops that had a game idea and pursued it with success that was usually local and most often short lived. The wooden card machines and spinning wheel formats were soon discarded as too delicate for everyday use, to be replaced by sturdier coin drop, baby bell, and other reel machine formats increasingly in aluminum cabinets following WWI.

The coming of Prohibition killed the card machine market, but led to a greater store and checkout counter market. For that the games changed. The Mills Novelty TARGET PRACTICE of 1918 and the Silver King BALL GUM VENDER of 1925 were landmark machines that were widely copied by other producers. They popularized the concepts of a skill coin flip and the miniaturized three reel baby bell format in simple cast cabinets with a minimum of mecha-

L. Horton's cigar stand at the corner of Golden Gate Avenue and Taylor Street in San Francisco shows off its Sittman And Pitt LITTLE MODEL CARD MACHINE introduced in 1894. Photo is dated September 10, 1898. The Mitchell saloon is to the left.

nism. Their very simplicity encouraged copycatting, setting the stage for a Golden Age of trade stimulators, only by this time they were beginning to be called counter games.

By the early 1930s the cast aluminum cabinet baby bell format, established by the J. M. Sanders Manufacturing Company with their BABY VENDER in 1928, and produced in great volume by the Buckley Manufacturing Company, with both firms in Chicago, took the trade machine and counter game market by storm. Buckley started producing the machine at the end of 1931, anticipating the coming of Repeal two years later. When Groetchen Tool &

Cigar makers in the 1890s often gave trade stimulators away to sell their cigars. The machines became so popular, some of the companies switched to coin machines. Chas. T. Maley of Cincinnati became the Charles T. Maley Novelty Company in 1893.

15

Grabbing pennies. The desk clerk in this Fulton, Illinois, hotel oversaw the awards offered by the United States Novelty WINNER dice machine of 1894 sitting at the end of the counter.

Manufacturing Company entered the market with their similar DANDY VENDER in 1930, along with the A. S. Douglis & Company with their own version (with Douglis becoming the Daval Manufacturing Company in 1932) the era of volume counter game production had arrived. Soon swarms of other manufacturers were in the business.

Other formats were introduced, staying away from the delicate machines of the past. Flat counter dice and spinning wheel games, and all sorts of specialty reel machines that looked like small slot machines, were added to the mix. Gone were the "Free Drinks" and "Free Cigars" of the past, replaced by gumballs with every play (if the player took them) and "Free Cigarettes," or counter payouts in cash or merchandise.

Counter games continued to gain in application up until WWII, with some sort of machine on virtually all fast food counters, tavern bars, food store and restaurant check out counters, and almost everywhere imaginable. They came back after WWII to seek out the same locations, with an anticipated boom market ahead. But it was not to be. For one thing, chance taking and gambling had become an onerous activity. Even the small chance taking of the counter game was frowned on. When the Johnson Act of 1951 became law, ending the open use of slot machines across the country and restricting them to only those states that allowed gambling, the counter game also disappeared.

But it wasn't the Johnson Act that put them out of business. It was business. Trade stimulators and counter games were created for the individually owned businesses of the past, where the owner counted the money and made the awards. The post-WWI world of the '50s practically wiped out the Ma & Pa stores and replaced them with chains and hired store managers. There was no way that any management would allow the cash flow of a counter game to go through the hands of a hired hand. By that time most of the food and sundry stores that had supplied the prewar populace had been replaced by supermarkets and the first of the malls, none of which allowed counter space for a game, much less tolerate the time lost in playing a game for gain. The reign of the counter game abruptly ended as America moved into the automobile as the primary means of going to the fast food outlet or the store. Progress made the counter game a useless vestige of the past.

The closest thing to a comeback are the electronic poker machines that are to be found in bars and restaurants—but only some bars and restaurants. These are amusement machines, and do not contribute to beer or product sales as their forebears did in the past. They may be non-payout and counter pay, but they are of a different tradition. Yet they retain a link to the past. In all likelihood they are the collectibles of the future, but are not a part of the scope of this guide to vintage trade stimulators and counter games.

The large John M. Waddell BICYCLE DISCOUNT WHEEL of 1897 at the far right offered customers a chance to get discounts on their merchandise purchases. They at least got their money's worth, or a break on the price.

1899 Decatur Fairest Wheel Works FAIREST WHEEL No. 3 with a visible cash box graces the cigar counter, far right, of a Galena, Illinois, general store.

Hanging with "The Boys" in a San Francisco saloon, circa 1902. Card machine at the far end of the bar is the Reliance Novelty Company IMPROVED RELIANCE drop card machine of 1898.

A merchant oversees his Waddell Wooden Ware Works THE BICYCLE trade stimulator wheel to apply its awards to cigar or candy sales. Northern Illinois, early 1900s.

San Francisco Barbary Coast saloon around 1908. Dice machine on the bar in the foreground is the Charles Fey ON THE LEVEL of 1907, made the year after the devastating San Francisco earthquake and fire.

Consumer general store in Stockton, Minnesota, gives center counter space to the Kelley Manufacturing Company THE KELLEY gum machine, circa 1903. Trade stimulators paid their way, and were known as rent payers.

There were few trade stimulator locations as active as billiard parlors. This Caille Bros. Co. QUINTETTE of 1901 stands silently at the right, awaiting the afternoon playing crowd.

Bigger and better. The Mills Novelty Company JOCKEY, first made in 1900, was a general and drug store favorite for the next twenty years. JOCKEY, at far right, has the prime counter corner spot in this general store on the Iowa side of the Mississippi River.

Drugstores, and the Mills Novelty JOCKEY, were made for each other. JOCKEY, far right, is on the cigar counter waiting for customers at the Economy Drug Company, Fargo, North Dakota, circa 1904.

The Caille Bros. Co. GOOD LUCK of 1904 was their version of the Mills LITTLE PERFECTION. It sits close to the cash register in the Axt Drug Company in the Mississippi River town of Madison, Iowa. The location was busy because of river traffic and visitors to the Iowa Penitentiary.

Mills Novelty COMMERCIAL card drop machine of 1904 sits on the checkout counter of the Ballou Latimer Drug Store in Boise, Idaho, in the late teens. It generally paid off in cigars or candy.

Trade stimulators were such common sights in stores of all kinds, they were accepted as part of the scenery. A nickel plated Hamilton Manufacturing "Diamond Top" DAISY sits in the center of the counter of the Great Northern Cigar Store in Hot Springs, Arkansas, circa 1916.

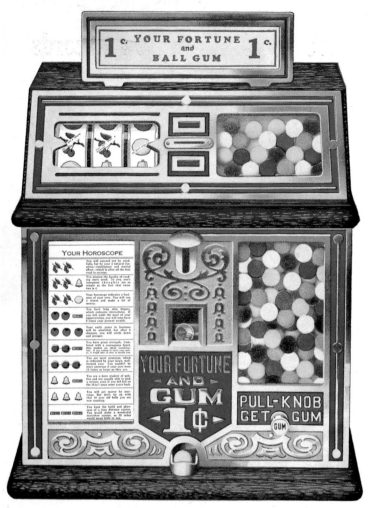

SUPERIOR
FORTUNE BALL GUM VENDER
All That The Name Implies

Size: Height 18", Width 14", Depth 10½". It only weighs 26 pounds.
Price for sample machine $45.00. Write for quantity prices.

Counter games were heavily promoted during Prohibition, and included the use of full color advertising. This ad for the Superior Confection FORTUNE BALL GUM VENDER appeared in the trade magazine *The Automatic Age* in June 1927.

After Repeal, counter games once again found their locations in taverns and bars. This is Vern And Jack's Jipp's Tavern in St. Cloud, Wisconsin, July 1941. An old Watling JACK POT "Blue Seal" slot machine is next to the wall while a brand new Groetchen IMP is on the bar.

How the games were sold. This 1938 advertisement for counter games includes, left to right from the top, Groetchen HIGH STAKES, 1936; Daval 1938 PENNY PACK, 1938; Groetchen ROYAL FLUSH, 1937; Daval DOUBLE DECK, 1937; Groetchen ZEPHYR, 1937; Groetchen SPARKS, 1938 and Groetchen GINGER, 1937.

Law didn't end counter games—it was changing business conditions that did them in. The arrival of the supermarket, arriving in the late 1930s, led to checkout counters that had no space for the machines. After WWII, with the big malls and supers, there was no room left for the games.

RACE GAMES

The easiest activity to achieve with a mechanical drive is to spin something. Released power can turn a shaft with greater ease that any other task. When stepped up, or down, by gearing, the shaft can create many forms of active and display machines. The result can be clocks or similar pointing arm devices when held in the vertical plane, with multiple arms for hour, minute, and second all geared to their respective rates of speed coming off the main shaft driven by a windup clockwork mechanism. Similar clockwork mechanisms were used for early trade stimulators, spinning a single shaft, often with multiple arms coming off the same shaft geared to the same or different speeds to create a variety of game opportunities when the clockwork mechanism is activated by the dropping of a coin. It was the basic original coin-operated trade stimulator game drive.

By laying the mechanism on its side, with the turning shaft facing upward, creating a game that spins its pointers or markers in the horizontal plane, or even small race horses at the end of the spinning arms, a whole new range of game opportunities are possible.

At the time this mechanical drive capability was available there were a number of common bartop games that allowed the patrons to bet on the outcomes to see who would buy the drinks, or win or lose the cash put down on the game layouts spread across the bar. The leading game among them was the "French Race Game," a smaller version of the large *Le jeu des Petits Chevaux* (The Race of Miniature Horses) horse racing machines operated by the legal gaming casinos in the resort areas of southern France. Many of these smaller pre-coin-op games were imported from France in the 1880s, becoming popular in saloons and clubs across North America. The inveterate entrepreneur Richard K. Fox, a New York man about town and the publisher of the lurid *The National Police Gazette* magazine with its massive saloon and barber shop circulation, was importing the games for sale through his growing direct mail sales organization promoted in the magazine's pages. Starting around October 1889 he was offering his non-coin operated FRENCH RACE GAME in company with the coin operated circular race track EXCELSIOR AUTOMATIC, produced in Chicago and also offered in the pages of the Rothschild's & Sons 1889 and 1890 saloon equipment catalogs out of New York City and Cincinnati, Ohio. They were the first commonly seen coin operated gambling machines in the United States, with the saloon, cigar counter, and billiard parlor proprietors getting a nickel a race while the players bet against each other for drinks and games.

Due to their massive rates of importation and production examples, many of these games are known and are in collector's hands, with a number luckily surviving the

The gambling palaces in southern France sported enormous horse racing games on which betters placed wagers of win, place, or show on any one or more of nine brightly colored horses spinning on a circular track. THE RACE OF MINIATURE HORSES (*Le Jeu des Petits Chevaux*) at Vichy was a popular gathering spot. The casino at Enghien-Les-Bains had three tables with betting layouts set up around each of their *Salles des Petits Chevaux* games. Postmarks on these tourist postcards are in the early 1900s.

years. Earlier games were made and played, with very few of them known to have survived, although it is believed that examples exist and just haven't made it to the hands of those who collect trade stimulators and counter games.

The earliest race games, seemingly coin operated, were created as toy banks, or at least they were promoted as such to make it possible to obtain exclusive machine patents. The leading toy bank maker, the J. & E. Stevens Company of Cromwell, Connecticut, introduced their RACE COURSE in August 1871, quickly followed by their BIG RACE COURSE and RACE AGAINST TIME, all of which found use as chance machines by allowing players to bet on their respective horses. These initial versions led to more sophisticated machines, including the Edmund A. Thompson RACE COURSE of July 1874, made in Amherst, Massachusetts; the Hawes, Butman & Company version made in Boston in March 1876; the coin operated race games made by James D. O'Donoghue followed by Josiah T. Marean's RACE COURSE, both made in Brooklyn, New York (O'Donoghue's in October 1883 and Marean's at the end of 1885); and the William H. Murphy race game made in the wide open western town of Brenham, Texas, in December 1886. None of these latter machines have ever been discovered, suggesting short production runs and a virtual dead end in American game development.

Just as this format was dying out, the British came to rescue of the genre by adopting the race game as a standard pub gaming machine. William H. Britain, one of the early British coin machine game designers, introduced his FOUR-HORSE RACE in 1886, followed by his THE MECHANICAL WALKING RACE in 1888. By that time Great Britain's genius pioneer coin machine designer William S. Oliver of London had introduced his landmark race game of February 1887, with others quickly following. By 1888 this new watering hole enthusiasm had leaped the Atlantic Ocean, with American coin machine designers quickly picking up further machine development. Nichols McManus of New York City created a floor standing game called RACE TRACK in June 1888, and had a counter model a few months later. By the end of the year and into 1889 improved models were being seen in New York and Brooklyn saloons. The New York City based American Mechanical Toy Company had their own version by the end of 1888, and Henry A. Behn of Union Hill (now Union City), New Jersey, had an advanced coin-op race game by May 1889. The respected Hudson Moore Company catalog was offering a race game on a cast iron pedestal in 1889, likely the American Mechanical Toy or Henry Behn machine.

Meanwhile, in far away Minneapolis, Minnesota, the National Automatic Device Company in Minneapolis was making their AUTOMATIC RACE COURSE in February 1889, quickly copied by the Flour City Manufactory making their own AUTOMATIC RACE COURSE a few months later. Other copycat producers were Universal with their AUTOMATIC RACE TRACK of 1889, the Chicago Nickel Works with their own AUTOMATIC RACE TRACK of 1889, the Lewis Manufacturing Company in Minneapolis, also in 1889, and elsewhere across the country from the Mississippi River east. This led to a round of lawsuits that never

Cie Jost RACE TRACK six-Horse, non-coin, 1902. French RACE TRACK (*De Course*) games were imported in great numbers for saloon and club use between the 1880s and prior to World War I. The major producer was Cie Jost in Paris, who also made the large French casino games.

H. C. Evans MINIATURE RACE COURSE, 1932. Makers lavished their cabinetry and mechanical skills on race games. This 18-inch cabinet "French Race Game" format by H. C. Evans and Company of Chicago is oak, with twelve hand carved horses. Betting was on a layout, often missing, as are manufacturer nameplates. The H. C. Evans name was sometimes stencilled on the front handle panel, sometimes not.

The RACE TRACK format became a generic. This is an English 9-Horse machine in a walnut case sold by Ambercrombie and Fitch in New York in the early 1900s. Players dropped their bets in the center cup, and spun the horses by turning the crank to set the horses in motion, which then coasted to a stop after a few revolutions.

found adequate settlement as the games came and went before the action was resolved. None of these machines have been found, either, and are eagerly sought by collectors. Possibly the most sought after are the complicated and larger AUTOMATIC RACE TRACK made by Kennedy & Diss in Brooklyn in October 1889 and the Hammond & Jones HORSE RACE made in Baltimore in 1888.

Other formats include oval race tracks, such as the Edwin J. Lumley machine made in Washington, D.C., in August 1889, followed by the elaborate Coyle & Rogers AUTOMATIC RACE COURSE of January 1890, also made in D.C. By the turn of the century the race game format had pretty much run its course, with the H. C. Evans & Company of Chicago practically the last holdout, making the machines as carnival games. By the 1920s H. C. Evans had added coin control, and the games found new life in the 1920s and 30s in a variety of models, the last of which were known as the SARATOGA SWEEPSTAKES machines made until the advent of World War II.

There was one other brief return of the earlier race games when nickelodeon pianos were facing their last days. By the mid-1920s the honky tonk pianos of the past had all but lost their charm to their speakeasy and roadhouse customers. In order to reinstate interest in the pianos, slot machines and coin-op target pistols were added to the machines, giving the customer a play for their nickel in addition to the selected piano roll. It was the Mills Novelty Company of Chicago that introduced the race game in a boxed windowed cabinet at the top of the piano, allowing the customers to bet on the outcome just as they had over the games of thirty or more years earlier. The Mills RACE HORSE PIANO was introduced in 1926, and was quickly followed by the J. P. Seeburg Piano Company GRAYHOUND RACE PIANO the same year. A few years later, after electrical amplification took the phonograph out of the shadows and created the automatic phonograph, later to be called the jukebox, Seeburg had their wholly owned subsidiary Western Electric Piano Company of Chicago introduce DERBY, another race horse piano similar to the Seeburg product of a few years earlier. But the time for both the race game and the coin operated player piano had come and gone.

A number of manufacturers created race game formats into the counter game era of the 1920s and 1930s, but few were made after that time. Most race games in collections tend to be older machines, with more of the original games still undiscovered than known. They have yet to be found.

J. and E. Stevens RACE COURSE, 1871. Coin control of race games began with the J. and E. Stevens RACE COURSE cast iron banks made in Cromwell, Connecticut, in the 1870s. Larger BIG RACE COURSE and RACE AGAINST TIME versions made it to barroom betting. Weight of the coin actuates the mechanism. It is the second design of leading toy bank inventor John Hall of Watertown, Massachusetts.

H. C. Evans and Company horse race wheel and layout, circa 1920. H. C. Evans started making race games at the turn of the century, upgrading them over the years to popular coin-op versions in the 1930s. The layout is typical of those that accompanied other countertop race games, with horse names and colors and 1, 2, and 3 betting squares.

Rothschild's AUTOMATIC RACE TRACK, 1889. Coin-op race games were bartop standards by the late 1880s. Customers bet against each other, or against the house, for drinks. They seem to be taking it all too seriously. This AUTOMATIC RACE TRACK was carried in the 1889 catalog of R. Rothschild's and Sons of New York City and Cincinnati, Ohio.

Chicago Nickel Works AUTOMATIC RACE TRACK, 1889. By the end of 1889 there were half a dozen race game makers, all copying each other. Almost identical to the Rothschild's AUTOMATIC RACE TRACK, with a shorter cabinet and different form of tri-color French flag, this version is believed to have been made by the Chicago Nickel Works, the major competitor to the Excelsior Race Track Company, also located in Chicago.

Excelsior Race Track EXCELSIOR, 1890. The race game classic. Between 1890 and 1894 these machines could be found on just about every other bar in America and Canada. They wore out, however, and were quickly relegated to the junk pile as new coin-op games came along. It is almost miraculous that any have survived at all. The Excelsior EXCELSIOR can be identified by its distinctive nameplate at the coin chute and play handle.

Richard K. Fox EXCELSIOR AUTOMATIC, 1890. The same game as the Rothschild's AUTOMATIC RACE TRACK (although the Fox version has a British flag, whereas Rothschild's has a French), the game was possibly produced in Chicago by the Excelsior Race Track Company, the leading maker of race games. Mechanism is clockwork, actuated by dropping a coin which allows you to turn the handle to initiate the spin. These machines are rare.

Excelsior Race Track EXCELSIOR, 1890. Glass domes and flags varied from game to game as production progressed. This example sold for $400 in an antique shop in 1950, almost half a century before this book was published. It was found in the basement of an old saloon building in a western state. Surviving domes are most often sold as antique cheese or cake covers.

Unknown manufacturer, circa 1890. Originally credited to the Excelsior Race Track Company of Chicago, the differences between this machine and the EXCELSIOR are so great it is likely made by another manufacturer. Surviving paper on the dome provides instructions but no maker name. Possibilities are the Flour City AUTOMATIC RACE COURSE or the Henry Behn race game, both of 1889. Perhaps it is the product of an as yet unidentified game maker. Time may tell.

National Table NATIONAL TABLE, 1928. The A. B. T. coin slide of the late 1920s made many new coin-op machines possible. National Table Company responded with a gambling race game that spins four colored race cars around an inset track, stopping at put-and-take panels (Put 2, Take 1, etc.) around the perimeter. They used the table top for local advertising for the movies, an ice cream parlor, Sunshine Coffee, and other products and stores.

Western Electric Piano DERBY, 1930. In 1926 Mills Novelty introduced their RACE HORSE PIANO with a race track under glass at the top to stimulate piano play. J. P. Seeburg Piano Company quickly followed with their dog spinning GRAYHOUND RACE PIANO. Western Electric Piano Company came back with their DERBY piano in 1930, using the Seeburg race game mechanism, detail shown here.

J. W. Whitlock THE DARBY, 1931. When their music machine business faltered during the Great Depression, J. W. Whitlock Company of Rising Sun, Indiana, turned to other products. THE DARBY horse race game operates by marbles. Players pick horses by name and color, and put a coin in the chute for their horse. Less than 150 were produced. Historian Q. David Bowers discovered a cache of them in the old factory building, and sold them off for about $250 each in the 1970s.

Peerless THE RACE OF HORSES AND MARBLES, 1932. A fairly simple kicker spins the horses and puts a lot of colored marbles in play. Peerless Products Company was located in North Kansas City, Missouri, and that's the area where these games have been found. The six marbles have point values when they land on their color. Score is added up. It looks like you can't lose, but the odds are overwhelmingly in favor of the house.

Western Electric Piano SWEEP STAKES, 1932. With pianos not selling, the wholly owned Western Electric Piano subsidiary of the J. P. Seeburg Corporation stripped out the DERBY piano race game units to create the Western SWEEP STAKES counter game, although it was pretty big for a counter so it generally came with legs. Sliding in a coin and pulling the handle sets the race in motion, with an electric "Results" scoreboard in the inside under glass flashing the name and number of the winner. The electricity could be had either battery or plug-in, to accommodate areas of the country that were not yet electrified. A payout version was also made in the same cabinet under the same name, with a horse selector on the front and a payout cup on the side.

H. C. Evans SARATOGA SWEEPSTAKES SPECIAL, 1933. The ultimate development of the "French Race Game," SARATOGA SWEEPSTAKES came in a variety of models. This is the top-of-the-line six-horse SARATOGA SWEEPSTAKES SPECIAL with a built-in gum vender. Player picks horse to win with the indicator at left, drops coin and pushes handle. Horses actually pass and repass each other. Every play gets a gum ball, but who cares.

Great States SANDY'S HORSES, 1936. The boss at Great States Manufacturing Company in Kansas City was Abe E. Sandhaus, and his nickname was "Sandy." So when their race game came along they called it SANDY'S HORSES. Fairly conventional, with a ring of spinning horses in an inset track. Six horses each have a name, color, and number. An odds drum at the top spins along with the horses, giving the final odds by the horse number. Coin chute for each horse to pick to win.

Rock-Ola OFFICIAL SWEEPSTAKES BALL GUM VENDER, 1933. Number and color dial selector, spinning horses, and a gum ball. It sounds like the SARATOGA SWEEPSTAKES, except the Rock-Ola OFFICIAL SWEEPSTAKES is a lot more exciting because everything happens so fast. A clever feature is the "moving ball of magic," or simply the "magic ball." While the horses spin, the ball rolls around the rim of the disk changing the odds as the race progresses. Hot game.

A. B. T. HALF MILE, 1939. Practically a replay of SANDY'S HORSES, the A. B. T. entry has seven coin chutes, spinning horses and changing odds. Drop the coin in the selected chute and depress the handle on the side. Not much different here, although the fact that it is an A. B. T. game that isn't a gun makes it unique and somewhat rare. The changing odds wheel is highly visible in the middle.

Maitland MISTIC DERBY, 1947. An absolutely sensational machine, and only one has been found to date. Play your horse in its own coin chute, and its number lights up on the backglass. Then a tone arm drops on a 78 rpm record and plays the race through the speaker. You hear it happen, and if your horse wins you win. The trick is there are five grooves on the record, and you don't know what race will be played. A race runs thirty seconds. This thing made money.

Daval DAVAL DERBY, 1937. Lots of action with this race game. The game takes up to eight coins to play the full field of horses on the track. Miniature horses spin at the bottom, changing the win, place, and show odds as the game progresses. Place your money in the coin chutes at the top—your choice of pennies, nickels, dimes, quarters, or even half dollars—and give the handle a twist. They're spinning. Large as counter games go, this was used as a big money gambling game—but the payback was over the counter, so it wasn't popular with merchants.

ROULETTE

They may look a lot alike, and they may even seem to play much like each other, yet the comparison between race games and the roulette wheel format for trade stimulators and counter games is strictly superficial. Both tend to be circular horizontal plane countertop games, similar in size and appearance. The difference is that race games drive mechanical arms and figures around a track to a winning or scoring finish line whereas roulette machines have no such mechanical components, instead spinning a free rolling ball (or balls) around a marked and generally pocketed scoring wheel which may or may not spin on its own, with the play action concluding once the ball has fallen into a scoring hole. In race games the machine provides all the action; in roulette the balls are wild and seek their own scoring finish. It is one-on-one action, with the player betting against the machine, rather than providing a machine that allows players to bet against each other. As such, the single player roulette format has had a much more enduring life than the race game and remains a viable game process to the present day.

Not that roulette format counter games are still being made. They seem to have had a series of vogue periods, followed by long dry spells of non-acceptance, after which they repeatedly returned again for another run. If counter games ever come back, you can expect that roulette format games will be among them, at least for part of their new tenure.

Roulette countertop games have two basic approaches. The early games had a metal ball, usually a steel ball bearing, loosely sitting on a plain or decorated disk. Once a coin is dropped and lever is pushed the disk is given a spinning kick which sends the ball careening around the circular wheel. As the spinning disk begins to slow down the ball seeks a hole, finally coming to rest. If the ball lands in a scoring hole, as evidenced by the award paper on the cabinet or in the center of the wheel, the player is awarded anywhere from one to a number of cigars or drinks depending on the award schedule. This win or lose approach is a clear indication of where the games were placed, for early roulette games tended to be saloon pieces. If you got a winning number you won. If you didn't, you lost your coin. The format encouraged a chance machine with "blanks," or non-scoring pockets or features that gave nothing in return. Although classified as a trade stimulator, the roulette games didn't encourage trade or provide the location with a selling advantage. It is a flat out gambling machine.

C. P. Young AUTOMATIC ROULETTE, 1893. Charles P. Young of York, Pennsylvania, was a creative and marketing giant, making a variety of trade stimulator, vending, and amusement machines in his back room manufactory. Young machines are very rare, with only single examples of two known. He probably placed them locally as both were found close to home. AUTOMATIC ROULETTE offers "Premiums For A Nickel" on its award panel.

Later roulette format counter games of the 1920s and 30s, and into the post-WWII years, often added a skill feature. An added mechanical component was a shooter that allowed the player to use a plunger to project the ball or balls onto the spinning disk once the coin was dropped and released the mechanism. It didn't make the play any easier, but it did give the players the feeling that they were involved in the end result. Later roulette counter games often eliminated the blank play feature by providing gum

Mansfield ROULETTE, 1893. If every other saloon had a race game by 1892, by 1894 the others most likely had a small boxed roulette game on their bar. Probably more, because by that time the EXCELSIOR race games had gone sour and were tossed out. The Mansfield Brass Foundry in Mansfield, Ohio, was a major maker. Early games had wood decorations on the sides. Mechanisms were easy to manipulate, and they jammed a lot.

31

balls or some sort of trade value for every play, making the games trade stimulators of sorts. But the straight gambling aspects remained with most roulette counter games and still made them a win or lose proposition with the scoring hole, or total score obtained by anywhere from two to more balls, determining if the player would get anything for their money, or nothing at all.

Western Automatic IMPROVED ROULETTE, 1894. The basic bartop trade stimulator of the 1890s into the early 1900s. Western Automatic Machine Company in Cincinnati, Ohio, was a large producer, as were Clawson in Newark, New Jersey, and Leo Canda and Charles T. Maley in Cincinnati. Sturdy mechanism. Neat touch is the added finger pointing graphic promoting "The Hub Bub" at Oakville, Iowa, September 13, 14, 1911.

Due to its basic simplicity, that of released energy spinning something, roulette format trade stimulators were among the earliest to be made. The Acme Novelty Works in Detroit made a countertop ROULETTE game in 1891, followed by the American Automatic Machine Company in New York City, New York, making their AUTOMATIC ROULETTE in 1892. The first of these small bartop machines were notoriously unreliable and could be manipulated by the players, so that genre ended fast. The first of the viable and secure crop of games seems to have been made by Charles P. Young of York, Pennsylvania, with his AUTOMATIC ROULETTE game of November 1893. Only one of these original Young machines has been discovered to date, and it was found fairly close to its original home. Young had a billiard parlor facing the town square in York, and made a variety of machines in the back room of the building. He was a customer of, and sold completed machines, to the almost legendary Clement C. Clawson of Newark, New Jersey, regarded as America's first multiple line coin machine builder. Young's development led to a rash of square boxed IMPROVED ROULETTE machines, including the Mansfield Brass Foundry ROULETTE made in Mansfield, Ohio, in 1893; New Yorker American Automatic Machine Company's IMPROVED AUTOMATIC ROULETTE of 1894; the Western Automatic Machine Company of Cincinnati, Ohio, IMPROVED ROULETTE of 1894; followed by the Atlas Manufacturing Company of Cincinnati with their IMPROVED ROULETTE in 1895; and fairly quickly by the Clawson Slot Machine Company in Newark; the Leo Canda Company in Cincinnati, and a number of others until the game lost favor around the turn of the century. These machines look very much alike, and it is difficult to tell them apart with only the cabinet front castings and award panels providing a point of difference for many of the games. These games aren't particularly exciting when played repeatedly, with their charm primarily based on the fact that at one point in time they graced the bar of a late 19th century saloon.

A number of variations on the theme were tried. The J. W. Stirrup Manufacturing Company of New York City made both square and circular versions with their FAIREST ROULETTE of 1896 and their WINNER ROULETTE and a AUTOMATIC CIGAR SELLER bartop roulettes of 1897, all three of which are rare and valuable. An interesting sidenote is that the first patent issued to a youthful Julius Roever of Brooklyn, New York (who soon had his own Roever Manufacturing Company), was for a countertop roulette with a number of complicated and interesting features, including a player ball release. Roever applied for his game patent in February 1894, coming back a decade later to invent and patent some of the earliest selectable recording (in this case Edison cylinders) automatic phonographs which he called the MULTIPHONE, leading to the development of the later jukebox.

Merriam Collins PEERLESS ADVERTISER, 1897. Cigar manufacturers promoted their brands at the sales level with dedicated trade stimulators. The PEERLESS ADVERTISER promoted "The Code" cigar. Maker is Merriam Collins and Company of Decatur, Illinois, the cigar distributor. This is one of the most elegant cigar trade stimulators ever made, one example known. 5¢ in the coin slot spins the roulette wheel below, with the steel ball stopping in a 1, 2, or 3 hole.

National Manufacturing LITTLE MONTE CARLO, 1897. Major advancement over the wooden box roulette games of the past. National Manufacturing Company of New York City encased the mechanism in metal on an oak base. Play was similar to a full size five-way slot machine in which the player picks the colors to play, with the yellow "00" paying fifteen cigars, with only one on a twenty-four-stop wheel. Bought out and subsequently made by Mills Novelty in 1898.

with a roulette version called SKILL ROLL. 1920 saw the simplified and smaller 36 LUCKY SPOT MIDGET dice and 36 ROULETTE versions, followed by the MIDGET 36 dice and MIDGET roulette games of 1926 and the NEW 36 GAME dice and PEE-WEE roulette counter games of 1927. Fey also licensed and sold the components of these games to half a dozen other "manufacturers" (who really only assembled the Fey games with their own cabinet castings) across the country who continued to produce them well into the 1930s. The A.B.T. Manufacturing Company in Chicago picked up the format in 1927 with their 36 LUCKY SPOT dice and 36 ROULETTE games of 1927. Few roulette format counter games have been made since the middle 1930s, with the more controllable reel and dice format games preferred for their more positive score display and lesser susceptibility to manipulation and accidental alteration of a delicate game result provided by a loose ball in a shallow flat wheel depression. Which doesn't alter the fact that a roulette counter game can be exciting as you have no idea where that ball will land. These games are a tantalizing entertainment treat on a modern rec room bar, particularly if you set up your own award schedule for drinks or, as the current fad increases, cigars.

A far more sophisticated roulette counter machine called the LITTLE MONTE CARLO was created by the National Manufacturing Company in New York City in November 1897. It was a hard core gambling machine patterned after the large floor slot machines of the period. The player had a choice of five colors to play, with varying odds, picking the appropriate coin chute. If the ball fell into the color played the barkeep would pay off in cash or drinks. In 1898 Chicago's Mills Novelty Company bought out National, including their building and the rights to the game, making their own version from that point forward. In 1902 Mills Novelty brought the game out in an elaborately trimmed cast iron cabinet as the IMPROVED LITTLE MONTE CARLO, followed by the 1903 model which sold for a number of years up until WWI, after which it disappeared from the Mills catalogs once Prohibition was nationalized after the war.

Roulette format trade stimulators and counter games found a new champion in the early 1900s and into the 1920s in Charlie Fey of San Francisco, the 1905 inventor of the three-reel automatic payout slot machine that led to the modern casino slot. Fey created a series of counter games using dice and roulette formats in the same cabinets, starting with his ON THE LEVEL dice game of 1907,

Mills Novelty (IMPROVED) LITTLE MONTE CARLO, 1902. Mills Novelty Company in Chicago kept making LITTLE MONTE CARLO in the same metal cabinet and wooden base as the National Manufacturing machine, all the while planning for bigger things. They popped their cast iron cabinet IMPROVED LITTLE MONTE CARLO on the market in 1902, and quickly dropped the "IMPROVED" part of the name. By 1903, with a newer model yet, the LITTLE MONTE CARLO was an established cast iron machine.

Mills Novelty LITTLE SCARAB, 1914. The most elaborate cast iron roulette game ever made, the Mills Novelty LITTLE SCARAB is an overgrown version of the LITTLE MONTE CARLO adding an internal payout mechanism and token payout cup to the side. There are five coin slots for colors, with a push of the crank starting the wheel spin. The LITTLE SCARAB name is in the nickel plated front casting. Only one was known in 1975. Since then two more have been found.

Wisconsin Novelty TRIPLE ROULETTE, 1923. If you were a friend of leading trade stimulator maker Charlie Fey of San Francisco in the 'teens and 'twenties, chances are you would soon be making a Fey machine under your own name, using Fey mechanisms and your own modified cabinet castings. The Wisconsin Novelty Manufacturing Company of Kaukauna, Wisconsin, made TRIPLE ROULETTE in a modified ON THE LEVEL cabinet with their name in the casting.

Mills Novelty LITTLE SCARAB VENDER, 1914. Some territories regarded trade stimulator machines that dispensed tokens as gambling machines, and outlawed them. To get around that ways were found to give value to every play while still avoiding a no blanks operation. Making a machine a gum vender often did the trick. This is the stick gum dispensing version of the LITTLE SCARAB with a side vender on the left. Only one known.

Fey TRIPLE ROLL, 1924. Chas. Fey & Co. in San Francisco was a creative and prolific small shop trade stimulator maker prior to the 1930s. Charlie Fey kept making new games in cabinets using old foundry patterns. He took his 1907 ON THE LEVEL dice game cabinet first converted to a roulette game as SKILL ROLL, kept the old names in the paper panel and front casting, and reissued the game in 1924 as the six slot TRIPLE ROLL. You can hardly tell the difference from its looks.

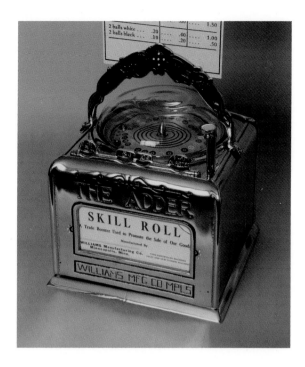

Williams Manufacturing THE ADDER, 1925. One of the Charlie Fey friends and machine makers was the Williams Manufacturing Company in Minneapolis, Minnesota. They made a modified version of the Fey TRIPLE ROLL and called it THE ADDER, offering four slot 5¢, 5¢, 10¢, Play Quarter "No Blanks" play. Confusing SKILL ROLL paper panel was retained. Cast iron dice symbol arch across the top says "Add Em," but was really placed there to prevent slapping. One known.

Mills Sales MIDGET ROULETTE, 1926. Sometimes an obvious Fey machine isn't obviously Fey. This MIDGET award card is unidentified, although headed 3 MUSKETEERS. As for the anti-slugging viewing window on the front, it's not there. In its place is a solid yellow panel. Fey? Atlas Manufacturing in Kaukauna, Wisconsin, L. C. Graham in Albany, New York, and Mills Sales in Oakland, California, also made and/or marketed the game as MIDGET ROULETTE in 1926.

Fey MIDGET, 1926. More spreading of Fey machines and names. MIDGET came out in 1926 with a Chas. Fey & Son (son Edmund joined his business at the time) award card and usually with a coin viewing window below the Indian head. The whole idea is that a merchant can see if a player is slugging a machine as the last coin played shows up in the window. If the player wins, and a slug is visible, no pay. It behooves a customer to be straight. 3 MUSKETEERS name of another game appears on the award card.

Michigan Novelty DETROITER, 1927. The Fey MIDGET rides again, only this time it's called the DETROITER, and is made by the Michigan Novelty Company in Detroit, Michigan. Other than the name in the casting, and the coin viewing window below the Indian head (where the copycat pattern has lost much of its detail), it could be a MIDGET clone. Including the endemic Fey duplicate name problem. The award card calls this machine THE PAWNBROKER.

Monarch Sales PEE-WEE ROULETTE, 1927. Looks familiar, doesn't it, with the familiar 3 MUSKETEERS name on the award card. It was the Fey way, to spread the largess out and about—and that included Indianapolis, Indiana. The initial maker was Monarch Sales Company, who made three models of Fey machines. Later in the same year the firm changed its name to Monarch Manufacturing and Sales Company, and then proceeded to make an additional three different Fey machines.

Keystone ROULETTE POKER, 1932. Few coin machines were made for outdoor placement as mostly venders and scales were placed in front of drug stores. ROULETTE POKER was announced in May 1932 as "Ideal for outdoor locations" because of its all metal construction. Maybe that's why these machines are so rare, because they certainly weren't made to resist the weather. Maker is Keystone Novelty Manufacturing Company of Chicago, out of business in less than a year.

Monarch Manufacturing BULL DOG GUM VENDER, 1927. Stretched Fey in Indianapolis. Take the MIDGET (or DETROITER, MIDGET ROULETTE, 3 MUSKETEERS, THE PAWNBROKER, PEE-WEE ROULETTE or whatever you call it), add gum balls behind a window, with a new award panel at the top and a gum ball dispenser below, and you get a new machine. The bulldog in the side casting gives the machine its name. But Fey duplication survives. The front paper calls it the 36 LUCKY PLAY.

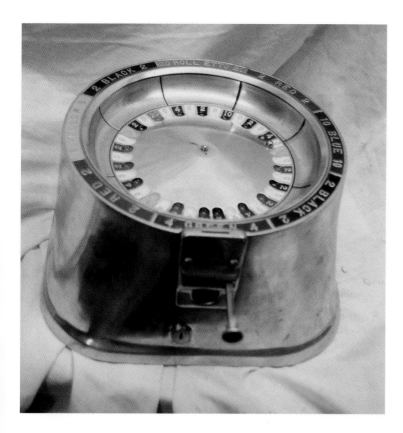

Roll-Etto Novelty ROLL-ETTO, 1933. Interesting graphic approach and clever design. The award card is cast into the outside perimeter of the circular cabinet, with a ball kicked into the spinning inside wheel by a plunger. A key selling point was that it didn't have a confusing color selector. What you shot is what you got. The four way coin chute takes pennies, nickels, dimes, and quarters. It apparently didn't sell well.

Jennings LITTLE MYSTERY, 1934. O. D. Jennings and Company, a major Chicago slot machine maker, added counter games to fill out their line and meet distributor requests. They created a line of flat countertop models in wooden cabinets with a metal "Ball Gum" gumbal vender on the side. All have the same inside mechanical kicker. LITTLE MYSTERY spins a roulette wheel, ARROW spins a pointer and DICETTE spins a felt pad to tumble dice. None are common.

Garden City DE-LUXE VENDER, 1933. The thick aluminum cast Fey look permeated the industry, particularly roulettes and dicers. It ended in the 1930s with the arrival of new cabinet materials. Last of the heavy aluminum cabinet roulettes is the DE-LUXE VENDER made by the Garden City Novelty Manufacturing Company of Chicago. It is a benign gum vending version of the WHIRL-SKILL, a fruit symbol model that also came with a jackpot.

WHEEL

We're back to spinning a shaft, the easiest form of mechanical drive, only this time the shaft is horizontal and spinning a wheel, disk or circular structure in the vertical plane. It makes a tremendous difference, and the early trade stimulator inventors quickly recognized the fact that they now had the most powerful and only consistent natural force working on their behalf: gravity.

M. O. Griswold WHEEL OF FORTUNE, 1893. The primordial counter wheel. Milton O. Griswold in Rock Island, Illinois, was already a successful coin-op game and peep show maker when he brought out his WHEEL OF FORTUNE in October 1893. The weight of the coin trips the escapement, unlocking the handle to start the spin of a cast iron flywheel. Thirty wheel divisions tell fortunes by number, and award cigars. The format endured for half a century.

This meant that a coin dropped at the top of a wheel would be pulled downward by gravity, and in the process could be made to impart a spin to the wheel. This led to the simplest and some of the most successful coin machines ever made. The first of this class appears to be the VOTING MACHINE created by Clarence M. Kemp of Baltimore, Maryland, in November 1889, although the exact method of drive is unknown as this machine remains to be discovered. The pioneer free wheeling trade stimulator wheel is recognized as the FAIREST WHEEL introduced by the Decatur Fairest Wheel Works of Decatur, Illinois, in April 1894. It was created as a cigar wheel that gave away from one to three or five cigars (you could order the wheel marked either way) for every nickel played, in the aggregate selling six nickel cigars for 25¢, giving the shopkeeper who operated the wheel a marketing advantage by providing a discount on cigars to those lucky enough to have the wheel land on a number larger than one as indicated by the numbers on the periphery of the wheel. If the wheel stopped on a one, the player got their 5¢ cigar for a nickel. If landed on a two, they got two cigars for their money, on

Lichty AUTOMATIC SALESMAN AND PHRENOLOGIST, 1893. A takeoff on the Griswold WHEEL OF FORTUNE, the AUTOMATIC SALESMAN AND PHRENOLOGIST spun two wheels with an indicator for each, with tallys indicating fortunes and awarding cigars. Des Moines, Iowa, druggist Norman Lichty patented and produced it in his Norman Lichty Manufacturing Company in town. Phrenology was a Victorian Age pseudo science that purported to read your character based on the bumps on your head.

up to five. There were only a few of the high numbers, but they brought in customers, in effect stimulating the trade. The wheel was so well received a DISCOUNT WHEEL version was made in February 1895 that gave discounts on any sale to those lucky enough to land on a number higher than zero, with the players always getting full value for their money, and maybe more.

CIGAR WHEEL, unknown manufacturer, circa 1894. The Griswold wheels led to an army of copycats, and variations. This forty-eight number boxed wooden wheel offers up to three cigars. This is the back, or merchant, side to confirm awards. So many cigar wheel machines showed up in 1893 and 1894, often in encapsulated geographic areas made by local artisans and cigar makers, that many still remain to be sorted out. Never advertised, and made in small numbers, they may never be identified.

The Decatur FAIREST WHEEL found its way into so many stores and cigar counters, and was so financially successful, it quickly generated competition with most of it taking place in the neighboring central eastern Illinois farming area. The Drobisch Brothers and Company, grocers and carpenters also located in Decatur, went in to the trade stimulator business in the summer of 1896 to briefly become the second largest in the country, with both of these leading firms located in the same town. To the west, about half a day's ride, a farmer named Monroe Barnes started making his own FAIREST WHEEL wheels in Bloomington, Illinois, in the summer of 1895, and was sued for infringement (by that time the Decatur Fairest Wheel Company had locked up a protecting patent), soon changing the name of his initial machine to CIGAR WHEEL. This was followed by similar and highly decorated Barnes BONUS (CIGAR WHEEL) and CRESCENT (CIGAR WHEEL) machines in 1897. A bit farther south, in Champaign, Illinois, Orin L. Percival began making his own CIGAR WHEEL in August 1896. With the whole country their market, and in spite of the keen competition, all of these Illinois farm country firms prospered in the heyday of the cigar wheel between 1894 and the end of the 19th century.

It was the original FAIREST WHEEL that continued to outsell all of the others, and by a wide margin. The firm was briefly known as the Decatur Fairest Wheel Company, and in 1896 was reorganized and enlarged as the Decatur Fairest Wheel Works, producing ball bearing models of Style 1 followed by smaller and less expensive wheels as the FAIREST WHEEL No.2 and No.3 until the end of the nineteenth century. The firm was again reorganized in 1899 and moved to nearby Pana, Illinois, as the Progressive Novelty Company, proceeding with THE FAIREST WHEEL and subsequent models into the early years of the twentieth century, finally becoming the Pana Enterprise Manufacturing Company in Pana, starting out with the NEW IMPROVED FAIREST WHEEL in 1907, with the trade wheel surviving into the World War I period. The wheels were still found on general, drug, and cigar store counters well into the 1920s and the advent of the counter game, making them one of the most successful forms of coin machines ever made.

CIGAR ONE WHEEL, unknown manufacturer, circa 1894. Stamped metal disk. Twenty-four stops include twenty-one 1, two 2, and one 3 to barely preserve house odds. There are often clues to manufacturers, with CIGAR ONE WHEEL having the same handle as the D. N. Schall and Company slot machines made in Chicago starting in 1896. Daniel Schall set up shop in 1894 to make "novelties." This may be his first product, but there is no known way of confirming that possibility. One known.

Direct competition to the FAIREST WHEEL came out of another small town, and the manufacturer of general store goods. The John M. Waddell Manufacturing Company of Greenfield, Ohio, was a significant producer of display counters, coffee grinders and rat traps, and was an early producer of cash registers that totalled their counts with colored marbles. The company took a turn in its product lines with the development of a cigar THE BICYCLE WHEEL in 1896 that was operated by dropping a coin and pushing a lever that gave the wheel a spin. Taking the wire spoked wheel form of the then popular fad of bicycling, Waddell put a miniaturized free spinning wheel into a curved top cabinet with etched glass faces that proclaimed the name of the machine, with a small white celluloid nameplate (which quickly faded to a cream colored panel) that identified the company in much the same manner as nickel plated brass nameplates were applied to store display cases. In the same manner as the Decatur Fairest Wheel Company, Waddell also made a DISCOUNT BICYCLE WHEEL with its own etched glass name panels. By 1897 larger wooden boxed BICYCLE DISCOUNT WHEEL merchandise discount machines were being made for their general and country store counter customers in two sizes, expanding their product line at a time when plus income was difficult to achieve. When the company name was changed to the Waddell Wooden Ware Works (known as the "Four Ws") in January 1901 production continued with new nameplates.

Decatur Fairest Wheel Works FAIREST WHEEL "Small Wheel," 1894. The preeminent cigar wheel, and the basic model that lasted the longest. This is the smaller version of the original Style 1 wheel made by the Decatur Fairest Wheel Works in Decatur, Illinois. Bronze coin chute at top right has company name on front, Decatur location on back. The "2" payout at eleven o'clock position on wheel has been lettered over the original "3" to reduce cigar payouts.

It was their next machine, a two wheel model of a miniaturized bicycle in a glass walled cabinet also operated by a lever push once released by a coin, that gave Waddell an unmatched product. Called simply THE BICYCLE, and introduced in July 1897, it became the next most popular trade stimulator in the country second only to THE FAIREST WHEEL in its many variations. Customers loved THE BICYCLE, and operational examples of the machine are known to have survived into the 1930s. Remarkably, because of their attractive appearance, the original boxed machine, or just the model of the bicycle removed from its cabinet, was sometimes saved as a child's toy. A number of these survivors have shown up at rural estate auctions over the years. The success of the machine led to imitation, with a former Waddell employee setting up his own BICYCLE and IMPROVED BICYCLE operation as the Sun Manufacturing Company in Columbus, Ohio, making the machines from 1898 into the early 1900s. Other nameplates on THE BICYCLE, specifically the Kelley Manufacturing Company and the Poole Brothers Manufacturers, both of Chicago, suggest additional production, although these machines may have been made by Waddell or Sun to be identified private label.

M. O. Griswold BLACK CAT, 1895. The next step in Griswold WHEEL OF FORTUNE development. By 1895 the Griswold wheel idea had been picked up by payout slot machine manufacturers, with fancy machine names. Griswold came back with BLACK CAT, with elaborate front trim. Play handle is on top, with instructions: "Drop nickel in slot, press lever up and let fly back." But watch it. The scissors grip handle can easily nip your fingertips. That might be why BLACK CAT is rare.

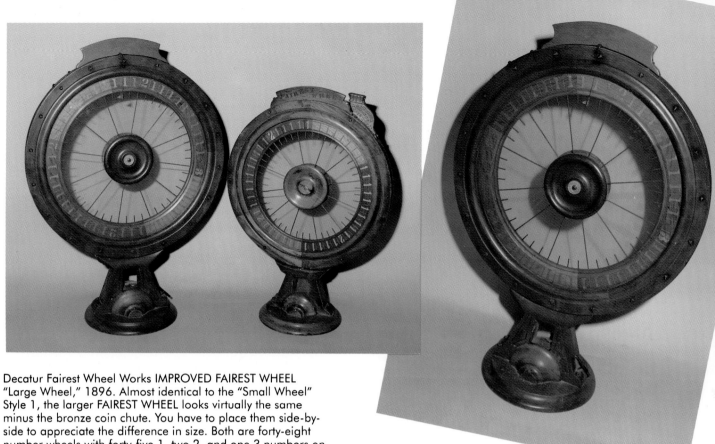

Decatur Fairest Wheel Works IMPROVED FAIREST WHEEL "Large Wheel," 1896. Almost identical to the "Small Wheel" Style 1, the larger FAIREST WHEEL looks virtually the same minus the bronze coin chute. You have to place them side-by-side to appreciate the difference in size. Both are forty-eight number wheels with forty-five 1, two 2, and one 3 numbers on the edge of a wheel which spins inside of an outer frame. The cash box at the bottom was quickly discovered to be too small.

A new wheel machine direction was taken with the ZODIAC made by the Wain and Bryant Company, a Detroit patternmaker, in February 1902. Inspired by the Caille Brothers Company BUSY BEE of a year earlier, the ZODIAC was a non-automatic over-the-counter payout miniaturized five-way color wheel slot machine made for bartop and counter usage. It was put into production by the Bryant Pattern and Novelty Company, which in turn was picked up by Caille Brothers. A modified version of the machine was continued in production by Caille as the SEARCHLIGHT, starting in April 1902. The Mills Novelty Company in Chicago got into the swing with their BULL'S EYE of 1902, with Caille Brothers following up over the years with similar machines called WASP, of 1904, and LINCOLN, a later version of the SEARCHLIGHT, produced in 1912. Even the Watling Manufacturing Company of Chicago got into the miniature color wheel action with their own BUFFALO JR. of 1905, placing all three of the "Big 3" slot machine makers in the trade stimulator color wheel field.

Generally operated under the watchful eye of a store proprietor (the player naturally wanted the storekeeper to see how many cigars, or what merchandise percentage discounts, were earned) the delicate wheel trade stimulators required the presence of a qualified person to make the awards. The growth of stores and the addition of hired personnel, as well as the rapid disappearance of the Ma & Pa grocery, cigar, and independent drug stores in the 1920s after World War I, doomed the wheels to extinction. And that just about ended that with a few post-1930 exceptions.

John M. Waddell THE BICYCLE WHEEL, 1896. The John M. Waddell Company of Greenfield, Ohio, produced general store goods such as coffee grinders, display counters, mouse traps, and cash registers. THE BICYCLE WHEEL was their first attempt at a coin operated trade stimulator offering cigars or other store merchandise. Primary feature is a delicately balanced wheel virtually identical to a large 1890s bicycle wheel, bike riding being a major fad of the time.

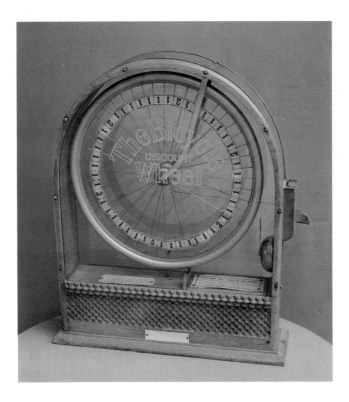

John M. Waddell THE BICYCLE DISCOUNT WHEEL, 1896. Same machine as THE BICYCLE WHEEL with a different etched front glass announcing it as THE BICYCLE DISCOUNT WHEEL. The idea was to offer discounts on store merchandise, but they were just as often used as cigar wheels. Numbers on wheel are on a printed paper strip woven in and out of the bicycle wheel spokes. Paper award panel inside of the cabinet offers "The Best 5¢ Cigar In The House."

John M. Waddell THE BICYCLE DISCOUNT WHEEL, 1897. For obvious reasons, this store counter model is called the "Large Square Wheel." The larger bicycle wheel is suspended on a stronger fork. Numbers are printed on cardboard, with two numbers per disk stuck on short spokes. The scoring numbers appear between spokes, but twisting sometimes led to disputes. This example is in the Greenfield, Ohio, museum as part of a Waddell display.

John M. Waddell THE BICYCLE, 1897. The second most popular cigar wheel next to the FAIREST WHEEL. The bicycle wheels are smaller than those on the single wheel machines, with numbers around their perimeter. Add numbers on both wheels to get the tally. Drop coin in at upper right, it rings the bell and you push the released handle down to spin both wheels. Generally placed over a cigar counter, primarily in drug stores. These sometimes show up at rural auctions.

Barnes CRESCENT (CIGAR WHEEL), 1897. The Barnes double printed paper wheel showed up on a number of models. The CRESCENT (CIGAR WHEEL) is a repackaged BONUS WHEEL with a wider coin entry at top, enclosed cash box at the bottom. The center disk was generally covered with a circular paper panel with the picture of an American Beauty, the generic pretty girl of the era. The Barnes wheel was patented. Only three known.

Barnes BONUS WHEEL, 1897. When Monroe Barnes Manufacturer of Bloomington, Illinois, first introduced this cigar wheel he called it the FAIREST WHEEL. A law suit by the Decatur Fairest Wheel Works in nearby Decatur, Illinois, brought that to an end. It was renamed the BONUS WHEEL and things went along swimmingly after that. BONUS WHEEL has a double printed wheel that can be used as a forty-eight number cigar wheel as well as an alphabetical letter wheel for special daily prizes.

T. J. Nertney COINOGRAPH SALESMAN, 1898. Thomas J. Nertney of Ottawa, Illinois, was a true character. He operated the leading THE SENATE bar in town, and made slot machines at his T. J. Nertney Manufacturing Company. "Coinograph" is the coined name for his line. He most likely had permission, or a license, to copy other machines. COINOGRAPH SALESMAN is the Nertney version of the M. O. Griswold WHEEL OF FORTUNE, sitting on a large cashbox, a Nertney innovation.

Brunhoff AUTOMATIC VOTE RECORDER AND CIGAR SELLER, 1898. Dropping a coin is just making a decision, not gambling, to get around anti-gambling laws. The twenty-four position cast iron AUTOMATIC VOTE RECORDER AND CIGAR SELLER made by the Brunhoff Manufacturing Company of Cincinnati, Ohio, is one of the most elegant cigar machines of the period. Vote for U. S. Grant or Napoleon, with coin windows and counters on each side. Another model shows two pretty girls. Two known.

Sun THE BICYCLE, 1898. The raging success of the Waddell THE BICYCLE spawned copycat production, with a former Waddell partner starting his own Sun Manufacturing Company in nearby Columbus, Ohio, to produce the machine. Most of THE BICYCLE trade stimulators found are Sun production, with lucky finds including nameplates etched in ivory or imprinted on celluloid and mounted on the front to the left of the handle. Inside award paper is also rare.

Bennett and Company COURT HOUSE, 1900. There were many ways to get around the FAIREST WHEEL design and accomplish the same thing. Bennett and Company in Kalamazoo, Michigan, made a series of similar boxed wheels for various Michigan and Ohio based cigar brands, who gave them away to their locations with a substantial order for cigars. COURT HOUSE, G. W. S., and LA CULTURA were sold by George W. Stattler of Charlotte, Michigan. Other examples have been found.

Bennett and Company SHENANDOAH UP-TO-DATE, 1900. Bennett must have been a terrific salesman, as his cigar wheels are certainly made for a lot of brands. At least three variations have showed up so far, and there surely will be more. SHENANDOAH UP-TO-DATE was made for the Stuckey Cigar Company of Lancaster, Ohio, active between 1899 and 1918. Stuckey was no slouch, as their name can be found on a number of other cigar machines, one apparently being their own.

Waddell W. W. Works THE BICYCLE WHEEL, 1901. The reorganized "Four Ws" of Greenfield, Ohio, continued to make the same machines produced by John M. Waddell Manufacturing Company. Celluloid nameplates are often found on these machines, identifying the Waddel W. W. Works with a single "L" in the company name, something not done on the company letterhead and business papers. The weaving numbered award paper often degraded, and wheels are usually found without it.

Decatur Fairest Wheel Works FAIREST WHEEL No. 3, 1899. The immense popularity of the FAIREST WHEEL, and its money intake, led to smaller, less expensive models with larger cash boxes. FAIREST WHEEL No. 2 has an enclosed wooden box, while No. 3 has glass walls to flash the money and show the level of play. The wheel is no longer a natural wood finish, with increased visibility numbers in black (for the 1) and red (for the 2 and 3) on a white background.

Waddell W. W. Works THE BICYCLE WHEEL, 1901. W.W.W.W. got around the degraded paper numbers by substituting printed number disks. This required a new wheel with half the spokes, with the intermittent spoke stubs holding the numbers. Disputes of the past were overcome by a single number per disk, creating a twenty four position wheel. The customer return is high on this altered wheel with nineteen 1, two 2, and three 3. Disk number wheels are rare.

Caille-Schiemer BUSY BEE, 1901. Designer Adolph Caille teamed up with businessman Jacob Schiemer to form the Caille-Schiemer Company in Detroit. Their sole trade stimulator was the LITTLE BUSY BEE of 1901, built to compete against the Mills LITTLE MONTE CARLO. The "LITTLE" name was soon dropped. Both are five-way machines spinning a wheel, one roulette, and the other a vertical disk. Pick your color and play a nickel. Black or red was two cigars; white was twenty.

Caille BUSY BEE, 1901. When The Caille Brothers Company was formed in Detroit in the summer of 1901, absorbing Caille-Schiemer, the BUSY BEE was their first trade stimulator. They produced five models; No. 1 for nickels, No.2 for pennies, No. 3 for pennies and nickels, No. 4 for "No Blanks" cigar play, and No. 5 as shown for chance and no-chance play with an added five-way coin head. These are rare and valuable cast iron machines. Dial has six bees in color, busy.

M. O. Griswold THE BIG THREE, 1901. Trade stimulators were not restricted to countertops. The seven foot THE BIG THREE is the largest known, and was built as a stand alone for cigar counters in hotel lobbies and other high traffic locations where counter selling space was more valuable than floor space. The wheel is almost identical to contemporary slot machines, with a four-way coin head. One known, found in 1980 in the effects of an old midwest hotel which closed in 1908.

M. O. Griswold STAR, 1902. Further development of the original Griswold WHEEL OF FORTUNE with no blanks play. The numbers on the wheel are jumbled and do not appear in sequence. Every number gets a cigar, with numbers 5, 10, 15, 20, 25, and 30 providing two. The return is based on selling six nickel cigars for a quarter, allowing the house to break even while stimulating trade. Stars on the wheel give the machine its name. Some STAR models have fully visible wheels.

Overton THE WHEEL, 1902. Another generic wheel promoting specific cigars. Clockwork mechanism spins a thirty position wheel when tripped by a coin. First patented and produced in 1901 as the THE HOO DOO by the August Grocery Company in Richmond, Virginia, a western producer adopted the machine and modified it as their own. Examples found so far promote Little Tom, Sir Lancelot, and Stickney Superbas cigars. Made by the Overton Manufacturing Company of Topeka, Kansas.

Wain and Bryant ZODIAC, 1902. A newer machine in the BUSY BEE tradition. Maker was patternmaking Wain and Bryant in Detroit, Michigan. ZODIAC put a five-way coin head on top of the machine in the same manner of the floor standing slot machines of the day. Signs of the zodiac are in the casting around the wheel. To distinguish it from a payout slot machine, the top casting carries the copy "For Trade Only." The Caille Bros. Company soon bought out Wain and Bryant.

Caille SEARCHLIGHT, 1902. The Caille Brothers adaptation of the Wain and Bryant ZODIAC. It has attractive Gibson Girl types in the center and a crowded fifty-six spaces on the wheel. As did the Mills BULLS EYE, the Caille SEARCHLIGHT came with two coin heads for single coin and five-way play. These machines are rarely found with both coin heads as they were operated one way or the other, with the extra parts discarded. We're lucky to have the machines.

Mills Novelty BULLS EYE, 1902. Not to be left out of the small counter wheel trade machine class, the Mills Novelty Company in Chicago created the similar BULLS EYE. The machine came with two coin chute plates. The single coin head as shown is for no blanks cigar machine play. When the five-way five color coin head is substituted, play is for 5¢ to one dollar in trade provided you picked the right color. Elegant cast iron, and a rare machine.

Brunhoff SLOTLESS (CIGAR CUTTER), 1903. Some territories objected to slot machines, covering trade stimulators in the same stroke of the legal brush. The way around that was to omit the slot. One of the most elegant of this class is the Brunhoff SLOTLESS (CIGAR CUTTER). It offered all the amenities of a cigar machine, except you paid the bartender to play. The Brunhoff SLOTLESS has one of the most elaborate cast cabinets of all cigar machines. Many sold, few found.

F. A. Ruff THE DEWEY, 1904. One of the lesser known makers of machines in the countertop LITTLE MONTE CARLO class was F. A. Ruff located in Detroit, Michigan. The knob-turn cabinet base was similar for a line of wheels, dicers, and possibly other variants. Frederick A. Ruff was in real estate in the 1890s, and then switched to electrical switch manufacturing, and then went back to real estate. Somewhere in there he also made coin-op machines. Rare.

Caille WASP, 1904. Further development of the single and multiple coin head cast iron single wheel trade stimulator theme. The Caille WASP of 1904 picks up the design features of the ZODIAC and the lively ladies and crowded wheel of the SEARCHLIGHT, and tosses in the color confirming coin viewing window of the Mills BULLS EYE. WASP came with two coin heads, single and five-way, the latter shown. Back of the machine has a glass anti-slugging window for viewing by the merchant.

Poole THE BICYCLE DISCOUNT WHEEL, 1904. Either Waddell W. W. Works Company made machines private label, which is likely, or other producers made their own versions of the many trade stimulators originating in Greenfield, Ohio. They are found with a variety of maker's names. Among them are the machines carrying the brass nameplate "Poole Bros. Manufacturers, Chicago," to be found on THE BICYCLE and the larger and imposing THE BICYCLE DISCOUNT WHEEL, shown here.

Griswold Manufacturing STAR, 1905. When M. O. Griswold and Company of Rock Island, Illinois, changed its name to Griswold Manufacturing Company in 1905, the STAR trade stimulator went along for the ride. It may have been the only one. Production continued with virtually the same cabinet and marquee, and a new Griswold Mfg. Co. decal on the case. Instructions say "Raise lever and let it fly back." This marquee awards five cigars for 5, 10, 15, 20, 25, or 30.

Albert Pick FAIREST WHEEL, 1906. The FAIREST WHEEL No. 3 with its smaller wheel and larger cash box went on and on into the 20th century, even after the producing company changed its name and location. A number of private label versions were produced, with perhaps the most notable the Albert Pick and Company model with the Pick name all over the inside wall of the cash box. The Pick version of the FAIREST WHEEL was cataloged in 1906 and 1907.

Pana Enterprise IMPROVED FAIREST WHEEL, 1907. Dead giveaways to the origins and history of the IMPROVED FAIREST WHEEL made in Pana, Illinois, in 1907 are the fact the FAIREST WHEEL name was used (without a law suit) and the copy behind the wheel saying "Over 250,000 In Use." That can only refer to the original Decatur Fairest Wheel Works FAIREST WHEEL of the 1890s. Same maker, with company name and location changed, and same white wheel, in a new advertising cabinet.

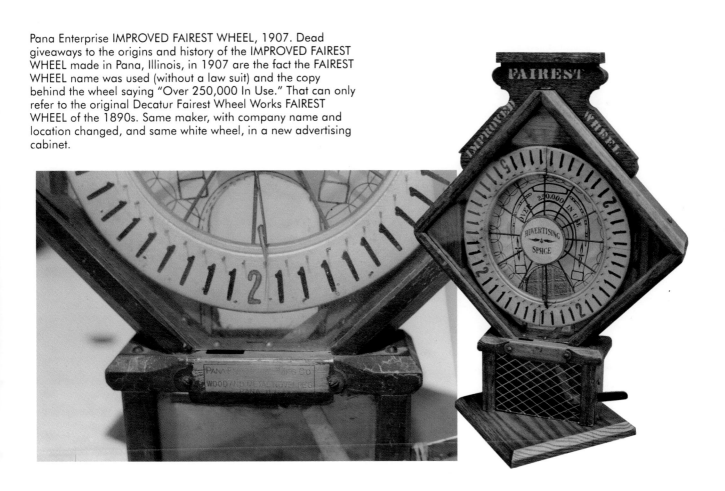

Caille LINCOLN, 1912. The Caille LINCOLN of 1912 is an upgraded SEARCHLIGHT in a new wood grained cabinet. Woody as it may look, that's cast iron under its painted skin. This time five Ulysses S. Grant faces have replaced the lovely ladies on the dial. At least he's not chewing a cigar. LINCOLN came with three interchangeable coin heads for single coin, five-way, and "For Cigars" no blanks play, operator's option. This is a rare find, with all three coin heads available.

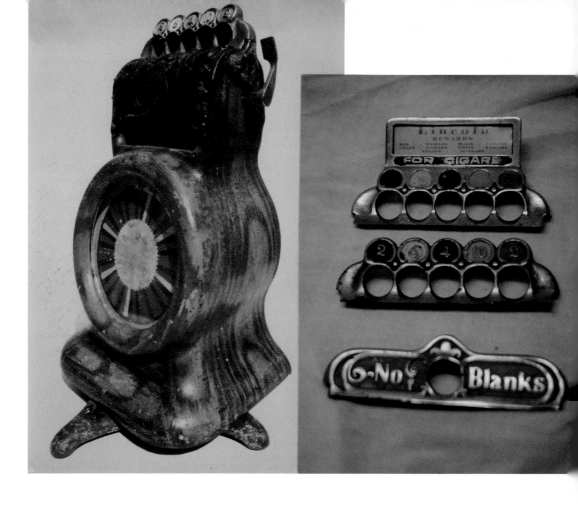

Griswold Manufacturing SELF PAY, 1916. Griswold took a dramatic turn from its WHEEL OF FORTUNE flywheel format with a token vending trade stimulator called SELF PAY that dispenses a token of varying amounts with every play. The press handle idea is retained, only this time it's press down and let the handle snap up. Griswold was always a finger snapper, so stay clear. The mechanism is completely different and has positive stops like a slot machine.

Smokers Supply CIGAR DICE, 1918. Hand made, home made and rough. Cigar wheels range from slick to shaggy, depending how and where they were made. Quite a number of local fabrications have shown up, and there are probably many more to be found. This carny dice wheel format has a primitive six-way coin head in wood. Drop the coin to show up behind the circular coin window you picked and physically spin the wheel. Match it and you can win two to five cigars.

Griswold Manufacturing NEW STAR, 1919. The Griswold STAR format was repackaged once again after World War I just before the arrival of the newer generation of counter games. NEW STAR remains the same in a more modern cabinet. Two coin chutes, with one in the back for the merchant to activate. Cigarette burns on the cabinet attest to the new post-WWI habit of smoking. The prizes are still cigars, and landing on 5, 10, 15, 20, 25, or 30 wins more than one.

Bally CUB, 1933. There isn't much difference between the Bally CUB of 1933 and an old cigar wheel machine of the 1890s, except the Art Deco trim and use of aluminum. In 1933, with repeal, and taverns opening up, everything old was new again. CUB came with four wheels; numbers, fortune, beer, and cigarettes. Usually only one wheel survived. The beer wheel is the one most wanted, and rarely found. The fortune wheel is shown, with its black cat printed blue.

Griswold Manufacturing NEW STAR, 1922. The last time around for the venerable Griswold Manufacturing Company NEW STAR, bridging the past with the future. The spinning flywheel is solid cast aluminum, the post-WWI wonder metal, with graphics and numbers painted in the recesses. 5, 10, 15, 20, 25, and 30 still pay off, in "5 Royal Buck or 3 Robert Mastell Cigars." Griswold closed its doors in 1922 to become Thomas-Kerns Company, not Inc., making vending machines.

Exhibit Supply SWEET SALLY, 1934. Set the selector to the number you are playing to win, slide in the coin, and wing-a-ding, it spins to a snap stop. Wheel has graphics on both sides; the front has cigarettes as a counter game, the back has numbers and multiple pictures of "Sweet Sallys" as a penny, nickel, dime, or quarter gambling machine. One of the best selling counter games ever made. Great entry level game to start a collection because it's fun and fairly common.

Stock CORK TIP, 1934. The five-way color wheel machine seemed to be a thing of the past by the 1930s, until H. J. Stock Company, founded by wholesale liquor dealer Harvey J. Stock, came out with a seven-way model that he sold to his tavern customers. The game looks thirty years older than it is. CORK TIP has a cigarette wheel. FLYING HEELS has horses. Starting in 1935 the games were being made by the Hub Manufacturing Company, also in Milwaukee. Not many known.

POINTER

In the Wild Wild West, or anywhere in the Americas of the 1880s for that matter, the travelling spindle game was a common sight. Travel by train or steamboat brought them out (the stagecoach was too bumpy, but some of the "Sportsman" gamblers probably tried it) and stories of losing the farm on the way to the homestead are ubiquitous in American folklore. When coin control came along the spindle, or pointer, game was one of the first to be mechanized.

The initial mechanical drive used in trade stimulators, or in vending and automatic payout slot machines for that matter, was clockwork, a dependable energy source already centuries old. Charged by winding a mainspring, the multi-geared clockwork stores energy for later release. When configured as a clock an escapement meters out the power at a predetermined slow rate, with a spinning "clock" timing and driving all of the moving components (hour, minute and second hands as well as phases of the moon graphics and whatever else was added) incorporated in the final machine.

As a trade stimulator energy source, a clockwork mechanism replaces the continuous drive escapement by a timing component that releases the exact amount of energy needed to drive the machine through a single cycle initiated by the dropping of a coin. Once the mechanism is so released, game activity begins. Often it was fast, with the gearbox imparting spins to single or multiple pointers journaled to the drive shaft, which then come to a quick snap stop. The clockwork trade stimulators required winding once or twice a day, or even more if the play exceeded the energy stored by the windings.

There is also another, even simpler form of pointer machine, more akin to the gravity driven wheel in the vertical plane than the power driven wheel or dial machines. In it, the dropping of a coin imparts a spin to the pointer or pointers, and a number, symbol, or score is identified by the pointer. Both formats, clockwork and gravity, characterized the early pointer machines, with clock timed mechanical drives and later electrical power continuing the format into the modern era. Pointer machines are basic and form one of the most numerically important classes of trade stimulators and counter games.

Gravity driven pointer machines constitute the very first chance devices operated by a coin. Edward S. McLoughlin,

Gambling layout, circa 1880s. Travelling by steamboat or train in the 1880s included "Sportsmen" who were quick to spread out a layout and set up a game. These portable gambling spindles were often encased in elaborate inlaid cases to look good and provide some assurance to the marks that things were on the up and up. But watch your pocket. The old adage was "Never play another man's game." Most wheels were fixed. Steamboat No. 999 cards enhance this layout.

a New York City house painter with grandiose schemes for financial success occupying his mind, got the first patent on a coin operated gambling machine with his BANKER WHO PAYS of 1876. McLoughlin purposely identified his machine as a bank to get around the anti-gambling laws and the machine restrictions of the United States Patent Office. Because of this subterfuge of over a century past toy bank collectors long ago adopted the machines as their own, and it has taken research and the finding of original examples to reveal these devices as chance machines. Dropping a coin at the top spins a pointer, which in turn indicates how many drinks or coins are to be returned to the player by the machine operator. McLoughlin went on to create a line of these chance machines, bringing out GUESSING BANK and DRINKS in 1878, followed by his masterpiece and artistic PRETTY WAITER GIRL in 1880. But success was not to be his with the machines, so he sold out his inventory and rights to a New England game maker, with buyer Smith, Winchester Manufacturing Company of South Windham, Connecticut, producing the GUESSING BANK in limited quantities starting in May 1877. They, too, missed commercial success with the machine, and the format languished, only to be rediscovered three generations later when a toy bank collector tracked down a number of surviving examples.

The free wheeling pointer was used extensively in the early days of cigar trade stimulators, with its initial application likely on the Kellogg and Company WHEEL OF FORTUNE made in New York City in 1888. The Standard THE STANDARD was made in 1892, along with the single pointer spinning ARROW of the Amusement Machine Com-

McLoughlin (Smith, Winchester) GUESSING BANK, 1878. First American coin-op gambling machine. Patented as a toy bank to avoid patent restrictions on gambling devices. Cast metal case states "Pays Five For One If You Call The Number." Who paid, kids? Not likely. Payoffs were in drinks or cash by a bartender. Inventor Edward S. McLoughlin, a New York City housepainter, sold his rights to Smith, Winchester and Company in South Windham, Connecticut, who made it.

pany, the New York City creator of their far more successful five-reel card machine. The Drobisch Brothers and Company of Decatur, Illinois, used the format on a number of machines made to compete with the Decatur THE FAIREST WHEEL, producing their VICTOR, STAR ADVERTISER and THE LEADER pointer machines all in 1897. John M. Waddell Manufacturing Company in Greenfield, Ohio, had a pointer in their line with their BOOMER of October 1897, and Comstock Novelty Works added a pointer at the bottom of the pinfield of their PERFECTION coin drop to create the PERFECTION WHEEL of 1898. The format even survived into the early 1900s with the MENU WHEEL made by the Menu Wheel Company of Cincinnati in 1903. But after that the powerless format all but disappeared.

Bypassing the eventual dead end of the early gravity driven pointer machines, it was the clockwork and other powered pointer devices that were destined to survive. They came early. Willard A. Smith of Providence, Rhode Island, developed a series of clockwork pointer machines, including the 3 DIAL FORTUNE of June 1892 and the STYLE A and STYLE B versions of his LITTLE JOKER in 1893. Murray, Spink and Company of Providence also marketed the 3 DIAL FORTUNE in the summer of 1892.

The Anthony Cigar Company of Cincinnati introduced its exciting coaxial twin arrow spinning ECLIPSE in 1892, and by 1893 had set up its own ECLIPSE manufacturing

M. O. Griswold LAYOUT, circa 1890s. When M. O. Griswold and Company set up shop in 1889 at the edge of the Mississippi River in Rock Island, Illinois, one of their first products was a mechanized layout for riverboat gamblers and the like. Generally known as "Cigar Wheels," the "Place coin where you choose" lettering on the layout board makes no such pretense. It's a flat out gambling wheel. Mills Novelty also made one called NEW IDEA CIGAR MACHINE.

McLoughlin (Smith, Winchester) PRETTY WAITER GIRL, 1880. If there is any question about McLoughlin machines being gambling devices and not banks, this lady should settle the dispute. Hardly a child's toy bank, the short skirted and busty PRETTY WAITER GIRL is perhaps the most explicit trade stimulator ever made. The casting says "Pays Five For One If You Call The Number." Only one known, in a "secret" collection. Collectors are scouring the country for more.

operations as Anthony and Smith. By 1894 the ECLIPSE was being made, or private label marketed, by the Grand Rapids Slot Machine Company of Grand Rapids, Michigan and the Charles T. Maley Novelty Company, Western Automatic Machine Company and Western Weighing Machine Company, with all three latter firms located in Cincinnati. The Mills Novelty Company was still making their own version of the machine as late as 1903, calling their model the ARROW (CIGAR SALESMAN). The introduction of new power sources, such as clock controlled lever and pull handle mechanisms and electricity, led to a rebirth of the format, with the Garden City Novelty Company SPIN-O of 1935 and the Bally SPINNERINO of 1937, both made in Chicago, typical examples.

There was one other pointer machine class that can be regarded as a trade stimulator, although the devices were not directly coin controlled. In a way they were, as they were trade stimulating devices located on top of a cash register which, when the sale amount was keyed into the register and the lever pushed (and coins or bills were put in the cash box) the wheel was kicked into a spin. If the pointer number matched the amount of the sale, the customer got the purchase free. Primarily located in saloons, it was a constant stimulant to beer purchases, with a lot of roaring and cheering when a patron got a free pass. The basic idea was initiated in Chicago in 1902 by the self

standing and complicated THE PROFIT SHARER by the Automatic Cash Discount Register Company. The unique system got a solid foothold on the market when the Page Manufacturing Company of Chicago introduced its line of cash register top SALES INCREASER pointer wheels in 1909, with "Drinks On The House," "Profits Shared," and "Free Merchandise" models differing by their names and graphics. Three years later the Mills Novelty Company picked up the line and cataloged it as their PROFIT SHARING REGISTER in STYLES A, B, and C. These register pointers had a limited service life, with Prohibition ending their tenure.

Because of the limitations of a dial, and the ever closer division of scoring numbers, pointer machines often led to disputes over winning or losing. That was enough to end their appeal. The other was the greater display afforded by symbols on reels, and the later solid state display machines. Pointers always looked old fashioned, and soon they were.

Murray, Spink 3 DIAL FORTUNE, 1892. Drop in a coin and the clockwork mechanism quickly spins three pointers to a total to tell a fortune. Don't you believe it. The so called fortunes indicated in the panels at upper left and right have numbers included, indicating how many cigars are to be awarded, if any. Few of these machines have been found. One showed up at the Chicagoland Show in the spring of 1991, having been in a family tucked away for years.

World's Fair COLUMBIAN FORTUNE TELLER, 1892. When this machine was first found it was thought to be from the 1930s it looked so modern. But the COLUMBIAN name in its stamped steel dish and the fact it was made by The World's Fair Slot Machine Company of Bridgeport, Connecticut, suggests something much earlier. The World's Columbian Exposition was in Chicago in 1893, a dated linguistic tie-in. Patents on the device proved the point: this is an 1892 machine.

Reliance Novelty THREE SPINDLE, 1897. First made in 1896 by Charlie Fey of San Francisco, California, THREE SPINDLE was based on an idea by Gustave F. W. Schultze of Berkeley, California, the inventor of the automatic payout slot machine. Fey added his own thoughts, and then sold the idea to Reliance Novelty Company in San Francisco. Drop a coin, push down the lever, and the three pointers spin to a stop to add up the score for a reward. Sometimes.

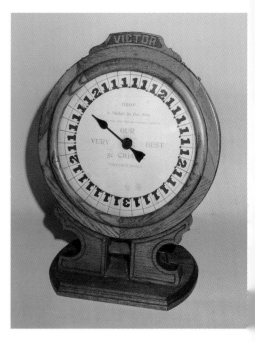

Drobisch VICTOR, 1897. Major competition to the Decatur FAIREST WHEEL, and made in the same town. Rather than spin a wheel, VICTOR spins a pointer. Made by the Drobish Brothers and Company woodworking shops in Decatur, Illinois. The thirty-five position no blanks dial is more liberal than the FAIREST WHEEL, offering four chances 2 cigars, one at 3, and thirty chances at 1. Sometimes found with the dial inverted to show the 3 at the top. Widely distributed.

United States Novelty JOKER, 1894. One of the truly great looking trade stimulators from the early 1890s. Dial is a plethora of color, showing a deck of playing cards around the perimeter. The clockwork two arrow mechanism and cabinet is virtually identical to that of the Maley ECLIPSE, except the face is entirely different. The dial design is attributed to George E. Stoneburner, which is splattered all over the front as part of the graphics. Two, maybe three, known.

Fey KLONDIKE, 1897. More vintage Fey along the same lines, KLONDIKE is a six-way machine. You pick the color in one or more of the selected six slots, turn a crank and this thing spins a pointer that snaps to a stop. Pick the right color you win. Pick wrong, try try again. These machines took in a lot of money on the Barbary Coast, and didn't give much in return. That's why the saloon proprietors, and game maker Charlie Fey, liked them so much. Old and valuable. Two known.

Drobisch STAR ADVERTISER, 1897. An advanced development of the Drobisch VICTOR idea, the STAR ADVERTISER has a thinner free spinning pointer and face display area for cigar advertising. In this case it promotes the Star Advertiser cigar made by the W. L. Kline Company in St. Louis, Missouri, with the cigar name on the face and the Kline name in the panel below the dial. A no blank machine, the front promises "You always get one, and sometimes more."

Drobisch THE LEADER, 1897. Similar to the STAR ADVERTISER, the Drobisch THE LEADER promotes a different cigar. This is a major blanks machine, with thirty-two 0 on a forty-eight position dial. There are 1, 2, and 5 pots, and a single 10, but the returns heavily favors the house. Lower panel says "Drop a nickel in the slot and try your luck." A greater variety of Drobisch cigar machines were made in a few years than by any other maker. Yet they remain fairly rare.

Ogden and Company DEWEY SALESMAN, 1898. First discovered in an 1898 catalog, and then in its patent. Invented by Joseph Nichols and John Davis of Chicago, and marketed by the Ogden and Company mail order house, also of Chicago. That was enough to start the search for an example. Miraculously, one showed up in the fall of 1977 with an Ogden decal—but only one. Drop a nickel, the pointer spins on the no blanks dial, bells ring, soldiers march, and the flag flies.

Watling THE FULL DECK, 1905. What a whirligig of action this is. The wonderful color graphics of a full deck of fifty-two cards are laid around the dial perimeter, with the center of the machine sinking into a deep bowl holding five independently shafted arrows. Put in a coin, push down the handle, and zip! All five pointers roar into action, spin furiously, and then snap to a 1, 2, 3, 4, 5 stop. You have a full hand of cards to check against the award card.

John M. Waddell THE BOOMER, 1897. Waddell developed their own simplified cigar pointer in THE BOOMER. This example promotes the Bear In Mind cigar. The dial is careful to note that THE BOOMER is "To be used only in the presence of the proprietor or clerks." That put it next to the cash register. Inside copy on the dial says "For each nickel dropped in the slot we will give you a 5¢ package of chewing gum or our best 5¢ cigar." You can win up to five.

Comstock PERFECTION WHEEL, 1898. Comstock Novelty Works of Fort Wayne, Indiana, made their classic PERFECTION coin drop cigar machine in 1897. To make the game more vigorous they added a wheel at the top of the pinfield in 1898 to create the PERFECTION WHEEL. Drop the coin at the top and it kicks the pointer before it works its way down the pinfield at the bottom. The wheel positions are mostly 1, with some 2 and 3, which multiply your score at the bottom.

Page "Drinks" SALES INCREASER, 1909. Cash registers with actuators and drawer openers gave merchants a new source of energy at the sales level. Page Manufacturing Company of Chicago tapped into the system with three SALES INCREASER headers: "Drinks On The House" for saloons, "Profits Shared" for cigar counters, and "Free Merchandise" for stores. Key in a sale, push the key and the pointer spins. Match the value of the sale and it's yours, free. One known.

Page "Free Merchandise" SALES INCREASER, 1909. The SALES INCREASER works by a rod descending from the top display into the cash register mechanism that kicks the pointer once the drawer opens for a sale. That makes it a form of coin machine. The "Free Merchandise" version was used in soft goods and general stores. The notice on the dial states "Free Merchandise if arrow stops at amount purchased. Mfd. by Page Mfg., Co., Chicago, Ill." These are rare. One known.

Keeney and Sons MAGIC CLOCK, 1933. Shades of the past, it's the Reliance Novelty THREE SPINDLE in a new set of clothes. Old game ideas kept getting new again in the 1930s when manufacturers searched desperately for a continuing flow of new products. The Keeney and Sons MAGIC CLOCK works the same way; coin in, lever down, and three fast spinning pointers clicking to a 1, 2, 3 stop for a total. The fruit symbol version shown scores like a three reel slot machine.

W. C. Steel Ball PUT AND TAKE, 1931. When miniature pool tables were all the rage in Texas in 1930, the W. C. Steel Ball Table Company of Fort Worth made them along with a dozen or more other manufacturers in the state. When the fad was over most of these companies folded. W. C. Steel Ball came back with PUT AND TAKE. It's a competitive game, with players putting and taking with every spin of the pointer. Score is kept by the markers across the bottom.

Advertising System M-R, 1936. The pointer-on-a-dial is one of the oldest forms of carny games, and among the easiest to coin control. The result is a large reservoir of simple and similar pointer machines that are difficult to identify and date. The way they work sometimes helps. The M-R Advertising System M-R machine with its top coin chute, aluminum arrow, and electric power dates from the 1930s. But how did it work? A layout is likely missing.

Jennings SWEET MUSIC, 1940. The rise of the jukebox at the end of the 1930s led to an array of counter games and wall boxes that added amusement and award features. SWEET MUSIC by O. D. Jennings and Company of Chicago gave the jukebox addicts what they wanted most; you could win tunes. Coin in, lever down, and the indicator spins to award plays of the tunes indicated by the pointer, played off by pushing the button. Rare. One known, but there must be more.

SWEET SIXTEEN unknown manufacturer, circa 1960s. A further development of the M-R and Mooney and Goodwin advertising games, electric but not coin-op. This seems to have been the ultimate development of the pointer games as solid state electronics and simulating video came soon afterward. The top spinner lands on the selected "horse" by number and red, white, blue, or yellow color while the bottom spinner indicates the odds. The layout was probably on the bar.

Mooney and Goodwin TRADE STIMULATOR, 1946. The post-WWII equivalent of the prewar M-R Advertising System game, this time enhanced by the addition of a neon ring around the dial. Mooney and Goodwin Advertising Company (even the name sounds similar; could the "M" in M-R be Mooney?) of Glendale, California, made this wall hanging TRADE STIMU-LATOR for bars at a time when amusements had to compete against the shuffleboard. Electric plug-in.

DUNSTAN'S CIGAR SELLER unknown manufacturer, circa 1900s. While cigar machines could well be expected to be regarded as an exclusively North American idea, it seems to have skipped the Atlantic Ocean. DUNSTAN'S CIGAR SELLER is an enigmatic machine that looks too new to be old, yet has an elder approach to cigar selling. Likely English, or even Irish, made, circa 1900s. Forty-four position dial has forty 1, three 2, and one 3. 5¢ play suggests Canadian usage.

CARD MACHINES

If there is a "first" original chance slot machine, it has to be the card machine created for saloon and cigar counter placement. Visualized at the end of the 1880s, with their primary patents in the 1890s, the classic, coin operated, sophisticated and simple card machines are the children of the late 19th century, with few new games of their class developed or primary patents issued after 1900. Starting with simple reel mechanisms, leading to more complicated drop card and hold and draw features, the interesting and clever pioneer trade stimulator card machines of the 1890s most often encased in wooden cabinets led to later cast iron versions, such as the Canda AUTOMATIC CARD MACHINE; Mills, White, Cowper, Watling and Royal LITTLE DUKE and TRADER models; and others, with wonderful new machines entering the market including the Schiemer-Yates, Caille Brothers and Mills HY-LO hold-and-draw machines; the Mills NATIVE SON and Caille GLOBE; the outstanding Mills SUPERIOR; the Portland Novelty OREGON; the enigmatic I. X. L. JR.; and the beginnings of the more modern counter games in the PURITAN, GEM, PILGRIM, MAYFLOWER, and other smaller cast iron (and later aluminum) counter games made by Puritan Machine, Caille, Mills, Watling and a host of others.

The first producer of the genre was the Amusement Machine Company of New York City, making their CARD MACHINE, the device described and illustrated in patent No. 442,683, with early production models carrying the notation "Patent Pending" prior to the December 16, 1890, issuance. The original patent papers show the machine in

Amusement Machine BABY CARD MACHINE, 1891. The small machines that just about started it all. The Amusement Machine Company of New York City initiated the genre, with others immediately jumping into the opportunity. Improvements were rapid. Black BABY CARD MACHINE at left matches the firm's 1892 advertising, with the later stained wood model at right moving the "Free Cigars" award card to a marquee on the top, adding designs to the front of the cabinet.

Card machines were practically everywhere after they were introduced in the United States in the early 1890s, surviving well into the new century. Border countries, such as Canada and Mexico, had them in their bars and shops during Prohibition after American saloon locations had disappeared. Here a Mills Novelty COMMERCIAL card drop machine of 1904, lower right, provides 1920s tourists with a round of fun in a souvenir store in Tijuana, Mexico.

a wooden case, with eleven full-size playing cards pasted on flats in sequence on a series of five rotating drums. For all of the language in the patent, the mechanism is simple, with a toothed rack imparting spin top a shaft with a ratchet at the lever end, with pawls at each of the five ratchet-fitted drums to stop the card-faced reels approximately lined up in back of a wide window on the front to show the hand of poker. This rack-to-ratchet mode was common machine practice in the 1880s, but it had never been applied to a coin released mechanism. It was this feature of coin-released lever actuation that permitted the spinning of circular drums that established new art and led to the issuance of the patent, becoming the basic document that led to all subsequent forms of coin-operated symbol reel machines. Produced in some quantity based on contemporary reports, it is probable that at some future date one of these initial large bartop card machines will be discovered. But as of this writing, none have. It remains the pioneer origin of all spinning reel coin machines, with none known to exist.

By the end of 1891 the firm had over eight machines in their line, and it was growing. The next in line, and even more successful, was the BABY CARD MACHINE of 1891. If surviving examples are any indication, the BABY CARD MACHINE had to be an even greater success than the original CARD MACHINE and POLICY models. Once started,

Card Machine Combinations

In ordering Card Machines mention if "liberal," "medium" or "strong" combination is desired. We recommend combinations Nos. 3 or 4, which will earn from 20 to 30 per cent. more than retail cost of premiums given out as rewards.

Card Machine Combination No. 1

Ace spade	10 spade	Queen spade	King spade	Jack spade
10 diamond	6 diamond	4 heart	4 diamond	3 heart
2 diamond	4 club	8 diamond	8 spade	7 diamond
9 spade	5 spade	7 spade	Queen club	6 spade
3 club	Queen heart	Ace heart	5 heart	Queen diamond
Jack heart	9 club	3 diamond	7 club	4 spade
8 club	7 heart	Jack club	10 heart	7 diamond
6 heart	Ace club	2 spade	3 spade	10 club
King club	5 spade	King diamond	Ace diamond	King heart
2 diamond	2 heart	6 club	2 club	5 club
5 diamond	Jack diamond	9 heart	9 diamond	8 heart

Royal Flushes, ALL.

Straight Flushes,	spade 6	club 8	heart 8	diamond 7

Four of a kind, ALL.

Card Machine Combination No. 2

Ace spade	Ace club	10 spade	Jack spade	Queen spade
7 heart	6 heart	Ace heart	4 heart	4 diamond
3 spade	3 club	5 club	Ace diamond	7 club
Queen diamond	King spade	2 diamond	Queen club	10 diamond
9 club	5 diamond	9 spade	8 diamond	2 club
2 heart	8 club	8 heart	10 heart	Jack heart
8 spade	2 spade	4 spade	6 club	6 spade
4 club	Queen heart	7 spade	5 spade	9 diamond
King heart	2 spade	Jack club	3 diamond	King club
10 club	Jack diamond	3 heart	5 spade	5 heart
6 diamond	7 spade	King diamond	9 heart	4 diamond

Royal Flushes, ALL.

Straight Flushes,	spade 4	club 4	heart 4	diamond 4

Four of a kind, ALL.

Card Machine Combination No. 3

10 spade	Jack spade	Queen spade	King spade	Ace spade
4 club	3 heart	4 diamond	3 diamond	10 diamond
Jack heart	5 club	5 spade	4 spade	2 diamond
6 spade	King diamond	10 club	10 heart	3 spade
2 heart	9 spade	7 heart	6 club	King heart
3 club	Ace club	Jack diamond	9 diamond	8 club
Queen diamond	4 heart	9 club	Queen club	6 heart
7 spade	Queen heart	Ace heart	5 heart	Jack club
8 diamond	2 club	6 diamond	Ace diamond	7 diamond
King club	8 spade	2 spade	7 club	9 heart
Extra 2 heart	Extra 5 club	8 heart	Extra 3 diamond	5 diamond

Royal Flushes, Four, All suits.

Straight Flushes,	club none	spade none	heart 1	diamond 1

Four of a kind, ALL.

Card Machine Combination No. 4

Ace spade	Ace club	Queen diamond	King club	Ace heart
10 diamond	7 heart	6 spade	9 diamond	7 spade
9 diamond	9 club	7 diamond	2 club	8 diamond
2 diamond	5 spade	3 heart	King spade	4 heart
3 club	4 club	Jack spade	4 diamond	Queen spade
Jack heart	6 diamond	8 heart	8 spade	9 heart
8 club	Jack club	5 club	Queen club	6 club
6 heart	10 spade	King heart	5 heart	King diamond
Ace diamond	Jack diamond	4 heart	7 club	2 spade
2 diamond	5 spade	8 heart	10 heart	Queen heart
5 diamond	2 heart	10 club	3 spade	3 diamond

Royal Flushes, 1 in Spades.

Straight Flushes,	heart 6	diamond 7	spade 6	club 5

Four of a kind, Nine.

Card Machine Combination No. 5

King spade	King club	Queen spade	Jack diamond	King diamond
4 diamond	5 diamond	7 diamond	10 club	7 spade
8 spade	9 heart	9 spade	4 heart	8 club
Queen club	2 diamond	6 heart	Ace spade	Jack spade
5 heart	3 club	Queen heart	6 diamond	3 heart
7 club	Jack heart	2 spade	10 spade	6 spade
3 spade	8 heart	3 diamond	2 heart	Queen diamond
Ace diamond	10 diamond	Ace heart	Ace club	4 spade
2 club	6 club	8 diamond	7 heart	King heart
9 diamond	5 diamond	4 club	9 club	5 club
10 heart	Jack club	2 spade	6 diamond	3 heart

Royal Flushes, None.

Straight Flushes,	spade 1	club 1	diamond 1	heart 0

Four of a kind, Eight.

Card Machine Combination No. 6

King spade	2 diamond	3 diamond	Ace club	10 heart
2 club	Jack club	6 club	10 spade	4 spade
4 diamond	5 diamond	7 diamond	7 club	3 heart
3 spade	8 heart	Queen spade	4 heart	King heart
5 heart	6 club	8 diamond	Ace spade	5 club
Queen club	10 diamond	3 club	2 heart	Queen diamond
9 heart	8 spade	6 heart	10 club	7 spade
4 club	5 spade	9 spade	2 heart	6 diamond
9 diamond	Jack heart	Ace heart	Jack diamond	Jack spade
2 spade	2 diamond	6 spade	9 club	8 club
Ace diamond	King club	Queen heart	7 heart	King diamond

Royal Flushes, None.

Straight Flushes,	diamond 1	heart 1	spade 0	club 1

Four of a kind, None.

The major manufacturers of card machines, including Canda, Mills Novelty, Caille Brothers and others, usually offered a selection of reel strips to provide "Liberal," "Medium," or "Strong" play, meaning a high return for the player (Liberal), a return of more than the retail cost of the awards (Medium), and an unconscionable high return for the house (Strong). Note that the recommended No. 3 combination provides players with a possible four royal flushes and four of a kind for all cards, whereas No. 4 only offers one royal flush and nine four of a kind. "Strong" No. 6 gives players no chance at either one.

---

the "Free Drinks and Free Cigars" trade stimulator card machines made rapid technological advancements, moving into large floor standing cast iron cabinets as SUCCESS machines, and into card drop, hold-and-draw, and a wide variety of other formats. Saloon floors and bartops were never the same again.

Card machine development after that was rapid, with the greatest changes based on out and out piracy. It was evident that The Amusement Machine Company had too much of a good thing, so others piled on the bandwagon. A dozen copy-cat producers suddenly showed up, surfacing in areas where the CARD MACHINE was barely being sold, or was hardly recognized. The pressure of competition changed the machines, as did the saloon placements themselves. The saloon customers roughly handled the wooden models, so increasingly greater use was made of metal and cast iron cabinets. The paste-on poker cards of the original multi-flat surface drums of the CARD MACHINE fell off, so the drums became circular with cards printed on reels instead of being individually applied. Drum spin was alternated, with every other drum spinning in an opposite direction to add excitement as the reels clicked to their stops. Cash boxes got bigger, and bigger again, to handle the excessive play. Changes occurred monthly, with the whole process of machine improvement taking only a few years.

Tibbils CARD MACHINE, 1891. Early card machine makers were in the east, with the Tibbils Manufacturing Company located in Rochester, New York. The idea of placing a machine in a store or at a cigar counter was brand new, so size considerations were hardly considered. The "footprint" of the Tibbils CARD MACHINE is enormous, but at least the customers couldn't lift up this heavy cast iron game and shake the reels.

The next big maker was the Leo Canda Company in Cincinnati, Ohio, home since 1880 to the United States Playing Card Company, with Canda ultimately becoming the largest dedicated producer of card machines both in terms of variety and volume. In the brief fourteen year period between 1893, when Canda made his first MODEL CARD MACHINE, until 1907, Leo Canda produced well over forty different interpretations of the card machine as The Leo Canda Company, and another dozen or more under the Leo Canda Manufacturing Company name. Canda's original entry into the market was a wooden cabinet variation of the BABY CARD MACHINE called the MODEL CARD MACHINE. The MODEL was quickly followed by the NEW CARD MACHINE with a horizontal spinning advertising drum, and the POLICY MACHINE, improved models with card and number reels respectively,

with a dice reels DICE MACHINE and tall floor standing "Giant" (meaning the large full size playing cards preferred by gamblers) GIANT CARD, GIANT COUNTER CARD, GIANT POLICY, COUNTER GIANT POLICY, GIANT DICE, and COUNTER GIANT DICE models.

In addition to its own ever expanding lines, Leo Canda Company also did private label work, making machines for the Novelty Manufacturing Company and the Charles T. Maley Novelty Company, in Cincinnati; and the Samuel Nafew catalog house in New York and Chicago. It was in 1897 that they hit their full stride, creating the card machine that exemplified the genre for the remainder of its history. After a continuing stream of models, including a wide variety of "iron card" machines on cast iron pedestals, Canda created a check vending trade stimulator model

Amusement Machine STANDARD "Iron Card," 1891. The Amusement Machine Company came out with design after design in rapid succession, all utilizing their basic card machine patent while incorporating mechanical improvements. The self-standing STANDARD created the cast iron pedestal "Iron Card" machine. Its modified play lever and internal return coil spring was an enhancement over the first bail spring model. Just about everybody copied it.

they called the FIGARO, a tall countertop machine with three number reels, that had a considerably modified and detachable (for maintenance, repair, and lint or hair removal from the shaft) reel bundle. So different was the approach, in fact, Canda applied for a patent soon after the first of the year, and got it before the year was out. The new format led to a revitalized vertical cabinet line with three or five reels which included the Canda COUNTER PERFECTION, UPRIGHT FIGARO, and FIGARO CHECK machines of 1898. By 1898 a smaller version was being produced, utilizing the same mechanism, which became the PERFECTION CARD, soon becoming the quintessential card machine. Its small counter footprint, curved wood "humpback" top, alternate reel spins (two spin backward, three forward), center cabinet coin chute and right sided

By the beginning of the 20th century the die was cast. Large multi-operational full line coin machine producers were the wave of the future, and the leading card machine producers were The Mills Novelty Company, The Caille Brothers Company and the Watling Manufacturing Company, the very same "Big 3" that produced the leading automatic payout slot machines, arcade machines, phonographs, and coin-operated scales of the era. The card machine had grown up, and they were now being sold by the tens of thousands.

There were other producers, but their markets and offerings were often limited to their own geographical areas. Hundreds of card machine makers produced copycat machines, or their own individualistic models, until the market dried up. Five reel counter card machine development really came to a halt in the teens as state after state went dry, and the saloons disappeared, finally knocked out by Prohibition at the end of 1919.

Sittman and Pitt LITTLE MODEL CARD MACHINE, 1894. First of the wildly popular card machines. Sittman and Pitt was a partnership foundry and pattern shop in Brooklyn, New York. LITTLE MODEL CARD MACHINE is the oldest coin-operated chance machine that can still be found in some numbers. Actual "Little Duke" size playing cards are affixed to drums which spin when the coin is dropped and plunger depressed. The wire bar at the top of the cards holds the show in place.

plunger established the classic look for card machines, and made Canda the leading producer of the genre.

The firm continued to produce the PERFECTION CARD machine with a variety of cabinet decorations, from filigree castings in the early years to colorful decals in the later production runs, until they folded shop as the Leo Canda Manufacturing Company, selling out to the Mills Novelty Company in Chicago, a company that had been buying their machines under private label as their own since 1898. At first Mills Novelty Company took delivery of the Canda machines, but they soon made plans to produce the machines themselves. So Mills Novelty took over the line and embraced it as their own, starting with the PERFECTION CARD, UPRIGHT PERFECTION CARD, CHECK CARD, CHECK FIGARO, CHECK POLICY, THE GIANT, THE JUMBO, and THE CHECK JUMBO between 1898 and 1900, with the JUMBO GIANT adopted in 1900, followed by the UPRIGHT CARD MACHINE, CHECK UPRIGHT CARD MACHINE, and the line leading LITTLE PERFECTION "Round Top" and UPRIGHT PERFECTION "Square Top" machines in 1902. Once the large Chicago producer had added the popular Canda card machines to their line, exposing them to the massive Mills mailing and customer lists, the machines took off on yet another wave of popularity.

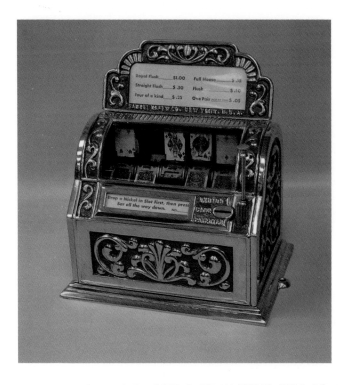

Samuel Nafew LITTLE MODEL CARD MACHINE, 1894. Private label Sittman and Pitt LITTLE MODEL CARD MACHINE made for the direct mail catalog house of Samuel Nafew Company, with sales offices in New York City and Chicago. Other names associated with this machine are "Monarch, S.F." in San Francisco and Ogden and Company in Chicago. Most examples of this machine are found with Samuel Nafew marquees. Sold in great numbers, these machines are even found in Australia.

The "Iron Card" pedestal card machines were the cheap, cost-effective and throw-away machines of their day. Once their usefulness had ended, and their barroom locations had closed their doors, or their cases were cracked, they were generally discarded. The wonder is that so many survived the years at all. No manuals have ever been found, and it seems unlikely that they ever will.

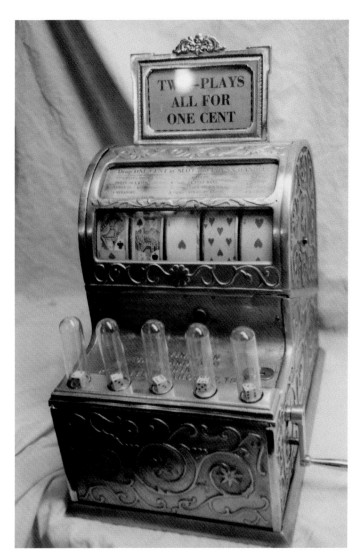

Watson COMBINATION CARD AND DICE MACHINE, 1894. Is it a card machine, or a dice game? Most of the cabinet is built around the five reel card machine, so that wins precedent. This machine was known from its patent No. 517,316 filed on January 15, 1894, and issued to John J. Watson of Buffalo, New York. Little hope was held out for finding an actual example, then one was discovered in the summer of 1989. Handle spins reels and pops the dice.

and a house button. Pace also made HIT ME. Buckley made their PILGRIM VENDOR in a five reel poker version, and the Groetchen 21 VENDER was a blackjack player. The closest thing to a true card player in this period was the DRAW POKER by National Coin Machine Exchange of Toledo, Ohio, made between 1934 and 1936. The Groetchen POK-O-REEL had such a good name it came back time after time in different cabinets and formats starting in 1932, finally ending its days in a modernized version in 1949, after which it was made by others.

The poker machine was far from dead in the 1930s and 1940s. It just wasn't as singularly important as it used to be. Its format showed up in the '50s and '60s and into the '80s and '90s as a countertop video game. As "Gray Area" games, the solid state poker and blackjack players can be seen most everywhere, operating both legally and illegally. If anything, the counter game collectibles of the future are here right now, with no one seemingly saving them as home games.

A new generation of card machines are projected to be commonplace into the 21st century, replacing the elec-

Canda GIANT, 1894. By the mid-1890s the Leo Canda Company in Cincinnati, Ohio, was the largest producer of trade stimulators in the country, ultimately earning its place among the top thirty makers of all time. They took their version of the MODEL CARD machine and stuck it high on hollow shafts to transmit the power and display the game where a whole saloon could see the play. To give you encouragement, this first example was found in New England in 1994. Another one has surfaced since.

While pre-Pro type card machines were still being made in the 1920s, with Mills Novelty the largest producer, it was with repeal in 1933 and the proliferation of taverns and bars (the word "Saloon" was outlawed in many states) in the 1933-1936 period that the games came back with almost as much fervor as they had in the beginning. Only this time they offered options beyond playing cards. Most counter games with three or more reels offered card symbol, fruit symbol, and number or cigarette reel models.

Few counter games were created solely as card machines in this period, although there were some notable examples. THE NEW DEAL made by Pierce Tool in Chicago between 1933 and 1936 was a poker player, with the machine also made by Pace Manufacturing Company, primarily a slot machine and scale maker. Pierce made HIT ME in 1934, a five reel blackjack version with two hits

tronic poker players on today's bars. The new TOUCHMASTER touch screen countertop poker and other games introduced by Williams Electronic Games of Chicago late in 1996 provide a glimpse into the future. Once in operation by a coin, the player prompts the game into action by touching the screen, dragging the object indicated by fingertip moves, or cancelling the action as desired. Game variations include ROYAL QUEST, a poker game; TARGET 21, a blackjack version; and TOURNAMENT SOLITAIRE, a solo card player. Other options are WORDZ, a word game; SHELL SHOCK, a computer version of the old shell game; and HOT HOOPS, a basketball game. But it's the card players that command the show. Clearly the card machine is facing a new beginning rather than an end, with the computerized versions owing as much to the machines of the past as to the technology of the present and what promises to be an exciting future.

Samuel Nafew MODEL CARD MACHINE, 1895. Growing up and out on the job. The designers and maker of the MODEL CARD MACHINE are Gustav Sittman and Walter H. Pitt of Brooklyn, New York; the west coast producer of the machine is the Monarch Card Machine Company of San Francisco, who called their model MONARCH CARD MACHINE; and the marketer of this version is Samuel Nafew Company of 392 Broadway in New York City. They all made this machine popular.

I. X. L. JR. unknown manufacturer, circa 1895. The growth of card machines was so rapid, it can never be totally tracked. Every year since trade stimulator collecting began in the late 1970s at least one unidentifiable machine, and often more, have shown up. Questions regarding maker, name, and date were resolved for many—but not for others. I. X. L. JR. has its name on the other side, and had this eagle below a coat of paint. It is unique, unlike any other machine.

T. F. Holtz Novelty CARD MACHINE, 1895. Theodore F. Holtz copied the machines of others when he started his T. F. Holtz Novelty Machine Works in San Francisco. His hired assistant was a young Charles August Fey, who soon went on to his own coin machine fame. CARD MACHINE seems to be a duplicate of the Amusement Machine BABY CARD MACHINE of 1891. What is likely is that Holtz took the older machine, re-routed the coin chute, and stuck it on a large cash box.

Reliance Novelty RELIANCE, 1896. With the wild and raucous Barbary Coast close at hand, and the gold mining and logging country nearby, San Francisco was a hotbed of coin-op chance machine design and placement. Reliance Novelty Company took the MODEL CARD MACHINE card flipping idea and made it a bigger, flashier game. RELIANCE cabinet is cast iron, nickel plated. Machines like these are generally found in the San Francisco area, usually in very poor condition.

T. F. Holtz and Company BROWNIE, 1897. As business grew, Holtz expanded. BROWNIE is an advanced development of the Sittman and Pitt MODEL CARD MACHINE, recreated in 1897 by the Monarch Card Machine Company of San Francisco. The elaborate new cabinet has Palmer Cox "Brownies" (an early "Mickey Mouse" type of character, only there were lots of them) and playing cards all over it. The Holtz version has lost a lot of detail, so he probably recast it from a Monarch.

Reliance Novelty STANDARD, 1897. TROPHY with the STANDARD nameplate in place. The back of this fabulous cast iron machine has an anti-slugging coin viewing window, and locked back door to the cash box. The operator, not the storekeeper, kept the key. The "Little Duke" cards inside sometimes came to a fairly sloppy stop, leading to disputes. Note the joker card with the TROPHY name. Only two examples of this machine have been found.

Reliance Novelty VICTOR, 1896. More of the same. The VICTOR is much like the RELIANCE with a different cabinet casting. A number of the eastern manufacturers had representatives in San Francisco to sell machines, and bring new ideas back to their home offices and manufactories. They did in droves, with the result that the Reliance Novelty RELIANCE was made by Caille Brothers in 1906 and the VICTOR was produced by Mills Novelty Company in 1907.

Reliance Novelty TROPHY, 1897. From the front it looks like the Reliance Novelty RELIANCE, but give it a one quarter turn and you see the name TROPHY in the side casting. There is a difference in the machines, with a much simplified and less delicate marquee on the TROPHY. Keep looking. The STANDARD nameplate at the bottom fits over the TROPHY name, making this one machine with two names. The award card is for Chas. Fey and Company, so Charlie Fey ran them.

anda JUMBO, 1898. It's a card machine
e any other—but BIG! The JUMBO name
mes from the full size playing cards
epicted on the reels. They are the size
eferred by gamblers, and not the "Little
ukes" used on most card machines to date.
JMBO was a prime saloon machine, and
so attractive for hotel lobbies. The Canda
dvertising said it could "be used for any
e, drinks, candies, notions, or jewelry, by
anging the reward card to suit the line."

Mills Novelty THE JUMBO, 1898. Newly formed Mills Novelty Company in Chicago started out with a few machines of their own, and acquired others. Canda card machines were built for them under private label. Often the Canda and Mills machines were identical, but sometimes small differences are apparent. Mills advertising states their machine was made of oak with "nickel plated trimmings." Pictures show the elongated coin window of this machine, so this is possibly Mills.

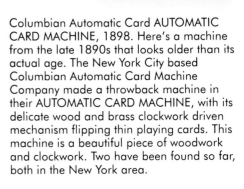

nda COUNTER JUMBO, 1898. After the
ended popularity of the Sittman and Pitt
 Samuel Nafew card flipping MODEL
d machines, it was the Leo Canda
mpany of Cincinnati, Ohio, that almost
gle-handedly re-popularized the five reel
d machine once again, with the cards
ted on reel strips rather than individually
erted into drums. The Canda machines
e much more durable and went a long
y toward avoiding conflicts.

Canda PERFECTION CARD "Round Top," 1898. The most significant Leo Canda contribution was their 1898 PERFECTION CARD machine. This seminal design characterized card machines for the next thirty years. The cabinet lid is reminiscent of a humpback trunk, earning the name "Round Top." Every major maker made it as LITTLE PERFECTION, PERFECTION CARD, or just PERFECTION; Mills Novelty, Caille Brothers, Watling, Berger, White, Cowper, and a host of others.

Columbian Automatic Card AUTOMATIC CARD MACHINE, 1898. Here's a machine from the late 1890s that looks older than its actual age. The New York City based Columbian Automatic Card Machine Company made a throwback machine in their AUTOMATIC CARD MACHINE, with its delicate wood and brass clockwork driven mechanism flipping thin playing cards. This machine is a beautiful piece of woodwork and clockwork. Two have been found so far, both in the New York area.

Mills Novelty THE LITTLE DUKE, 1898. Mills Novelty didn't mince words. If the machine flipped "Little Duke" cards, its name was THE LITTLE DUKE. This is said to have been the third new machine developed and produced by the Mills Novelty Company in Chicago after official formation of the firm in 1897. It is a pace setter with an elaborately trimmed copper oxidized cast iron cabinet. Only two or three of these machines are known, and they are highly desirable.

Clune VICTOR, 1900. When things got hot for trade stimulators in San Francisco, they warmed up in Los Angeles. In the time honored tradition of coin machines, L. A. copied S. F., and William H. Clune Manufacturer was in business. His "Free Drinks" award card indicates his customer base was saloons. The case is a direct steal from the Reliance Novelty VICTOR. Clune's second similar machine was called COMMERCIAL, and Mills Novelty in Chicago copied it.

Canda CANDA CARD MACHINE, 1900. Leo Canda Company in Cincinnati, Ohio, created the largest card machine of all in the CANDA CARD MACHINE. When the company changed its name to Leo Canda Manufacturing Company in 1902, production continued as QUINTETTE. The name comes from the encasing of five Canda UPRIGHT CARD MACHINE machines on a pedestal. Once tripped it sounds like a jet taking off. Caille Brothers made their own version in 1901 as QUINTETTE.

J. Salm Manufacturing, name unknown, 1899. Once the format for the cast iron pedestal "Iron Card" machine was established by Amusement Machine, Tibbils and Canda, more than a dozen other makers began to churn them out. It was known that J. Salm Manufacturing Company made "slot machines" in Cincinnati in 1899, but what kind? Then this machine came out of a barn in central Illinois in the summer of 1996 with a Philadelphia address for the firm. Bingo!

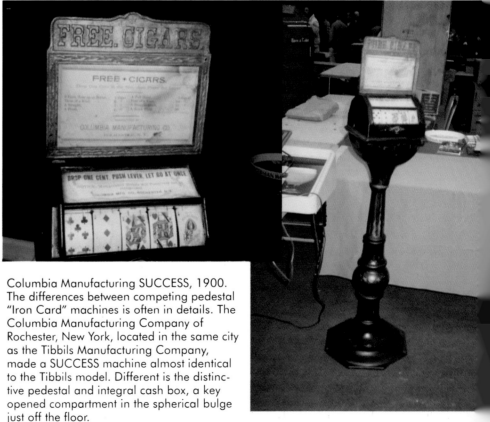

Columbia Manufacturing SUCCESS, 1900. The differences between competing pedestal "Iron Card" machines is often in details. The Columbia Manufacturing Company of Rochester, New York, located in the same city as the Tibbils Manufacturing Company, made a SUCCESS machine almost identical to the Tibbils model. Different is the distinctive pedestal and integral cash box, a key opened compartment in the spherical bulge just off the floor.

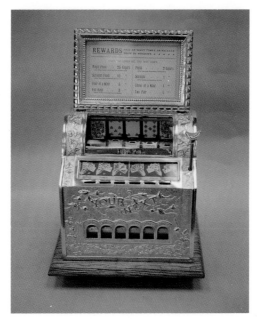

Mills Novelty JUMBO SUCCESS No. 2 "Big Success," 1900. Mills Novelty takes over as the largest producer of trade stimulators. JUMBO SUCCESS came out of their association with Canda, "Jumbo" meaning full size card reels. "Pays Cash Only" is rare. Operator is the Mills agent located in Manhattan Billiard Parlors in Spokane, Washington. Marbleized finish is typical for early JUMBO SUCCESS machines. No. 2 in the name identifies the pattern and production run.

Mills Novelty YOUR NEXT, 1900. The name probably should have been YOU'RE NEXT, but the turn-of-the-century populace loved to mess with the language. Even the Mills Novelty Company catalogs called it YOU'RE NEXT, but the truth is in the front casting. This is a cast iron Mills interpretation of the wooden cabinet Canda AUTOMATIC CARD MACHINE. The machine graphics are elegant, with a choice of six coin chutes to play for the artistic fanned poker hands over each slot.

Mills Novelty JUMBO SUCCESS No. 4 "Big Success," 1901. SUCCESS and JUMBO SUCCESS machines were so popular, they were made in large production runs by their many manufacturers. Mills Novelty numbered their foundry runs, making changes each time they went back on the line. "Automatic Dump" window holds coins in view to entice play, and dumps them in the cash box when too heavy, resetting to start all over. No. 4 features a marbleized cabinet with copper scrolls.

Mills Novelty JOCKEY, 1900. A true classic, and one of the most desirable card machines ever made. JOCKEY was made by a dozen manufacturers, including Automatic Machine and Tool, Charles Felitor, Watling, Caille Brothers, F. W. Mills, and Mills Novelty, each differing slightly from the other. Examples of the first three have never been found. JOCKEY took card machines out of the small countertop class and made them important. Still being found.

Mills Novelty UPRIGHT CARD MACHINE, 1901. By 1901 Mills Novelty Company in Chicago had acquired most of the Leo Canda Company line, and proceeded to make the machines in Chicago. They didn't look much different, with the exception that Mills added some interesting details. The operating lever below the coin chute sports a devilishly intricate casting, and the double gold medal decals of the Mills awards at the recent Paris World's Fair are added at top center.

Caille QUINTETTE, 1901. Here's a machine that stands up to take its place in any room. QUINTETTE is the Caille Brothers Company version of the Canda CANDA CARD MACHINE and QUINTETTE, plus a lot of castings and plating to make this the largest and flashiest card machine ever made. You would think that they have all been found, yet these marvelous monster machines keep showing up once in a while. More than one have come out of auctions of the effects of old hotels.

Wheeland Novelty PERFECTION, 1901. Here's a completely new idea with an old name. The playing cards pop up in a full hand of five, and when you drop another coin and push the lever they are apparently shuffled and pop up again. The inventor was Charles E. Wheeland of Seattle, Washington, who established the Wheeland Novelty Company to make them. He probably picked the name PERFECTION to be sure people knew he made a card machine. Rare and valuable.

Portland Novelty OREGON, 1901. With the wild wild west in full sway at the turn of the century, an indigenous slot machine industry arose in the western Pacific states of California, Oregon, and Washington, mostly making card machines. The Portland Novelty Works in Portland, Oregon, made one called OREGON. They used actual full size cards with great visibility on their reels, providing big cards on a small machine. Only one of these is known.

Royal Novelty ROYAL TRADER, 1902. Takin a step forward, the six-way ROYAL TRADER made by the Royal Novelty Company in Sc Francisco substitutes reels for the flip cards of the YOUR NEXT and OAKLAND, while retaining the elegant fanned hands of carc panel immediately above the six coin chute to graphically demonstrate the bet. ROYAL TRADER adds another feature, an absolute beautiful cast iron cabinet. The wood of th past wasn't out, but iron was in.

Wheeland Novelty CALIFORNIA, 1901. The Wheeland card poppers were regarded as all but impossible to own rarities when they were first discovered in patents and none were known. By the mid-1980s a number of the machines had shown up, and reached collectors. Examples were also made by Mills Novelty, Caille Brothers, and Watling. CALIFORNIA was the next Wheeland development, with these machines remaining in the west. This one was found in Nevada City, California.

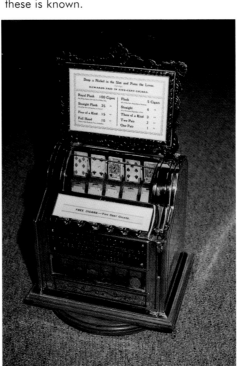

Oakland Novelty OAKLAND, 1902. The San Francisco area remained the center of California coin machine development. The Oakland Novelty Company, located at 1006 and 1008 Washington Street in Oakland across the bay from San Francisco, made a machine that was very much like the six-way Mills Novelty YOUR NEXT, with "Rewards paid in five-cent cigars." One wonders if the building is still there. It is likely the original machine had the colorful fanned hands of cards panel.

Mills Novelty LITTLE PERFECTION "Round Top," 1902. Mills Novelty Company of Chicago totally embraced the Leo Canda LITTLE PERFECTION "Round Top" (others called it "humpback") and ran with it. The made more of these machines than all ot makers combined. Mills started out with colorful decals on the front and sides, bu the years went by they dropped this prete at art and just banged out the boxes. Mo widely produced card machine of all time

Mills Novelty UPRIGHT PERFECTION, 1902. Looking much like the Mills UPRIGHT CARD MACHINE of a year earlier, including the Paris Exposition award decals above the reels, later examples of the Mills UPRIGHT PERFECTION starting at the end of 1902 incorporate an improved mechanism. Taking the Leo Canda original, Mills Novelty re-engineered the working parts and refined the reel release, adding a gum vending column in the process. UPRIGHT PERFECTION utilized the mechanism.

Mills Novelty IMPROVED JOCKEY, 1902. Mills Novelty Company continually upgraded their machines year by year to simplify manufacturing processes. IMPROVED JOCKEY (Mills called it "Improved style for 1903") substitutes stamped steel parts for some older castings, has interchangeable reels and adjusts for nickel or penny play by moving an inside lever. The physical expression of the improved machine is a simplified carved wood front scroll.

Mills Novelty COUNTER SUCCESS No. 6 "Little Success," 1902. Rarely seen bartop model of the SUCCESS. The top of the machine is the same as the pedestal model, mounted on a four-legged swivel base for the bartender to turn and confirm a win. Award card is for operator W. R. Bartley of Butte, Montana. Mills Novelty created these special award cards for their customers in their in-house print shop in Chicago, provided you bought enough machines to warrant the effort.

Mills Novelty BEN FRANKLIN, 1902. Take an UPRIGHT CARD MACHINE, clean up its working parts and add a gum column and you have a legal vending machine. Mills Novelty patented their gum slide improvements, which they stuck on the left side of their BEN FRANKLIN, adding old Ben in a casting on the front. It's the only member of the boxy PERFECTION lines made by anybody that is readily identifiable by name. That wonderful gargoyle push lever casting is still there.

Mills Novelty SUCCESS No. 6 "Little Success," 1902. Another year, another pattern, another production run, and SUCCESS No. 6 is born. No. 6 had the biggest production run to date and was a prime saloon piece. The award card offers "Free Drinks." Some examples had the Mills Paris Exposition decals, seen on front, and the Mills "Owl" logotype on the side. Sometimes these graphic elements are seen where they never were in real life as both had been reproduced.

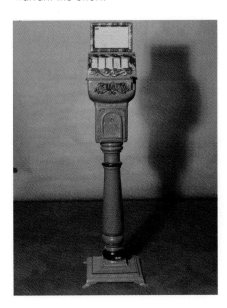

Hamilton THE HAMILTON (SUCCESS), 1902. "Iron Card" machines are usually grim, mostly basic black, with maybe some trim. But not the Hamiltons! The Hamilton Manufacturing Company in Hamilton, Ohio, in coin machine country just north of Cincinnati, made many things, including trade stimulators. They made a SUCCESS and stuck the name THE HAMILTON on the painted pedestal and case. This is the locked cash box model. They also made "Automatic Dump" and counter models.

Canda Manufacturing JUMBO GIANT, 1902. When Leo Canda Company became Leo Canda Manufacturing Company in 1902, they kept the machines that were tops in their line, JUMBO GIANT included. It is an advanced development of their earlier GIANT CARD of a decade earlier, with jumbo size cards stuck high. Mills Novelty also sold it. Nameplates are often missing. Most were found in original locations, one in an old saloon turned antique shop along the Ohio River in 1978.

Mills Novelty KING DODO (Three-Way), 1903. Named after a briefly successful Broadway musical comedy production, with well endowed showgirls cast into iron on the sides looking all the more alluring once plated in copper. Mills KING DODO was made in two models. This three-way model allows you to select the size of your bet—nickel, dime, or quarter—to make it more substantial if you feel lucky. Award card has different payout schedule for each coin.

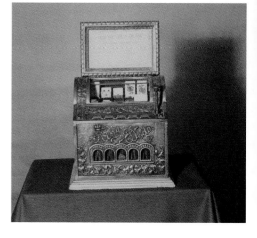

Mills Novelty KING DODO (five-Way), 1903. The five-way version of KING DODO has the same elegant cast cabinet, only this time it plays for nickels or pennies with the player picking a hand to win by selecting one or more of the five coin chutes. On both machines the last coin played is visible in the windows below the coin chutes. KING DODO is highly desirable as a collectible because of its castings. Only a few have been found.

Mills Novelty RELIABLE, 1903. The most elaborate and possibly the best looking car flipping poker machine. Play nickel in the "Straight Poker" chute at right. If you lose b still have a good hand to fill to in the five cards flipped into viewing position, play another nickel in the "Draw Poker" chute a left to spin a reel with a hand pointing to th card that was drawn. If you can make a winning hand with the draw card, you get cigars. Only one known.

Canda Manufacturing AUTOMATIC CARD MACHINE, 1903. Canda version of the ROYAL TRADER and the last Canda card machine. With trade stimulators going cast iron, Canda stuck to their sheet metal guns. What they had done over the years was convert their wooden cabinets to metal ones, eschewing castings, only to face the tough competition of classical plated cast cabinets. Canda decorated their flat surface machines with decals; birds, bees, and logos.

Mills Novelty SUCCESS No. 8 "Little Success," 1903. Last time around for the SUCCESS, suggesting Mills made so many No. 8 machines the inventory lasted well i the 1920s and the beginning of Prohibitior Mills literature states that SUCCESS "...ree can be changed very quickly and easily so to give different percentages of profit on th goods given out" with multiple award card to match. Greatly simplified from the mode of the 1890s.

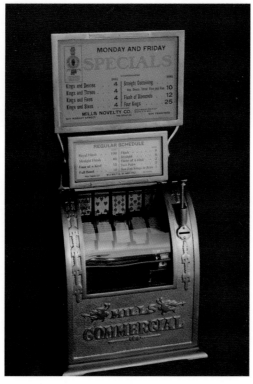

atling THE CLOVER (three-Way), 1903. ltimate development of the pedestal card achine. THE CLOVER was developed by anda, sold (and probably made) by awson Machine in Newark, New Jersey, d D. N. Schall in Chicago. When brothers m and John Watling bought out Schall, ey added THE CLOVER to their Watling anufacturing Company. Three row show s three players play, an idea also used on e JOCKEY. Watling also made THE LOVER PINOCHLE, yet to be found.

Mills Novelty COMMERCIAL, 1904. One of the most popular card flipping machines of all time. Mills Novelty picked it up from W. H. Clune in Los Angeles, and proceeded to take over the market for the COMMERCIAL, soon selling them back to Clune for his own operating routes. Some are found with W. H. Clune award cards. Many have Mills Novelty San Francisco branch office award cards with its 907 Market Street address. Sold all over the country up to World War I.

Mills Novelty SPECIAL COMMERCIAL, 1904. There are probably more varieties of Mills COMMERCIAL than any other single cast iron card machine. It was simple enough to make a new version. Just print a new card for the top frame, and make a special award. Mills Novelty also printed them in wads for their operators. This one offers a pint of wine or $2.50 in trade if you got a full house of ten or better. The "Special" part is the added cash drawer, often on a stand.

A. Feinberg SUCCESS, 1904. All the "iron card" machines made in Rochester, New York, look a lot alike. They must have passed the patterns around, or the foundry ran off batches for numerous customers on the sly. Award card says this machine was "Manufactured by The A. Feinberg Company." Its pedestal looks similar to a Columbia, or even the earlier Tibbils SUCCESS, although the Feinberg version is taller with a different bulging case box in the base.

Mills Novelty THE TRADER, 1904. The Mills Novelty version of the six-way Royal Novelty Company ROYAL TRADER. Mills kept the great looking cast iron cabinet, and even made it look better by adding a well shaped lady with what appear to be outstretched wings across the front in place of the company name lettering on the California version. Mills also changed the name to THE TRADER. The final touch is color: Mills plated the machine in antique copper.

Mills Novelty DRAW POKER, 1904. Mills Novelty made DRAW POKER first. Design is simple, with flowers in the lower front casting, numbers below the pushbuttons and a Mills logo owl in the header. "Two Plays For One Nickel" award panel explains the game: "1. Drop nickel and pull lever down. 2. Hold any cards you wish by pressing down corresponding button on top. 3. Prizes paid only on SECOND PLAY WHEN NICKEL SHOWS." Mills name is in the base casting.

Caille HY-LO, 1904. Another version of the draw poker theme, in a great looking Caille cast iron cabinet. Play a coin in the far right "Play All Slot" and pull down the handle and all five reels spin. To improve the hand you can play any one or more of the slots well above each of the visible five cards and get one more handle pull. Only one. Caille Bros. got a patent on the game. Mills Novelty copied it as the MILLS HY-LO but completely changed the cabinet.

Mills Novelty SUPERIOR, 1904. Many regard the wood cabinet Mills SUPERIOR trade stimulator as the most elegant card machine ever made. When a few showed up just as collecting got started in the late 1960s they were thought to be fairly common. They aren't. Few have shown up since. SUPERIOR also has an integral printing press. Winners get a preprinted fortune ticket with the award imprinted as it is dispensed. Troublesome, it is usually stripped out.

Caille GOOD LUCK, 1904. The "Round Top" LITTLE PERFECTION gets a new lease on life from Caille Brothers Company in Detroit. Caille started out with their version of the Canda as PERFECTION in 1901, and then redesigned it in 1904 and called it GOOD LUCK. It has refinements the Mills and Canda versions never had; a tongue-and-groove cabinet, typical elaborate Caille top casting and marquee, and a new GOOD LUCK front decal showing the name over a horseshoe.

Mills Novelty HY-LO, 1904. It is interesting to see how the various manufacturers handled similar machines to get around patents and someone else's proprietary products. The Mills Novelty HY-LO is essentially the same machine as the Caille HY-LO, except everything has been turned upside down. On the Mills the coin slots are below the reels, coin windows below that, just the opposite of the Caille. The Mills handle is low, the Caille high. Same, but different.

J. L. Foley DRAW POKER, 1906. Chicago machinist John L. Foley re-engineered the Mills Novelty DRAW POKER, and made it in his shop at 69-71 W. Jackson Boulevard. All the cast parts carry the Foley "F" mold identification. The new cabinet has scrollwork at the top and bottom and a seashell device in the header. The header is offset to the left, centered over the pushbuttons. He seemingly sold the idea to Watling, a block away, and Caille in Detroit. Few found.

Caille GLOBE, 1906. One of the last heavy cast iron card machines, and what a beautiful swansong it is to a departed class of trade stimulators. The cabinet castings alone are a work of art. Machines like this should be in museums of contemporary and early twentieth century art. GLOBE was also one of the last machines to offer its awards in cigars. The GLOBE name is in the front casting, and the Caille "CB" logo on the sides and back of the marquee.

Caille DRAW POKER, 1906. Shades of the Foley DRAW POKER, even to the "F" markings on the castings. Watling DRAW POKER also has them. Name panel at the bottom blank. Header and award card are incorrect here. They are Mills parts, with mixing and matching occurring at the time, and by later restorers. The Caille version was also sold by Silver King in 1917, and Buckley in Chicago made it without a header in 1934. Neither of the later, latter two have ever been found.

Mills Novelty VICTOR, 1907. As Mills Novelty swept up many of the California game manufacturers their machines joined the Mills line under their same names in new cast cabinets. The difference between the Reliance Novelty VICTOR and the Mills version is primarily the cabinet castings, with the "Mills Novelty Co." name on the front. These card flipping "card drop" machines rarely endure the years well, and generally need a complete restoration.

75

Caille MAYFLOWER "Style A," 1910. First made by the Puritan Machine Company of Detroit in 1905, the MAYFLOWER has a clever split cash box that tosses every fifth (or seventh, depending on how specified) coin into a separate toke for the operator, with the merchant having a key to the larger box to take the money as earned. Caille made it in four models, with "Style A" having card reels. "Style B" has fruit reels, and "Style C" and "D" have numbers and colors.

ROYAL CARD MACHINE, unknown manufacturer, circa 1920. Revamp, rebuild, from scratch or however, there are a number of card machines from the teens and twenties that have no identification, leaving them nameless. The production processes are advanced, and suggest new cabinetry utilizing old mechanisms, or new machines built from the ground up. As of now a number of them are "Mystery Machines," and await the discovery of examples with names to clarify.

Mills Novelty LITTLE PERFECTION "Flat Top," 1926. The Mills penchant for continual upgrading and manufacturing simplification finally caught up with the LITTLE PERFECTION "Round Top" in 1926. This time it is in a flat-sided boxy oak cabinet replacing the bent wood veneer used on the older model. An incredibly heavy marquee of cast iron holds the award card, with prizes "In Trade." This example was found in a Hart, Michigan antique shop in 1994.

Mills Novelty NEW JOCKEY "Plain Jockey," 1928. Time caught up with the Mills JOCKEY in 1928 with a completely redesigned cast aluminum front showing a horse racing scene in color. The greatly simplified straight lines cabinet, paper, and marquee styling is late 1920s semi-Deco. Also called "Aluminum Jockey." Inexplicably Mills retained the curved glass, an expensive production feature. They must have had inventory. Rarer than the older models, but not worth as much.

F. W. Mills JOCKEY, 1917. JOCKEY had one of the longest tenures of any card machine. It was made by half a dozen producers. One of the best looking was by F. W. Mills Manufacturing Company of Chicago just before the United States entered World War I. The F. W. Mills (Frank Mills, the brother of Herbert S. Mills of Mills Novelty) JOCKEY shows a hand of cards on the front, stresses three-way play with large numbers in color. Also found with Caille decals.

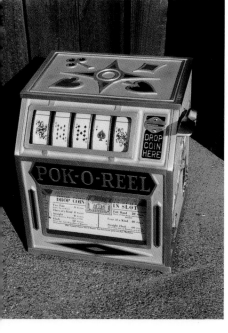

etchen POK-O-REEL, 1932. Translating card machines of the past into the minum Deco counter games of the 1930s 't take much more than new cabinets flashier display. Naming a game was of the process, with POK-O-REEL a ect handle for a poker playing five reel hine. Groetchen Tool POK-O-REEL with ig name across the front was a runaway ess and the pattern for many games to e. The name was retained for years in er models.

Daval JACKPOT CHICAGO CLUB-HOUSE, Model No. 3, 1933. For all the success that Chicago based Groetchen Tool had with their POK-O-REEL, competitive Daval also located in Chicago did far better with their remarkable CHICAGO CLUB-HOUSE. The games hit the public fancy just as Repeal brought back beer. The game was made in multitudinous models. Model No. 3 has a deep coin divider base and a jack pot, and came in poker or fruit reels.

Daval JACKPOT CHICAGO CLUB-HOUSE, Model No. 5, 1933. Model differences are a matter of add-ons and features. Model No. 5 is the same as Model No.3, except it doesn't have the coin divider base. If a player hit the three deuces that earned the jackpot, the merchant used a key provided by the operator to unlock the pot with the lock on its side. The coin viewing window on the side tells the merchant what coin was played; penny, nickel, dime, or quarter.

Happy Jack Company HAPPY JACK, 1932. The Depression 1930s were years of great experimentation in counter games, with some models far too complicated to survive. HAPPY JACK made by the Happy Jack Company in Glendale, California, is one. It shuffles the cards and deals two hands for two player play. It was known from only one ad that appeared in August 1932. Then miraculously one was found in a California garage sale in January 1991.

Buckley PILGRIM VENDER, 1933. In their own quiet way the Buckley Manufacturing Company of Chicago knocked out more "Plain Vanilla" counter games than all other makers combined. You can confirm it by the survival rate. If it's a plain PURITAN or PILGRIM type with no maker's name, it's a Buckley. The production runs were enormous, with private labels for other makers included in the runs. PILGRIM is the five reel card machine version of the PURITAN.

Daval "Gold Medal" CHICAGO CLUB-HOUSE, Model No. 7, 1934. In a clever marketing ploy, Daval gave themselves a Gold Medal "For Engineering Achievement" in the CHICAGO CLUB HOUSE, ostensibly awarded by their New York Distributors in November 1933. The award stated that "The machines manufactured henceforth should bear a replica of the Gold Medal." The initial match plates were made with a circular area below the coin chute. The Gold Medal emblem decal quickly wore off.

Groetchen 21 VENDER, 1934. From meager beginnings counter games quickly got sophisticated. 21 VENDER is a blackjack player, and complicated enough to get a patent. Drop coin, push handle down and the first two card rows show up. The other three are covered. Get blackjack with an ace and face card and you win big. If you are low, press down the third reel lever and get hit. Wins are 21, 20, and 19. Get up to three hits, but go over 21 and you bust. Fun game.

Pierce JACKPOT THE NEW DEAL, 1933. Pierce Tool and Manufacturing Company of Chicago came into its own in the mid-1930s with a series of well designed and somewhat unique counter games. THE NEW DEAL was named after a line that president Franklin Delano Roosevelt used in a 1932 campaign speech, offering a "New Deal" to the American public. Window display masks came with the game to convert to three reel fruit play or four reel "Mystery." Plain and jackpot models.

Daval "Gold Medal" JACKPOT CHICAGO CLUB-HOUSE, Model No. 6, 1934. "Gold Medal" CHICAGO CLUB HOUSE counter games started with 1934 deliveries. Not much else changed in the line, with the machines looking much the same except for the front casting. The exception was the JACKPOT CHICAGO CLUB HOUSE which replaced the lash-on side jackpot with a better looking pot that also held many more coins to stimulate play. Remains of the Gold Medal emblem can be seen.

National Coin Machine Exchange DRAW POKER, 1934. Shades of the past, the DRAW POKER was made by the National Coin Machine Exchange of Toledo, Ohio. Game maker Charlie Jameson was an old time operator from the early 1900s, and copied the features of some of the early Mills Novelty poker players. Put in a penny in the chute at top right that says "Play first coin here." To draw and improve the hand you add pennies to the five slots above the respective cards.

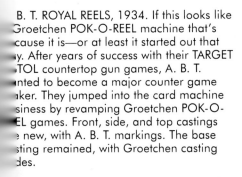

B. T. ROYAL REELS, 1934. If this looks like Groetchen POK-O-REEL machine that's [be]cause it is—or at least it started out that [wa]y. After years of success with their TARGET [PIS]TOL countertop gun games, A. B. T. [wa]nted to become a major counter game [ma]ker. They jumped into the card machine [bu]siness by revamping Groetchen POK-O-[RE]EL games. Front, side, and top castings [ar]e new, with A. B. T. markings. The base [ca]sting remained, with Groetchen casting [co]des.

A. B. T. ROYAL REELS GUM VENDER, 1934. In matters of cabinet design, A. B. T. outdid Groetchen. The Groetchen POK-O-REEL GUM VENDER has its gum vending column built into the left side of the cabinet, with a plain design across the panel. ROYAL REELS GUM VENDER does the same thing, except A. B. T. went crazy with the poker motif with fanned cards across the side to make as elaborate a casting as you'll ever see on a 1930s card machine. Wow!

Jennings CARD MACHINE, 1934. Talk about "Plain Vanilla," O. D. Jennings and Company of Chicago made card machines so plain and wooden boxy, collectors first thought they came out of the early 1900s rather than the 1930s. Even the name of their first venture is plain: CARD MACHINE. Jennings really didn't want to be a counter game maker, but they needed the machines to fill out their line and keep their distributors in line. Nothing unusual here, just the maker.

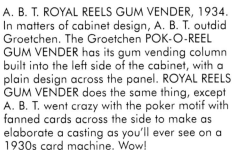

Pierce JACKPOT HIT ME, 1934. Pierce Tool had high hopes for HIT ME before they were slapped with the Groetchen gripe, and even produced a JACKPOT HIT ME model before they were shut down. They obviously took shortcuts, because the starry American flag cabinet originally projected, while it never showed up on the front, was at least made in side panels. It can be seen on the side of the JACKPOT HIT ME, in colors non-suggestive of the flag.

Pierce HIT ME, 1934. Advanced blackjack development of the poker player Pierce THE NEW DEAL with hit features. Actually, it was a direct copy of the patented Groetchen 21 VENDER format. Pierce advertised it in a star studded American flag cabinet, but when it finally came out it looked a lot like THE NEW DEAL. Groetchen was so riled over the copycatting and shuttered reels, they sued, and won. Pierce Tool had to cease and desist. Rare game.

Pace THE NEW DEAL JAK-POT, 1935. Pierce Tool seemed to lose heart for their card machines once they were reprimanded (maybe it was part of the Groetchen settlement), at which point Pace Manufacturing Company of Chicago picked up the pace and proceeded to make the same games, only a little bit different. The Pace NEW DEAL has a lower molded nameplate on top (the taller Pierce version is removable), a swing down gum ball dispenser and some inside mechanical changes.

Pace CARDINAL JAK-POT, 1935. The Pace jak-pot models ape the Pierce versions, only the jackpots are different in both name and action. On the Pierce machines the JACK POT has a trap door drop, but on the Pace machines the JAK-POT swings up and out. Both are key operated by the merchant to pay the player, with the lock on the bottom. The side panels on the Pace CARDINAL JAK-POT are virtually identical to the short-lived Pierce HIT ME.

Jennings 21 BLACK JACK, 1935. Back to th simplicity of the Jennings wooden card players. 21 BLACK JACK is made in the same tradition as the Groetchen 21 VENDE and Pierce HIT ME, but it didn't blow on Superman's cape by directly copying the patented Groetchen machine. It is simpler. The last three reels are indeed shuttered, and are released one at a time as selected by a single push down lever at the far righ of the reels. Penny version has a gum vender.

Pace CARDINAL, 1935. At last, the American flag cabinet sees the light of day. From Pace, not Pierce. THE CARDINAL is a conventional poker player that you could almost mistake for the Buckley PILGRIM if it wasn't so flashy. The front casting is a marvel, and if it were in red, white, and blue you'd want to salute. One interesting variation between the Pierce and Pace machines is that the Pierce brakes are located below the reels, whereas the Pace are above.

Reel Profits YOUR DEAL, 1967. These machines have amazed their owners. They often believe they have something made in the thirties, and are surprised that the date is 1967 and a few years thereafter. Insert coin, pull handle, and then push levers up one at a time to stop the reels for poker hand, high cards, whatever. Hard to do. Maker is Reel Profits, Inc. of Englewood, Colorado. Cast aluminum cabinet is color coded by coin. Nickel play is red, dollar green.

National Coin Machine Exchange DRAW POKER GUM VENDER, 1935. A gum vender was added to DRAW POKER a year later to create the DRAW POKER GUM VENDER. Marquee at the top was also added to show players where to put their coins for the draw cards. Maker Charlie Jameson was a member of Detroit's Purple Gang and a bootlegger in the 1920s, and built a summer home on Stony Lake at Shelby, Michigan. After he died, I bought it. It's the ultimate coin-op collectible.

A. O. Buchanan THE AUSTRALIA, 1937. Y are not looking at an artifact from the turr the century. THE AUSTRALIA is a 1937 product of game maker A. O. Buchanan Sydney, Australia, and an almost exact duplicate of the Mills Novelty THE TRADER 1904. Buchanan made the games off and on until new poker machines were made i Australia after World War II. The West Belconnen Leagues Club museum has one on display if you're ready to go "Down Under" to see it.

80

DIAL

The dial trade stimulator and counter game is a variation of the wheel machine, but rarely runs free wheeling without a powered start. The flat display side of the wheel, or wheels in the case of multiple dials, is generally decorated with display symbols or award markings on the face, with these markings often placed at the edge near the periphery of the wheel. The player sees the action, and near misses, as the wheel or wheels spin to a stop. The added weight of the graphics, or wheel coordination needed by a multiple disk machine, requires a mechanical drive. So dial machines generally accept a coin to release their mechanism, with a lever or handle pushed or pulled to set the wheels in motion. Prior to the introduction of electronics and solid state displays, the dial machine remained a leading format right up to the end of the active life of the mechanical counter game.

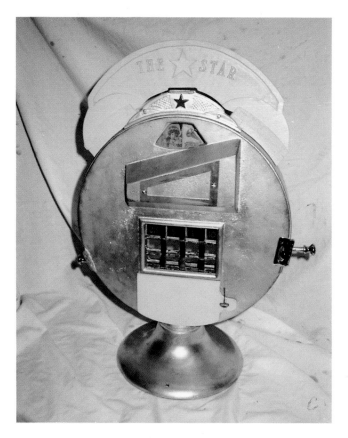

Pearsall and Finkbeiner "FIRE ENGINE", 1898. First THE (RED) STAR found, long remaining unidentified. Graphics on the wheel behind the viewing window are very old, yet it didn't seem possible the machine could be dated from the 1890s. One picture is a horse drawn fire engine, which gave the machine a tentative name as "Fire Engine." The fact it plays penny, nickel, dime, or quarter in a very advanced looking coin acceptor suggests a 1930s counter game. Nope. 1898.

Pearsall and Finkbeiner THE (RED) STAR, 1898. Collectors knew it was old, but it looked modern, even in cast iron. At the same time a fairly complicated number wheel patent credited to Pearsall and Finkbeiner in Syracuse, New York, made the rounds among collectors to see if anyone could identify the machine. Collector Tom Gustwiller, not paying particular attention, just happened to look at the patent one day, then his machine, and put it together.

An initial form of dial machine was the partially visible dial WHEEL OF FORTUNE machine introduced by the M. O. Griswold & Company of Rock Island, Illinois, along the Mississippi River, in October 1893. Griswold was one of the first of the broad line trade stimulator makers and can be compared to its contemporary Clawson Slot Machine Company in Newark, New Jersey, and in fact copied or produced modified versions of a number of Clawson machines. The Griswold approach was to mount a free spinning cast iron flywheel on a shaft, given a spin by quickly depressing a front lever once a coin was dropped. Controlled by a stop, the wheel would display a color or a number in the front display area which would determine the award, if any. Variations of the Griswold WHEEL OF FORTUNE were made for more than a quarter century, with BLACK CAT introduced in 1895, followed by THE BIG

Watling GOOD LUCK RACE HORSE, 1931. Dial machines are very 1930s. The later game makers liked to hide the wheels, in direct opposition to the early machine makers who liked to show off the honesty of the wheel. Watling counter games are rare as they only made a few. Set the pointer at the horse you're betting to win, put in a coin and down with the crank. The odds are six to one. Win, and you get your penny back. And whatever else the merchant offered for a win.

THREE in 1901. When the firm was reorganized as the Griswold Manufacturing Company, the machine came back in 1905 as the STAR with a fully visible dial wheel, and continued to be produced as a wheel format trade machine well into the 1920s.

The format returned to the market as a number dial counter game in 1933 in a cast aluminum cabinet machine called SOLITAIRE made by the Groetchen Tool (and Manufacturing) Company of Chicago, Illinois. The Groetchen SOLITAIRE and its fruit symbol gambling GOLD RUSH version of 1934 were both temporary aberrations as the Chicago firm sought new ideas out of the past, literally reproducing the earlier Griswold machine. Somewhat similar machines were created by the Watling Manufacturing Company of Chicago in 1931 with the introduction of their GOOD LUCK RACE HORSE and GOOD LUCK GYPSY countertops. None of these later versions were very successful, keeping these Groetchen and Watling machines in short supply.

The dial machine came into its own in the 1930s with the introduction of multiple dial models that acted much like the reels of the Baby Bells and novelty reel machines. The beginning of the genre was the FAIR-N-SQUARE made by the Western Automatic Machine Company of San Francisco in 1933. Basically a Fey machine, the Charles Fey and Company of San Francisco soon revealed its birthright with the PLAY AND DRAW of 1934, SKILL-DRAW of 1935 and THREE CADETS of 1936. Fey enthusiasts and licensees also made the games, with A. B. T. in Chicago picking up THREE CADETS in January 1936 and SKILL DRAW later in the year, followed by the Exhibit Supply Company, also of Chicago, making SKILL DRAW in 1937, and adding their own similar RED DOG and TURF TIME games in 1937 and 1938 respectively.

Multiple dial machines became a standard counter game format in the 1930s due to their action characteristics and varied display capabilities. Formats were both vertical, such as the Pierce Tool and Manufacturing Company FOUR LEAF CLOVER made in Chicago in 1934, and horizontal, such as the lengthy flat plane COMMERCE machine made by Sheffler Brothers, Inc. of Los Angeles, circa 1932, with a wide variety of model variations between. Machines can be simple single stop, or programmed 1-2-3 or more multiple stops to extend the suspense of the final score. They endured as long as counter games were popular, with later machines such as the 1940 MATCH COLOR and DIZZY DISKS machines made by Acme Game Company of Benton Harbor, Michigan. With the demise of counter games after WWII the mechanical dial machines

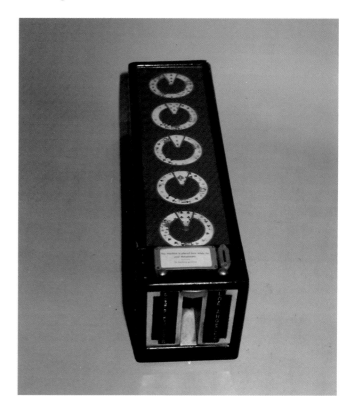

Sheffler Brothers COMMERCE, 1932. A full hand of five cards, in five viewing dials. A truly unique game, with nothing else like it ever made again. Sheffler Brothers, Inc. were Los Angeles, California, coin machine distributors and operators that went into making their own counter games and scales. That idea didn't last long, and they were out of the counter game business in a hurry. That makes this machine a rarity. This machine is 20" long. A bar filler.

didn't come back, although their display formats have reappeared in some of the solid state poker display and bartop card games that have started to reappear in pouring spots across North America. The format is so basic it will probably never disappear.

But specific games have over the years. One of the most sought after trade stimulators of all time is a vertical dial machine in a standing box-like cabinet. Called FUNNY FACES, it was created in January 1898 by Levi W. Yaggy of Lake Forest, Illinois. Yaggy was a local doctor who was also interested in mechanical contrivances. FUNNY FACES was created as a cigar machine that spun a series of circular dials with facial components, clicking to a 1-2-3 stop to create a face from a wide variety of offerings, usually quite distorted and funny. If the right rings lined up and a contiguous face was created, the highest cigar awards resulted. For less than perfect countenances, you got nothing. Not so if someone were to find FUNNY FACES today. As yet undiscovered, it is estimated that the machine will sell in the low five figures. With luck, someone will find it at a yard sale some day. I for one hit a lot of yard and garage sales in Lake Forest, a nearby Illinois suburb of Chicago, in hopes of facing FUNNY FACE. It hasn't happened yet, for me or anyone.

Groetchen SOLITAIRE, 1933. Groetchen Tool and Manufacturing Company of Chicago made a big thing out of their SOLITAIRE as the return of an incredibly profitable trade stimulator. Guess which one? It's the 1893 Griswold WHEEL OF FORTUNE in an aluminum cabinet, with only a few numbers showing through the window at the top. That makes it a dial. A SOLITAIRE GUM VENDER was also made with gum column at far right. Four different award cards came with the game.

Pierce Tool WHIRLWIND, 1933. When Chicago slot maker O. D. Jennings introduced their LITTLE DUKE payout with three disks replacing reels, it was revolutionary. No slot maker copied it. Pierce Tool and Manufacturing Company of Chicago did in both mechanical (shown) and electrical Deco counter games. To their dismay. It was a constant problem, quickly changed to three reels with the same name. Electrical version shocked players. Jackpot paid by unlocking pot.

Groetchen GOLD RUSH, 1934. Groetchen took SOLITAIRE a step further with a single symbol window in front of the wheel with a drum spinner above with its own, smaller window. The effect is two spinning elements behind the windows. For a chance symbol machine called GOLD RUSH they put two fruit symbols on each position of the wheel and one each to the drum, giving a three symbol show. Rarest is the GOLD RUSH CALENDER, day on top, date below, match and win.

Rock-Ola RADIO WIZARD, 1934. Griswold WHEEL OF FORTUNE type machine with a display window that shows a full hand of cards. The ribbed wheel drum that clicks to a stop can be seen at the bottom of the window. Seemingly simple, there is a lot of patented mechanism behind the plain RADIO WIZARD front. Penny, nickel, dime, quarter. Looks like a contemporary radio so it won't attract the attention of the law, and sounds like a carnival wheel when it's played.

Fey THREE CADETS, 1936. A transitional machine. Original Fey square cabinet format for a disk spinning counter game with the substitution of the A. B. T. coin slide for a push down handle. THREE CADETS works the same as SKILL DRAW, except there are only three spinning disks and no hold-and-draw feature. What you spin is what you get. It's a match color game, with yellow the bad guy. You want white, with the target. Also made by A. B. T. who added a fruit symbol version.

Fey PLAY AND DRAW, 1936. All five disks in line in a fashion similar to the Sheffler Brothers COMMERCE of 1932, except they are left-to-right to keep the show close to the player. Coin in, plunger down, dials spin, and hold buttons re-spin what you want. Only one found, mechanism only minus the case, possibly the prototype. Collector Bill Whelan made a case in keeping with contemporary Fey design. Date is guessed although it may be earlier as it matches 1932 formats.

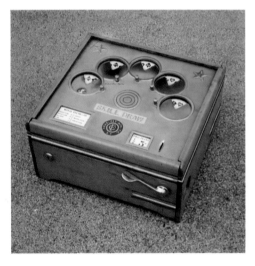

Fey SKILL DRAW, 1935. Charles August Fey, San Francisco inventor of the LIBERTY BELL slot machine, established a long enduring counter game format with SKILL DRAW in 1935. Drop coin in, and five disks spin to click 1, 2, 3, 4, 5 to a five dial display. It's a poker hand with the spot symbols only, with the card numbers in the suit graphics. Keep the handle down and push buttons to draw better cards. Simple and exciting. Completely new hold-and-draw format.

Fey SKILL DRAW, 1936. Fey quickly commercialized SKILL DRAW by modernizing it with an A. B. T. coin slide. Now it's coin slide in, spin the dials, and push the hold buttons. Fey got away from the older square shape and opted for a horizontal rectangle. The A. B. T. coin slide is a far more reliable and durable coin actuator than the dated handle, with the result that other manufacturers clamored for the Fey product to add to their own lines.

Exhibit Supply SKILL DRAW, 1937. Exhibit Supply Company in Chicago cleaned up the Fey SKILL DRAW design by moving the coin slide to the center, and making the whole top a single aluminum casting with an award card in a circular frame. The coin view and tilt windows are also placed elsewhere. The play is more exciting as the graphics don't give away the secret of spinning disks behind the windows. A. B. T. of Chicago also made the game.

Exhibit Supply RED DOG, 1937. The Exhibit Supply Company RED DOG has the same basic format as their SKILL DRAW, except no skill and no draw. RED DOG is a hard hitting card game that was (and probably still is!) favored by a younger crowd that drank beer at home and played poker on the dining room table. Push slide in to deal "House Card" at left. Pulling out spins a hand of four. Get two higher cards in the same suit as the house card and you win. One known.

Acme Game MATCH COLOR, 1940. Cast aluminum cabinet, two inside disks, dial window, coin slide (pushed in from the top) and an anti-slug window. The simple lines of MATCH COLOR made collectors think that these were '50s games, yet they are pre-WWII. Most of them have been found in Michigan where they were made in 1940 by Acme Game Company of Benton Harbor. Match the upper and lower dials and you win cigarettes or trade. Various dials and award cards determine play.

Star Amusement SPARKY, 1952. Fey card spot spinners return years after their prime into a market that had little or no place for counter games. The elegance of hold-and-draw or playing against the house is long gone. In its place is sudden death playoff. Coin in, slide in, and pulled out with a poker hand spun to a 1, 2, 3, 4, 5 stop. What you see is what you get. No more, no less. Star Amusement Company of West Columbia, South Carolina, made this game for years.

DICE

If you think that human nature has changed over the years, and that different cultures always react in different ways, think of dice. Seven-Come-Eleven, Little Joe, whatever. Dice. If you see them pictured on a large billboard for a modern riverboat casino or a land based gambling hall they move you in a certain way. Taking chances. Winning money. Beating the tiger, all of that.

So don't think you're different. Those same dice in a banner over a gambling table in the Wild Wild West, and painted on cloth at a desert oasis or a gambling den in North Africa thousands of years ago, or maybe in Rome at its height, moved people just like yourself the same way since time immemorial. Dice have been with mankind for a long time, and have spurred tens of thousands of players since they sprang upon the world stage.

Schloss and Company DICE TOSSER AND CIGAR CUTTER, 1889. Practically a nuts on duplicate of the 1889 Clawson Slot Machine Company DICE TOSSER No. 1 made in Newark, New Jersey, as the first bartop device made by the firm, Schloss may have been an agent for the Clawson dicer . It has "Schloss & Company, 157 Hanover Street, Boston, Mass." painted in gold gilt on the case, so the name isn't part of the casting. But who knows for sure this long after the fact.

M. E. Moore DICE BOX, 1887. Dice machines showed up on bartops long before coin control. Often they served more than one utilitarian purpose, such as the M. E. Moore DICE BOX and cigar cutter made in Chicago in 1887. The Moore machine anticipated the elaborate cast iron cabinets to come. Up to three cigars could be clipped at once with the lever at the right, which then released a steel band to pop the pair of dice under the glass dome. It was patented in 1877.

That is one of the reasons that they were used on some of the very first trade stimulators over a hundred years ago, in the 1880s. The other reason is that they are simple to mechanize. All that is needed is a closed container that pops the base, making the dice jump, and land in a way that changes their show every time. That gave the players something to bet on, and the coin-operated dicer was born.

The fact is, there were bartop dicers long before there was coin operation. For about thirty years before coin control was applied to dice machines, non-coin versions were available for a mechanical dice toss. They usually held their dice in a glass chamber, or a number of dice in separate glass columns, so that they popped under control and provided their display in a way that could not be manipulated (unless you picked up the usually small machine and shook it and then put it back down again). All sorts of bartop dicers made their appearance in the 1870s and into the 1890s, with the Keane Novelty Company SQUARE DEAL made in Chicago in 1891 and the Bernard Abel & Company SQUARE DEAL of 1893, made in New York City, classic examples. By the late 1880s dicers increasingly came under coin control. That way the barkeeps and saloon proprietors could make a few dollars here and there as players bet against each other, and finally with award cards that offered "free" prizes in cash, cigars or beer, make the play contribute solely to the coffers of the house.

By 1890 there were dozens of formats for coin-op dicers, ranging from dice boxes and bartop dice poppers

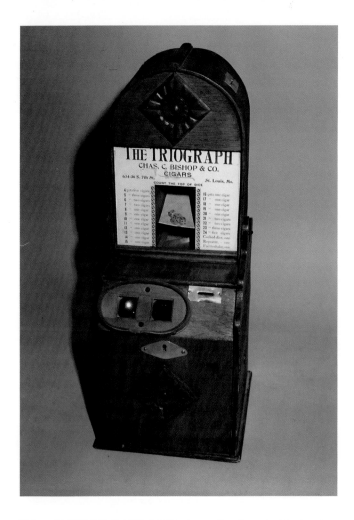

Bishop THE TRIOGRAPH, 1889. An early attempt at trying to put too much into one cabinet. For a nickel you got a shot at some cigars, always getting at least one for the throw but having a chance at up to five. You also got a peep show of a racy "girlie" picture. Maker is the Charles C. Bishop and Company of St. Louis, Missouri. Another nickel a different picture. So who cared about the cigars. But don't take your time or share with others; it flicks on and off in a hurry.

to large cabinet machines and games with an unusual degree of articulation. Numerous models, such as the John Lighton DICE SHAKER made in Syracuse, New York, in 1892, and the Charles T. Maley AUTOMATIC DICE SHAKING SLOT MACHINE made in Cincinnati in 1893, even used flat bottomed bar glasses turned upside down as the chamber, making replacement of the frequently broken dicer glass tops as easy as reaching down for a clean glass. Other machines were far more complicated, with wooden cabinets that held dice tossing cups under glass. Examples of this genre are the Clawson AUTOMATIC DICE and AUTOMATIC FORTUNE TELLER dice cup shakers of 1890, and the Griswold DICE MACHINE of 1892. One machine even had an automaton of a monkey free tossing the dice in a glass box, although the Roovers EDUCATED MONKEY of 1889 remains to be found. The same for the Coyle & Rogers ELECTRICAL DICE made in 1888, in which a simulated human hand scoops up the dice, shakes them in the hand, and free tosses them under a curved glass dome. When these machines are finally discovered they will be worth thousands as the most unique articulated dice tossing coin machines ever made.

But until that day comes we have to make do with the many exciting and unique dice tossers that have been discovered and are available to us from the 1880s up through the 1950s and beyond. As the years went by the dicers became less complicated, less expensive, and more profitable to run. The flat base wood cabinet coin slide Kalamazoo Automatic Music Company KALAMAZOO and KAZOO-ZOO dicers of the mid-1930s, and the virtually identical Quality Supply Company of Sioux Falls, South Dakota, multiple model aluminum cabinet dicers of 1949 generally paid for themselves in a few days, and after that they made nothing but money for their owners. The TWICO dicers made in 1957 carried on this tradition, and there is little question that if mechanical coin-op counter games are made in the future, dicers will be among them. They are fun to play, and their features are often unique and very clever.

GOOD LUCK, unknown manufacturer, circa 1890. Dice aficionados are aware of the vicissitudes of the game. Dice isn't a game. It's a collective. Hundreds of dice games can be played with the dies, maybe thousands. Some are gambling, some aren't. A variety of game machines were made for the variations. The attractive cherry wood GOOD LUCK was made as a non-coin poker player, tossing five dice under glass and giving the player a chance at a draw to fill a hand. One known.

Clawson AUTOMATIC DICE, 1890. America's first serious coin machine maker was Clement C. Clawson of Newark, New Jersey, starting out with an enormous musical scale. Using his 1888 coin viewing patent, he created AUTOMATIC DICE in 1890, forming the Clawson Slot Machine Company to make it. It is remarkable. Drop in a nickel and a prewound clockwork mechanism brings two red dice cups down to pick up, shake, and then free toss the dice on trays right before your eyes.

DeGrain DICE GAME, 1892. Bartop games to promote a whiskey or a cigar (rarely beer as it didn't make enough money to cover the promotional expenses until the national brands came along) were common. The 1890 three side window non-coin DeGrain DICE GAME made in Washington, District of Columbia, was replaced by this 1892 model with five displays on top, and a hold-and-draw feature. Sponsor of the game is "Soelerjenlen 5 cent cigars."

Bucyrus DICE BOX, 1891. Hundreds of dice games were submitted for patents, with more rejected than approved based on their presumably base usage. The art was all but brought to an end when saloonkeeper Frederick W. Mader of Bucyrus, Ohio, got a patent for his DICE BOX in March 1892 after an October 1891 application. He applied for and got five uses: this simple hand dice popper, variations in a table, on a table, electric, and in a plunger tripped box.

Griswold DICE MACHINE, 1892. An attempt to share some of the popularity of the Clawson DICE MACHINE, this semi-copycat version was made by M. O. Griswold and Company of Rock Island, Illinois, in 1892. Every toss a winner, ranging from one package Pepsin gum, two fish hooks, three marbles, and up to ten packages of gum on this drug store model. Saloon models offer cigars. Not quite as articulated as the Clawson, six dice are popped under two inverted glasses behind glass. One known.

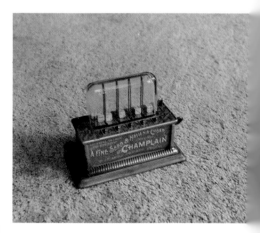

Keane Novelty SQUARE DEAL, 1891. One of the better looking dicers of the early 1890s, the Chicago based Keane Novelty Company made the elegant but quite small SQUARE DEAL as a bartop game. It may look like one of the coin-ops to come, but isn't. The graphics on the front say "A SQUARE DEAL/ Manufactured By The Keane Novelty Co. 36 LaSalle Street/Chicago" in a terrific Victorian lettering that you wish you could get on a computer font. Cast iron, hold-and-draw.

Keane Novelty SQUARE DEAL, 1892. Once the ornate Keane Novelty SQUARE DEAL took hold as a national hit it was a major piece for side panel advertising and product promotion. Designer is Frank H. Smith of Chicago, who also made coin-op trade stimulators for the Ideal Toy Company of Chicago. Colorful side advertising panel on this SQUARE DEAL promotes Champlain Havana Cigars. Turn of the side knob pops the dice. Hold levers let you pop again to improve the show.

Colby COMBINATION LUNG TESTER, 1892. Toss five aces and you get ten cigars for a penny. You always got at least one, and a chance to show the "Boys" in the bar that your lungs were stronger than theirs by blowing in the tube. They had to try, too. Colby Specialty Supply Company of Chicago was one of many companies founded by inventor John Colby to make scales, strength test, and other machines. A number of these were found in an old Michigan hotel in the late 1980s.

Lighton DICE SHAKER, 1892. Already established as a slot machine maker, and originator of floor saloon lift test machines, the John Lighton Machine Company of Syracuse, New York, expanded their lines by adding this dice shaker. It appears to be based on the original Pope design. The flat topped glass is an innovation. Dome breakage was a dicer problem, so Lighton made its replacement an ordinary beer glass stocked by most saloons, placed upside down.

American Automatic AUTOMATIC DICE, 1893. William R. Pope continued to invent dice poppers. His Automatic Manufacturing Company LUCKY DICE of January 1893 has a far more complex mechanism than the POPE DICE MACHINE. Machine parts and cabinet are made of cast metal. Actuation once a coin is dropped is by a knob on the side, sturdier than a flimsy plunger. This is the AUTOMATIC DICE version made by the American Automatic Machine Company that also produced Pope products.

Automatic Manufacturing POPE DICE MACHINE, 1892. Also AUTOMATIC DICE. This seminal dicer set the standard for the plunger dice poppers to come. Created by William R. Pope, prolific early coin machine designer, and made by his Automatic Manufacturing Company, New York City. Automatic Machine Company version has two glass domes, one protecting the other. Patent applied in February 1892, issued in July. Most dice poppers that followed owe their existence to the POPE.

Maley AUTOMATIC DICE MACHINE, 1893. Practically a duplicate of the Lighton DICE SHAKER of the year before, the Charles T. Maley Novelty Company AUTOMATIC DICE MACHINE may have been produced private label for Maley in Cincinnati, Ohio, by Lighton in upstate New York. A different treatment of the dome is apparent, with a wire bail over the top of the glass to prevent player slamming to jog the dice. Brass corners protect the cabinet.

American Automatic AUTOMATIC DICE SHAKING MACHINE, 1893. Probably the most frequently found dicer from the early 1890s, they are usually discovered in poor condition with their award paper severely degraded or gone. These machines are major candidates for restoration. AUTOMATIC DICE SHAKING MACHINE is made by American Automatic Machine Company of New York City, and was cataloged and sold by numerous other "manufacturers," agents and mail order houses. Loop plunger.

"Triangle Dicer," unknown manufacturer, circa 1895. It has the characteristics of half a dozen known dice poppers, yet its cabinet is absolutely and thoroughly unique. Only one of these great looking machines has ever been found, and its fairly good paper gives no clues to its origins. Collectors both love and hate machines like this. It's rare. It's valuable. And it remains a mystery. Collectors call this the "Triangle Dicer," and you can see why.

U. S. Novelty WINNER. One of the most long lived coin machines ever made. Cast iron cabinet and knob actuated dice popper. Created by the United States Novelty Company of Chicago in 1893, with easily assembled six piece cabinet that could be had nickel plated or in an alluring antique copper. When U. S. Novelty folded in the mid-1890s, other companies picked up the machine. Caille Brothers Company in Detroit made it as WINNER DICE between 1907 and the early 1920s.

AUTOMATIC DICE, unknown manufacturer, circa 1894. Who made it, and what is it? Practically identical to the American Automatic AUTOMATIC DICE SHAKING MACHINE with the difference being a plunger rather than a loop actuator. Otherwise these are virtually the same machines, although the bezel holding the glass dome and coin chute differ. AUTOMATIC DICE has been found in a variety of modes, with surviving paper suggesting a number of sources, none distinguishable.

"Spittoon Dicer," unknown manufacturer, circa 1894. Coin-op dicers arrived in such profusion in the early 1890s it is likely that examples will be found for years, with many never to be identified by maker and date. Without nameplates, award paper panels and corroborating advertising in the saloon and tobacco media, the origin is anybody's guess. That is the case with this orphan. The circular brass base reminds collectors of a spittoon, so that's its nickname.

"Hex Dicer," unknown manufacturer, circa 1895. One of the most marvelous coin operated dice machines ever found, this hexagonal cast iron cabinet beauty remains an enigma. At one time it was nickel plated. The cabinet is eight slab pieces; six sides, a top and a bottom, all in iron. There is nothing else like it that would suggest other products or a maker. For some reason there are more unidentified dice machines than any other class of trade stimulators.

Samuel Nafew AUTOMATIC DICE MACHINE, 1898. A John Lighton touch with the flat top glass. The unmistakable mark of the William R. Pope LUCKY DICE side knob with the coin chute in the same spot. One is about ready to identify this as an Automatic Manufacturing Company machine, except it appears in the Samuel Nafew Company catalog of 1898. Either way it is a New York City machine. It is likely that Automatic Manufacturing made it for Nafew.

E. A. M. EAGLE, 1899. One of the great classics of trade stimulator collecting, both for beauty and the mystery surrounding it. The cabinet is marked "E.A.M.," with nothing on the paper. Who? Theory has it that the maker is "Eagle Amusement Machine" and the machine is called EAGLE, which seems logical based on its casting artwork. But that's an educated guess. Player sees "One Cent" and "Nickel" coin chutes in front. Bartender sees anti-slug coin windows in back.

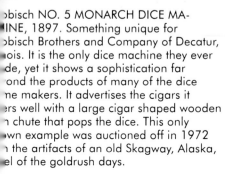

Tobisch NO. 5 MONARCH DICE MACHINE, 1897. Something unique for Tobisch Brothers and Company of Decatur, Illinois. It is the only dice machine they ever made, yet it shows a sophistication far beyond the products of many of the dice machine makers. It advertises the cigars it offers well with a large cigar shaped wooden push chute that pops the dice. This only known example was auctioned off in 1972 with the artifacts of an old Skagway, Alaska, hotel of the goldrush days.

Erickson LOG CABIN, 1898. Another discovery from the goldrush days in the Pacific Northwest. This sole example of the Erickson LOG CABIN made in Portland, Oregon, was found in the effects of an old Portland saloon when it was dismantled and sold off in the mid-1980s, just about a hundred years after the machine was made. LOG CABIN is a very substantial, with a screw assembled cast iron cabinet and a lever actuated, dice pop. Only example known.

Mills Novelty I WILL, 1900. A true classic of trade stimulator collecting, regarded as number one by some. It was called "three machines in one" because of coin chutes for each of three dice columns. Early models were dime, nickel, quarter, with later models penny, nickel, quarter. Sold for years. Collector Ed Smith found one in 1985, mint condition, still in its shipping crate, as a result of a coffee shop conversation with strangers. Its papers were dated July 7, 1912.

Ennis and Carr PERFECTION, 1904. One of the last uses of the by then trite name PERFECTION for a coin machine. Ennis and Carr was located in Syracuse, New York. When these machines were first found they were thought to be a lot newer. In spite of the modern 1930s appearance, the cast iron nickel plated cabinet gives away its age. Other period giveaways are the cigar awards, and the three holes on the push down front. This dicer doubles as a cigar cutter.

United Automatic 6-WAY DICE, 1905. One of half a dozen cities noted for coin machines, Kansas City, Missouri, primarily produced trade stimulators and counter games by numerous manufacturers over the years. United Automatic Machine Company produced 6-WAY DICE as a battery operated dice popper. Play a nickel in any number in the six-way coin head and it shows up in the glass. When the dice pop, you get as many cigars as the number that shows up, or twelve if you are lucky to get all six.

F. A. Ruff CRAP SHOOTERS DELIGHT, 1900. Similar to the F. A. RUFF cigar wheel THE DEWEY in a small counter cabinet with a side knob much like the Mills Novelty LITTLE MONTE CARLO. The Detroit, Michigan, maker first used the cabinet on this CRAP SHOOTERS DELIGHT dicer. Play a penny in the slot you select out of five possibilities, push down the lever you are playing for, turn the knob to pop the dice and hope to win from one to three cigars.

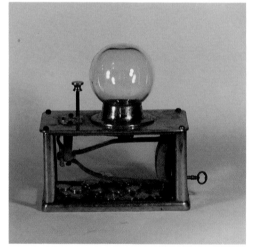

Dunn Brothers WRIGLEY DICE, 1905. Another machine that looks a lot younger than its years. Paper states "You Can't Lose" with advertising for "Juicy Fruit" or other gums and the name "W. K. Wrigley Jr. & Co." Winners get a package of gum or a cigar every pop, or two of either if the two dice match colors. Shopkeepers got the machine free when they stocked up on cases of gum. Dunn Brothers were located in Anderson, Indiana. Mechanism behind glass shows action, and cash.

Bradford Novelty THE LARK, 1907. When Bill Whelan entered the gloomy, flooded basement of the old saloon building in the Russian River area of California's gold country, his eyes adjusted to the darkness. There was a pile of rusty machines. Rusty? That's cast iron. One chunk was brown with five rust colored tubes. They turned out to be etched glass. He found, and restored, the first San Francisco Bradford Novelty Machine Company THE LARK. Two found since.

McClellan BOARD OF CRAPS AND GUM MACHINE, 1907. The dice machine that has a name that is practically a book. You drop the nickel in the slot right in front of the globe where the two red fingers are pointing, then turn the wheel on the right side to pop the dice. This is a game that easily stimulates play as a result of the largest award card ever made for a dicer. Maker William McClellan of Danbury, Connecticut, travelled around New England selling his own game.

Fey ON THE LEVEL, 1907. Charlie Fey of San Francisco was a man with a multitude of ideas, making a rash of similar machines with different play formats. ON THE LEVEL added a new level of risk taking with a single throw of the dice. The rewards were consistent with the level of risk, with the six-way head taking 5¢, 5¢, 5¢, 10¢, 10¢, and "Play Quarter" bets. This machine was all over San Francisco where a Mills Novelty rep saw it, and swiped the idea for Mills.

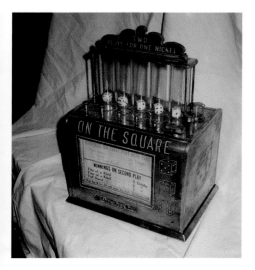

Fey ON THE SQUARE, 1907. Another example of a San Francisco five tube dicer, the Fey ON THE SQUARE may have been the first and the style leader. The Fey version adds hold-and-draw and promotes "Two plays for one nickel." Plunger is at the side. Player sees possibilities of winnings on both plays on the award panel on front; bartender sees anti-slug coin window and a brief repeat of the award schedule on back. Pushing hold buttons down holds dies during second toss.

Mills Novelty ON THE LEVEL, 1909. Mills Novelty finally ran out of the quickly modified CRAP SHOOTER cabinets (with the awkward blank spot at upper left where the single reel used to be and the plugged token payout cup on the side) and made a new cabinet for their six-way dicer and go with the original Fey name: ON THE LEVEL. The PILOT influence is still evident in the cabinet, but now it looks made for the game. It differs from the Fey in having a lever, not a plunger.

Mills Novelty CRAP SHOOTER, 1908. Mills Novelty took the single reel PILOT cabinet patterns acquired from Paupa and Hochriem, cut them down, copied the Charlie Fey ON THE LEVEL to put dice on a spinning disk under a top dome, and created CRAP SHOOTER. Play nickel, dime, or quarter with rewards based on the level of risk. Numbers, not spots, are on the die faces. Reward schedule on back is reversed so bartender can pay off based on coin played in window below.

Mills Novelty MILLS PIPPIN, 1909. Mills Novelty copied every good idea they could find, most coming from an active San Francisco rep named Matt Larkin. The Fey ON THE SQUARE and the Bradford Novelty THE LARK came in for the duplication treatment in the MILLS PIPPIN. Then Mills did something neither of the others could possibly do, creating a monumental nickel plated cast iron cabinet that shows three African-Americans playing craps and the Mills factory building.

Fey AUTOMATIC DICE BOX, 1909. You can't keep a good Fey fettered. When the San Francisco lawmakers clamped down on coin machines, Charlie Fey cranked up a non-coin version of ON THE LEVEL. Everything was the same. The cabinet, the "Read 'Em" header, the plunger, the dice and the award card, only this time the card was stuck up high. The only things that changed were the plugged coin head and the new AUTOMATIC DICE BOX cabinet paper with large words "No Slot."

Specialty Machine PORTULA, 1909. San Francisco, that hotbed of chance coin machine development, produced this unique dicer. Five ten sided poker dice deliver a hand of cards with a pair awarding "1" (cigars, drinks, or cash) up to "100" for a Royal Flush. The really unique feature is the fact there is no coin slot. In 1909 San Francisco outlawed slot machines, so Specialty Machine Works in town, among others, made slotless machines where you pay the bartender to play.

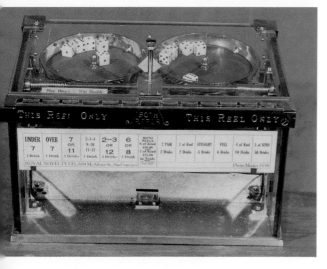

Royal Novelty DICE, 1912. 13-way play! That's right, thirteen coin chutes to pick your pockets. Two dice on the left side always paid at least one drink on under or over 7. Wins went up to two drinks for 7 or 11. On the right side five dice awarded from two to fifty drinks depending on the show, but you could also lose. But the center chute is the killer: 25¢. If you get six of a kind on both disks you get $30; seven of a kind, or all dice the same, pays $75. In 1912 dollars.

Imperial IS IT ANY OF YOUR BUSINESS, 1924. A gum vender and dice popper, all for a penny. Imperial Manufacturing Company of Chester, Pennsylvania, first made a wooden version as the IMPERIAL sometime after the company changed its name from the Automatic Pruner Company in 1911 with prizes on 3, 9, or 18. Success being what it is, they produced an updated version around 1924 with an aluminum front and header holding an award card. Prizes when dice total 3, 6, or 18.

Keystone Novelty WINNER, 1924. Perhaps the best known and most often found version of the post-Caille WINNER DICE is the Keystone Novelty and Manufacturing Company WINNER made in Philadelphia in 1924, and for at least another six to eight years afterward, stretching the game into the 1930s. The cabinet moved from cast iron to aluminum, making it lighter and even easier to assemble. Penny, nickel, and dime models were produced.

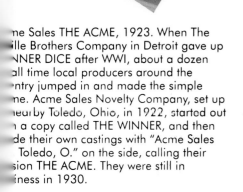

Acme Sales THE ACME, 1923. When The Caille Brothers Company in Detroit gave up WINNER DICE after WWI, about a dozen small time local producers around the country jumped in and made the simple game. Acme Sales Novelty Company, set up near Toledo, Ohio, in 1922, started out with a copy called THE WINNER, and then made their own castings with "Acme Sales Co., Toledo, O." on the side, calling their version THE ACME. They were still in business in 1930.

Star Novelty WINNER DICE, 1925. Acme Sales Novelty Company wasn't all alone in Toledo, Ohio, in the twenties. They weren't even alone in making their freebooted WINNER DICE knock off, with the competing Star Novelty Company in town making their own called an unimaginative WINNER DICE. It is a far rougher copy than the Acme Sales version, or most of the others, with the cast iron details filled in from making a direct copy of the Caille castings. Bezel is plain.

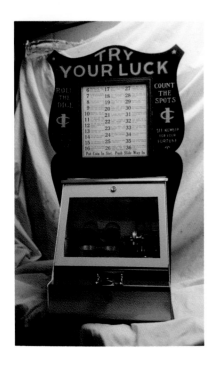  

Exhibit Supply DICE FORTUNE TELLER, 1925. Ostensibly a fortune teller, and theoretically an arcade machine, the Exhibit Supply Company of Chicago duplicated the Clawson AUTOMATIC DICE machine of 1890, right down to the clockwork and the two cup dice toss. The principal difference is that the Exhibit Supply machine is made of newer materials, with the cups of aluminum whereas the Clawson is cast iron. Machine came back in 1926 in a taller cabinet as OPERATORS DICE MACHINE.

Fey MIDGET WITH SALESBOARD, 1926. Whenever Fey made a dicer after 1920, he made a corresponding roulette version. Or the other way around; when he made a roulette a dice machine was sure to follow in the same cabinet. This is the early MIDGET cabinet with "Chas. Fey & Son SF" in the top casting, plus the coin window. Salesboard adds to the play, attached to the machine by a hinge that lifts it up for punching. Rare machine with the hinge and board. Two known.

Southern Novelty CHUCK-O-LUCK, 1926. The most widely advertised knock off of the WINNER DICE. The Southern Novelty Company of Atlanta, Georgia, took full page ads in the trade media that explained the use of the machine in detail, calling theirs the "New 1926 improved model (with) satin-finish aeroplane aluminum (and) crystal-like Lumite glass globe." How's that for making a silk purse? The Southern Novelty name is included in the bezel ring casting.

Mills Sales 36 LUCKY SPOT MIDGET, 1926. Mills Sales Company in Oakland, California, across the bay from Charlie Fey in San Francisco, was a marketer of Fey machines apparently made private label. Some of the Mills Sales machines had coin viewing windows, and some didn't. And without paper to back up the origins or owners, it is difficult to tell these machines apart from those made by Atlas Manufacturing in Wisconsin, or L. C. Graham in New York state. Fairly common.

Fey 3-IN-1, 1927. Pure Fey, yet it doesn't seem to have been sold by him at all. The Fey version was advertised by his distributor in San Francisco, Jno. R. Moore & Son. The same machine is advertised by L. C. Graham in Albany, New York, as the manufacturer. That's what Atlas Manufacturing Company in Kaukauna, Wisconsin, also said. So it was made in three places around the United States. The maker of the one you find depends on where you find it. Not common.

Monarch Sales 36 LUCKY PLAY PEE-WEE, 1927. The dice counterpart to the Monarch Sales PEE-WEE ROULETTE, both games created by Charlie Fey of San Francisco and made and sold by Monarch Sales Company of Indianapolis, Indiana. The days of drinks and cigars were waning, with 36 LUCKY PLAY PEE-WEE positioned as a trade machine, nickel play. Of six dice, two are one color and four another to shot for trade score with four while playing craps on the side with the two.

Keystone Novelty CHUCK-O-LUCK, 1927. The game didn't change, but the name did. Keystone Novelty and Manufacturing Company WINNER of 1924 developed quite a following in the mid-1920s. By 1927 it was being carried in distributor catalogs as CHUCK-O-LUCK. The rewards for play were progressive, with the nickel machine paying five times as much as the penny, and the dime ten times as much. Keystone has crisp, new patterns, with the "Keystone Novelty" name in the bezel ring.

Superior Confection AUTO-DICE AMUSEMENT TABLE, 1930. Experimentation in counter games led to making the counter itself in the AUTO-DICE AMUSEMENT TABLE. Columbus, Ohio, inventor Henry K. Renz came up with the game in 1930, found it tough to make on his own, so he sold the idea to the Superior Confection Company in town. The coin slide actuates the dice popper in the center of the table. Superior also made shuffle and bridge versions. Few found, all in Michigan.

A. B. T. 36 LUCKY SPOT, 1927. The A. B. T. Manufacturing Company in Chicago wanted to get in a Fey way in the worst way, but wanted it their own way. So Fey gave away the play of his matching roulette and dice way, while A. B. T. had their own say on the way the games would stay. A. B. T. created completely new cast aluminum cabinets with sturdy plungers actuating Fey mechanisms that stood up better on the counter. 36 ROULETTE and 36 LUCKY SPOT were the result.

Exhibit Supply JUNIOR, 1927. It looks like something from the deep, dark past of thirty years earlier. But it isn't. Promoted as a fortune teller, percentage award cards came with the machine for use as a merchandise trade stimulator. With luck you'll find them behind the fortune card on the front. Another practice was to pencil in the awards on the fortune card, as shown here. Standard format penny play, but was also made in nickel play on special order.

Keystone Sales KEYSTONE, 1931. Looking much like the top of an oil drum with dice under glass, the KEYSTONE was the first counter game attempt by the Keystone Sales Company of Chicago. The case is wood grained metal. The award card is held by a clip, sitting upright over the back of the game. Nothing fancy, just a clip. Same for the coin slide, very simple, almost primitive. Keystone Sales became Keystone Novelty Manufacturing Company, who made a roulette version.

Bowman DIXIE DICE, 1932. The Fey penchant for dice and roulette versions of the same game permeated the industry. Bowman Specialty Company in Cleveland, Ohio, created DIXIE and proceeded to roll it out as DIXIE DICE and as a roulette. The appearance is a step back in time, a wooden cabinet (with a bit of Deco trim) with a paper front and aluminum castings forming the features of the game. Set the pointer to select die number, push lever to flip to get as many as you can.

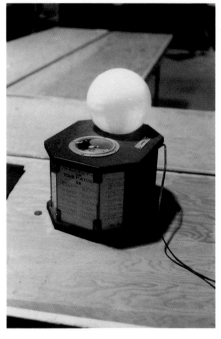

Ad-Lee FOUR WAY FROLIC, 1932. Plug-in electricity was a new stunt for counter games in the early 1930s. North American cities were electrified by 1905, but it was another quarter century before rural electrification was a reality, and a lot of stores were in rural towns. Drop in a penny, and the dice automatically pop. Electrically. Dice total adds up to a fortune on the front paper, with numbers for awards. Crystal glows all the time to attract attention.

Chicago Coin SHOOTEM, 1933. Never a big name in counter games, SHOOTEM was the first effort by Chicago Coin Machine Exchange of Chicago, a firm deeply committed to pinball games. A unique game. SHOOTEM came in standard dice or fruit dice as shown. At 15 inches long it is bigger than it looks. Suction cups hold cast aluminum cabinet to the counter. When the firm changed its name to Chicago Coin Machine Company SHOOTEM continued as their first counter game product.

Ad-Lee CRYSTAL GAZER (Fortune), 1932. A truly alluring counter game, CRYSTAL GAZER made the player feel like a fortune teller. The light glows and makes you want to try this gadget. First promoted by Automatic Coin Machine and Sales Company of Chicago in late 1931, by early 1932 it was solely an Ad-Lee machine. The game is entirely automatic with plug-in electricity lighting the globe and popping the dice. Cigarette version is 5¢ play.

New Era JUMPING JACK, 1933. The games made by the New Era Manufacturing Company of Chicago were never prize winners for beauty, probably designed by the pragmatic mechanical engineers that came up with the games. But workable they were. The first JUMPING JACK games awarded cigarettes. The IMPROVED JUMPING JACK added the machine name in the cast aluminum dice frame, with awards in beer. "Free Drinks" and "Free Cigars" were back, only now it was beer and cigarettes.

Buckley BABY SHOES, 1933. Buckley Manufacturing Company of Chicago only built three dice machines. BABY SHOES of 1933 was first, and Buckley dice games didn't come back for three years. They did need to. BABY SHOES was a big hit, and sold for years, paving the way for a wide variety of flat box coin slide dicers by other Special dice have colored dots. Standard dice were also used. Tilt the game and a "Tilted" button flips out over the dice field.

Bally BOSCO, 1933. When the Bally Manufacturing Company of Chicago made their big run for counter game domination after the success of their BALLYHOO pinball game, they turned out three different dicers in 1933 alone. BOSCO was the first. The game is a plug-in that automatically pops the dice when a nickel is dropped. One award card came with the game, cigarette payoffs on one side and beer on the other. Paid four packs for three "6." Later models paid two.

Bally DICETTE, 1933. Bally finally made a reliable dice game with DICETTE. Wooden box, coin slide, dice in deep hole and a tilt mechanism to avoid the game shaking that players tried when no one was looking. The tilter is a pedestal that stands upright in a hole. Tap or shake the box and it falls over, which voids whatever score is showing. New coin, slide in and tilter is reset, and you play again. Cautiously. Dice shaker shimmies to avoid dice pop and self tilting.

Pioneer Games BIG BONES, 1933. One of the oldest dice gambling formats known. BIG BONES is a wooden box Chuck-A-Luck cage with a coin releasing the handle to turn the cage over and drop the dice into new positions. Pioneer Games Company was off the beaten track in Minneapolis, Minnesota. Play penny, nickel, dime, with the same odds for all. The award panel is on both sides so you can play the game from either side. Not many known.

Bally BALRICKEY, 1933. Bally joined the coin the flat box dicer set with BALRICKEY, a very simple game. Bally marked it "Patent Pending," yet no example has yet been found with a patent date or number to indicate it was issued. Nor has a patent. It's something to look for. The award schedule is listed in black at the top of the gold playfield paper. Award paper scores by points with six of a kind valued at 1,900, or "Paid in trade." Uncommon game.

Ad-Lee CUBA, 1933. In the 1920s and early '30s mention of Cuba suggested open drinking and gambling, with Havana one of the offshore paradises for Americans suffering from Prohibition at home. The Ad-Lee CUBA had it all as five games in one. It came with dice, a horse race spinner, extra disk fields, balls for roulette, rubber pins, and score cards to set up for any of the five games. It was said the location could convert in minutes, but it must have been a pain.

Palamedes Sales LOTTA DOE, 1933. Everything old is new again. The top looks like a late 19th century dicer while the cabinet is a wonderment of Art Deco design and color. Palamedes Sales Company, Inc. was located in Fort Wayne, Indiana, not exactly coin machine country. The origin of the name is interesting. Palamedes is the mythical ancient Greek credited with inventing dice and the first dice games. The only game the company ever made, and not for long.

Groetchen DICE-O-MATIC VENDER, 1934. Groetchen Tool and Manufacturing Company in Chicago got into dice games late, but they did it up big with their DICE-O-MATIC VENDER. It came four ways, with standard, poker, fruit, or number dice with award cards for each. This game is not seen very often. Groetchen may have cut it short, for they never made another dice game as long as they stayed in coin machines (before they made the first post-WWII suburban outdoor grills).

Stone Brothers KENTUCKY DERBY (non-coin), 1934. In 1932 Nathan M. Stone Company in Chicago made a flat countertop non-coin dicer. Not exactly non-coin. Put your money on the horse and number 2 to 12 played on the top glass layout. Slide the lever and tumble two dice on a spinner to a total. Reformed as Stone Brothers, Inc. they numbered up to 56 with six dice and made it a coin-op. They used a variation of the new glass on this non-coin 2 to 12 two dice version.

A. J. Stephens FLIP, 1934. The perfect Rep piece, and an early attempt at an interacti game. And we do mean interactive. Make A. J. Stephens and Company of Kansas C Missouri, who churned out bartop games galore for the new taverns. FLIP looks ancient in its simple wooden box made to look like a beer case. Buy a bottle of beer stick in holder at upper left, invert your gl over the dice, put in coin, and pop. Then remove glass, fill with beer, and enjoy.

Garden City CHERRY JITTERS, 1934. Colorful game, and rare. Garden City Novelty Company of Chicago seems to have been an auxiliary of Pierce Tool and Manufacturing Company as they had the same address. They even made a number of the same games. CHERRY JITTERS is one, with Garden City making theirs in 1934 while Pierce Tool introduced theirs in 1935. Drop in coin, pull down handle, and a 6v battery pops the dice in their respective chambers. Valuable.

New Era NEW ERA VENDER, 1934. The New Era dice machines didn't get any better looking as time went by, although they did upgrade. This is virtually the same game as the JUMPING JACK of 1933, except it has a coin slide in place of the drop coin chute. The anti-slug window let the storekeeper see the coin before a payoff was made, avoiding the slugging problems of the earlier games. NEW ERA VENDER came in regular dice cigarette or fruit dice trade award.

Exhibit Supply SELECT-EM, 1934. The yea beer really began to flow was the year of dice game. Small, easily mechanized and fun, they were endemic to taverns and ba Exhibit Supply Company of Chicago stake out a claim to the genre and turned out more than anybody else in the thirties. SELECT-EM seated the format for their dicers; flat, coin window, delicate tilt, 1 to number selector, and distinctive graphics. They were all the same but looked differe

FAIR-N-SQUARE, 1934. By 1934 Charlie was an old guy, but he kept cranking out stuff with a better idea, or two, or three. FAIR-N-SQUARE "Blue Eagle Model" you [put] the coin in and an electric motor spins [and] tumbles the dice. Then you use the side [knob] to select your bet (in this case two pairs), [pull] out the slide and the action stops. The casting says: "The electric motor in FAIR-[N-]SQUARE permits your own personal [con]trol."

Kalamazoo Automatic KAZOO, 1935. It took half a century for this bartop dice game format to go away. Some bars still have them. National Coin Machine Exchange in Toledo started their line in 1934 with HAZARD. Kalamazoo Automatic Music Company in Kalamazoo, Michigan, made them larger in walnut cabinets, including KAZOO, KAZOO-ZOO, and KALAMAZOO. Camco Products Company, Inc., Grand Rapids, Michigan, started in 1939 with HI-LO. Most are common, but not KAZOO.

Pacific Amusement P.D.Q., 1935. When Pacific Amusement Manufacturing Company of Los Angeles moved to Chicago to be in the heart of the coin machine business, they expanded beyond pinball games to everything else, including consoles, arcade machines, and counter games. MARBLO, a roulette, flopped in 1934. They did okay with the electric P.D.Q., with letters on three dice and spots on the fourth. Spell P-D-Q and the spot die tells you how much you win.

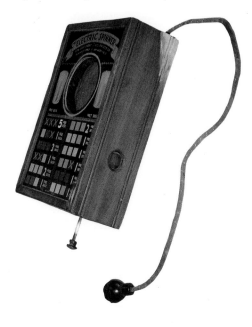

[Fort] Wayne LITTLE JOE, 1934. If this dicer [looks] familiar to you, go back to the [Archi]medes Sales LOTTA DOE of 1933. Here [it is] again, in the same town, now made by [the] Fort Wayne Novelty Company of Fort [Way]ne, Indiana. Not exactly the same. [See]ms they had a lot of slug problems with [LOT]TA DOE, not uncommon in industrial [tow]ns. Three more models were made by [Fort] Wayne, with this the definitive model [num]ber three. Anti-slugging window is on [the] left side, unseen.

Pierce PIX-IT, 1935. No mistaking what this dicer is promoting. Cigarette packages are all over the top, including Omar, Fatima, and pre-WW II "Lucky Strike Green." Nothing extraordinary here, except the dice. They are 90% of the value of the game. You select the brand you're playing for by the knob on front, slide in a coin and an electric kicker pops four dice with cigarette packages on them in color. Get two you get a pack; get all four you get six.

Pacific Amusement ELECTRIC SPINNER, 1935. Same game, different name, new glass, spinning, and tumbling three colored dice. Unique dice was a Pacific Amusement Manufacturing Company specialty, tough to replace if missing. Great Repeal piece as payouts are in "Free Beer," not counting the dough dropped on the game. Pacific's last counter game. For years they were "Mystery Machines" as no names are carried on the games and they weren't advertised.

101

National Coin TIA JUANA, 1936. Already making small dice games with HAZARD, National Coin Machine Exchange in Toledo, Ohio, took a step upward with a larger counter dicer that added an intriguing feature. Just like state lottos of years later, the jackpot was increased "Daily Up" a little each day until it was won, then started over. A single TIA JUANA example was known for years, with no manufacturer identified. It was finally connected to National Coin.

Gottlieb INDIAN DICE, 1938. Five column dice game with hold-and-draw feature. Inside dial behind center dice column sets "house point" as coin is dropped. Pull handle to shake dice. You want to get five of a kind, or at least four, and get a second handle pull to re-pop the dies not held in the hold-and-draw. Aces are wild. It takes five sixes or five aces to beat house point 5. Games are found in red and blue cabinets, depending on production runs.

B. A. Withey IMPROVED SEVEN GRAND, 1938. B. A. Withey Company of Chicago made of a series of classic dice games in large cabinets, spinning the dice to a substantial tumble with a handle push. IMPROVED SEVEN GRAND was top of the line. Came with two award cards and set dice; plain, and cigarette dice. The second set is sometimes found inside the machine. Later made by Withey Manufacturing Company in 1940, finally by Bradley Industries in 1948 as 7-GRAND.

Gottlieb DAILY RACES JR., 1938. D. Gottlieb and Company, well known pinball game makers out of Chicago, knew how to spin out a game. They made a large DAILY RACES JR. payout pinball, and miniaturized it to a counter dice game of the same name. Game top glass is lighted like the pinball backglass. Push in a coin and a horse or two are lighted up to win. Pull out slide and two dice pop to name winning horse and win, place or show position, with odds up to 30:1.

Exhibit Supply 36 GAME, 1938. After years of flat dicers, Exhibit Supply Company of Chicago stuck one on end to create 36 GAME. One of their last counter games. The idea was to make it look like a radio to avoid local laws against chance taking for money. Plug-in electricity lights the inside of the cabinet. Drop the coin and push down the plunger and the dice cage spins to a tumbling stop. Twelve losing numbers and nineteen winning make it look easy. It isn't.

Mikro-Kall-It, Inc. MIKRO-KALL-IT, 1938. Produced at the height of the radio boom, Mikro-Kall-It, Inc. of New York City made MIKRO-KALL-IT live up to its name. Put coin in and push down the lever, and the mike activated by two flashlight batteries. Forcefully speaking into the mike makes the ten dice jump, spin, and shake. Long only known from advertising, the first one showed up at the March 1985 Chicagoland Show and was snapped up fast. Only two known.

Norris ROLL-EM, 1939. The flat countertop dicer all growed up, as they say. The Norris Manufacturing Company of Columbus, Ohio, was long known for their outstanding MASTER peanut, gum ball, bulk, and novelty venders. In a break with their tradition they made reel counter games, and in 1939 made a dicer. It was their last counter game. ROLL-EM is electric, with an enormous cash box and a cabinet that makes it look like a miniature PACES RACES payout console.

Keeney JITTER-BONES, 1939. A fantastic gadget, and a great toy for a basement barroom. Electrical plug-in, the game is always on. All you need to do to make the dice hop is place your palm over the glass. Bang! Up they go. Is it magic? Is it heat? Is it the shutting of light? It's not the latter as they'd be popping all night in a dark room. Whatever it is, you do it. Non-coin. Plays 26 and all favorite bar games. Losers buy drinks. That's how the bar makes out.

aker Novelty PICK-A-PACK, 1939. Baker ovelty Company of Chicago revamped rge floor model PACES RACES automatic ayout consoles into BAKERS PACERS. As at well ran dry, they made counter games arting with PICK-A-PACK, a direct steal of e Gottlieb INDIAN DICE. The difference is ree columns. Cigarette packages are on e dice. "House point" dial indicates brand for win. Match three for a penny, with two andle pulls hold-and-draw, and you win a ck.

Victor Vending ROLL-A-PACK, 1941. A new player entered the scene just before America's entry into World War II. Victor Vending Corporation of Chicago made a machine similar to the Baker Novelty PICK-A-PACK in a much simplified wood and glass cabinet, calling it ROLL-A-PACK. No hold-and-draw this time, just play a penny and pop three dies with cigarette packages on their faces. Suspense was added by popping the dice one at a time 1, 2, 3 in sequence.

Dependable DICE-MAT (10-Column), 1948. One of the first counter games after World War II, DICE-MAT (10-Column) was made for bars, service affiliated lodges, and bowling alleys that rose like weeds when veterans came home. DICE-MAT was made a decade earlier in a five-column format by Dependable Enterprises, Barberton, Ohio. Miraculously, the firm still existed after the war, and brought it back bigger. Hold buttons permit games of from three to ten dice. Non-coin.

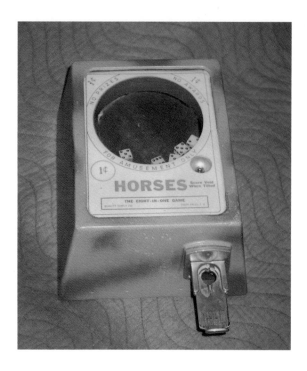

Quality Supply HORSES, 1949. The National Coin Machine Exchange HAZARD and its variants found new life after WWII when the idea was picked up by Quality Supply Company in Sioux Falls, South Dakota. Under the primary name HORSES, the Quality Supply countertop was called the "Eight-In-One-Game." Each game came with eight award cards that could be put on top: HORSES, HI-HAND, FOUR OF A KIND, ADD 'EM, WIN YOUR SMOKES, BEAT THE HORSE, WIN A BEER, and HI-LOW SEVEN.

Jess SHAKE, RATTLE AND ROLL, 1954. When Bill Haley and the Comets recorded "Shake, Rattle and Roll" in 1954, it became the first million seller for Haley's Comets. The expression became part of the language, and became a perfect copycat theme. As a dicer it started showing up on bars. Although counter games had long reached their nadir, smaller makers around the country made short runs of various counter games, mostly dice, including the Jess Manufacturing Company.

Quality Supply HI HAND, 1950. Special dice go with the eight award cards with each "Eight-In-One Game." HORSES shown uses five standard spot dice, HI-HAND has five poker dice, FOUR OF A KIND has four spot dice and one numerical die, ADD 'EM is five spot dice, WIN YOUR SMOKES is four cigarette dice and one numeral die, BEAT THE HOUSE is three numeral dice, WIN A BEER three dice of any kind, and HI-LOW SEVEN seven spot dice. The special dice are rarely found with the games.

Twico TWICO, 1957. The death knell of vanished locations never made counter games completely go away. A number of small dicers, dial and skill machines were made in the '50s, '60s, '70s, '80s and into the '90s with the arrival of solid state poker players and other games. The Twico Corporation, Chicago, Illinois, was one of the last dice game makers, with TWICO offering "Ball Gum 1¢ Read Your Fortune" play. Rewards are still based on up to five of a kind. Lotsa luck!

COIN DROP

Even simpler in construction than the free spinning wheel, the gravity driven coin drop trade stimulator is likely the oldest form of coin operated chance machine offering a predetermined payback, with its origins going back almost two centuries. By creating a vertical playfield studded with pins, coins dropped at the top will bounce back and forth among the pins as gravity brings the disk ever downward into a series of scoring pockets. In addition to gravity other principles with chance applications are evident.

Maley NICKEL TICKLER No. 2, 1893. Charles T. Maley Novelty Company, Cincinnati, Ohio, offered three NICKEL TICKLER models in 1893 and 1894, produced private label by Western Weighing Machine Company, also of Cincinnati. NICKEL TICKLER No. 1 was in a simple, oak cabinet. No. 2 adds an elaborate cast iron and copper plated header, sides and stand. Simple stuff. In No. 1 there were no blanks, offering 1 or 2 cigars. No. 2 was tougher; win 2, 4, 6, or nothing.

Canda TRADE VENDING MACHINE, 1893. Temptations in trade stimulators were enormous. Imagine seeing this pristine railroad watch (railroads theoretically ran on time, and the watch was the timekeeper) sitting in a box on a cigar counter, and all you had to do was play a nickel to get it. Early machine by Leo Canda Company, Cincinnati, Ohio, and also marketed by Samuel Nafew Company, New York, New York. This example showed up at a rural auction in 1994, sans watch.

This was first theorized, and then proven, by Sir Francis Galton, an Englishman that applied his fertile imagination to the solution of any and all sorts of natural and scientific problems that fascinated the Victorian mind. It was known that clay balls played on the tilted playfield of a late 18th century bagatelle game came to rest in scoring channels at the bottom in a pattern that suggested that more than random placement was at work. Game makers were able to indicate high and low scoring troughs based on long term experience on where the balls would eventually rest after completing their journey over the pegged field. Galton had an absolute belief that all phenomenon, no matter how trivial, was quantifiable. He created his late 19th century Galton Board experiment with a sloped playfield studded with a triangular pattern of pegs. When marbles were fed from the top they bounced around the pegs and repeatedly collected in the troughs at the bottom in a characteristic pattern, with the troughs left and right centers

Western Weighing IMPROVED NICKEL TICKLER, 1894. Western Weighing Machine Company of Cincinnati "improved" the NICKEL TICKLER by adding a coin diverting wheel to top center of pinfield. If skill was ever involved it ended here, as you have no idea where that coin will go. If wheel pointer is on 1, 3, 5, 7, you get one cigar; on 2, 4, 6, you get two. Match number on the bottom with same number on the wheel and you get five cigars. Mechanism incredibly simple.

collecting the most balls, while those at the far right and left and in the center had the least. The same results were obtained time after time, with the Galton Board proving that making the center left and right troughs the lowest scoring, with the far left, right, and center providing higher scores, provided a controllable return. The same principle applies to coins dropped on a vertical playfield, with the disks at rest at the bottom conforming to a predetermined profile. Changing the pinfield pattern will alter the results, but every pinfield will provide a distinctive scoring pattern of high and low, enabling the game makers to test their playfields over a period of time, after which they could establish where the high and low scores should be placed. Judicious placement of a number of pins all but blocking the entrance to the highest scoring trough insured that this score would be difficult or impossible to make, offering a seemingly high prize while preventing its accomplishment.

By this time Galton was on to other things, not the least of which was an idea that every human being had a different pattern of skin whorls on their fingertips, leading to his discovery of fingerprinting and the first use of this individual form of human identification by England's Scotland Yard. For this, and his other contributions to natural science, Galton was knighted as Sir Francis in 1909 with his thoughts already on something else and far away from the seemingly insignificant returns of a coin drop trade stimulator.

It was the Galton Board experiment, and the fact that coin drop trade stimulator makers and operators could bank on the returns of their machines, that led to one of the most enduring and reliable forms of coin operated devices. The first identifiable maker of a commercial coin drop was the Little Giant Manufacturing Company of New Haven, Connecticut, whose LITTLE GIANT paid off in cigars or drinks. The idea was picked up by the Clawson Slot Machine Company of Newark, New Jersey, with an enclosed playfield cigar selling game of September 1892 called FAIR-SELLING MACHINE on which the playfield covering front panel provided advertising space. Clawson also introduced a series of completely unique coin drops called LIVELY CIGAR SELLER, of June 1893, followed by HEADS AND TAILS and LIVELY CIGAR SELLER NO. 2 of October 1893, in which the coins bounce over a series of bent wires into scoring positions. These latter games were never copied by other makers over the years, although THE LIVELY CIGAR SELLER NO. 1 was reproduced in the 1960s with these aging replicas sometimes selling at shows and auctions as original games.

By 1893 the coin drop format was a standard trade stimulator configuration, and appeared in a wide variety of models. The Leo Canda Company of Cincinnati, Ohio, introduced their TRADE VENDING MACHINE that offered a pocket watch as a prize, and their straight gambling THE EAGLE machine with repays over the counter. The pioneer Western Weighing Machine and Western Automatic Machine Companies, also of Cincinnati, introduced their venerable NICKEL TICKLER (with one of the greatest trade stimulator names ever conceived) models in 1893, quickly picked up by the Cincinnati cigar maker turned game marketer Charles T. Maley Novelty Company who offered their NICKEL TICKLERS in three models, followed by a PENNY TICKLER and a coin drop game called CASHIER in 1894. Across the continent, Frederick W. Bishop of Los Angeles, California, offered NINE POCKET and TEN POCKET counter coin drops in the summer of 1893, both of which are rare and highly desirable machines that are difficult to identify as maker nameplates and identifying papers are nonexistent. An even rarer machine, that remains to be discovered, is the remarkable tall bartop coin drop of 1893 with scoring electric lightbulbs and ringing bells made by Skeen and Farmer in St. Louis, Missouri.

THE COMBINATION, unknown manufacturer, 1894. Reminiscent of the Consolidated Coin 200 made in Chicago, THE COMBINATION is a more complex game. You keep feeding coins to the game until you get a combination of 1, 2, or 5 on both sides for the payout, with most coins shooting right straight down the middle. Two examples known, with one of them showing up in California suggesting a west coast manufacturer. Patent applied for but apparently not issued.

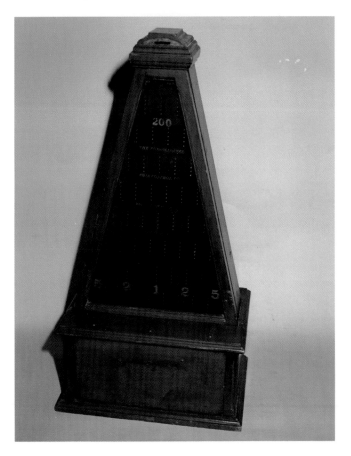

Consolidated Coin 200, 1894. A leading payout slot machine of the day was the TWO-DOOR BANK, also called PYRAMID, which dropped coins down a pyramidal pinfield that sometimes popped open two "bank" doors at the base that dumped out coins. Consolidated Coin Control Company located at 287 South Canal Street in Chicago, amid the railroad stations, made a trade stimulator version that paid up to five cigars, but no cash bank. Only one known, with paper missing.

By the later 1890s the coin drop had solidified into a conventional platform that was plain in appearance and highly successful in operation. The Comstock Novelty Works of Fort Wayne, Indiana, contributed THE PERFECTION in December 1897, with the Drobisch Brothers ADVERTISING REGISTER of June 1896 and the Friedman and Company SONG DICK made in Chicago in 1898 displaying cigar advertising on their playfields, greatly enhancing the graphics of these antique machines. An even more elaborate playfield was provided by the Jonas D. Bell DEWEY coin drop made in Chicago in 1899, with Spanish American War hero Admiral George Dewey's portrait spread across the playfield behind its brass pin studding. The game was quickly offered as a store premium with case orders of Wrigley's Gum, surviving as such into the first decade of the 20th century.

Much simpler and plain playfield coin drops were offered by Mills Novelty Company of Chicago and The Caille Brothers Company of Detroit in the early 1900s, with the Mills LITTLE DREAM coming out in 1907, and a Caille version appearing in 1910, with the latter soon changing its format and name to INDIAN PIN POOL in 1911. By the time Mills Novelty was offering its uniquely shaped inverted

V-shaped SILENT SALESMAN in 1916, the coin drop had become recognized as the most common and least expensive trade stimulator on the market. Much smaller drug and cigar store DAISY coin drops, made to be placed next to cash registers without taking much counter space, were made by the Hamilton Manufacturing Company of Hamilton, Ohio, starting with the DAISY NO BLANK "Bread Top" of 1907 to be followed by the DAISY "Diamond Top" of 1910 which continued to be made for many years. These small counter drops received great placement employment when they were featured in the Albert Pick and Company saloon equipment catalogs out of Chicago, and the games were still to be seen on cashier counters well into the 1930s.

The Jackson Company of New Castle, Indiana, with their ubiquitous NEVER LOSE, CAN'T LOSE, and HAVE SOME FUN drops of the late '20s and early '30s. World War II led to the Baker Novelty Company, Chicago, BOMB HIT and Runyan Sales Company, Newark, KEEP 'EM BOMBING coin drops of April 1942 and thereafter, among others, with anti-Axis graphics making these simple games popular during the war years, to be tossed out *en masse* afterward, making the surviving machines both interesting and desirable. The format survived the war years to show up in the late 1940s and into the '50s, with the Pioneer Coin Machine Company of Chicago contributing their SMILEY in April 1946, right up to the as yet unidentified BOUNCERINO of 1952.

If counter games are ever made again, coin drops will surely be among them.

Canda EAGLE, 1894. One of the most attractive coin drops ever made, if not the unquestioned leader. Only one has been discovered so far, and it was in terrible shape, although the paper survived under glass. Enough of the finish survived to show it was mottled copper plate over cast iron with a copper sheet pinfield. Coin chute is in the casting at the top under the eagle casting. This example was in the Stardust Slot Museum in Las Vegas until the museum closed.

Due to their simplicity of fabrication, wooden coin drops became the preferred style for regional distributors and trade stimulator string operators who made their own games and were able to assign production to local woodshops and thereby offer exclusive machines for their routes. The leader in this cottage industry expression was the Banner Specialty Company of Philadelphia, who came out with their generic LEADER in 1922, with dozens of THE LEADER models produced by other distributors for their contiguous markets soon showing up across North America in the 1920s. The basic format remained unchanged for years afterward, with game graphics becoming the major points of difference. A major producer of the genre was

Drobisch ADVERTISING REGISTER, 1896. Perfect machine to be turned out by a woodshop. Drobisch Brothers and Company in Decatur, Illinois, converted their cabinetmaking business into trade stimulators to give the Decatur Fairest Wheel Company in town some competition. ADVERTISING REGISTER is their first product. The idea was to sell cigar companies on using the pinfield as an advertising billboard. Weimer's PERFECTO is shown. Other versions have been found.

S. A. Cook HOWARD'S FAVORITE, 1896. Cigar sponsorship was typical of simple 1890s wheels, pointers, and coin drop machines. A lucky manufacturer would sell their machine to a number of brands. HOWARD'S FAVORITE shows distribution by S. A. Cook and Company in Medina, New York. More than likely it was made by someone else, with Cook the cigar distributor. A good producer possibility is Medina Manufacturing Company, a coin machine and specialty manufacturer.

Comstock THE PERFECTION, 1897. One of the most successful trade stimulators of the late 1890s, made for general and country store placement rather than saloons or cigar counters. Comstock Novelty Works in Fort Wayne, Indiana, was the outgrowth of a wholesale grocery supply business. THE PERFECTION was patented as a "Bonus-Indicating Machine." An advantage from the grocer's point of view is that a coin in a bottom award column can be seen through glass from the back.

Klondike THE KLONDIKE, 1899. Some machines may never be identified by name, maker, date, and location, even though there are clues. THE KLONDIKE GUM MACHINE was found in the late 1970s, but has always been suspicious. Nothing else like it ever showed up again. Klondike Slot Machine Company existed in Cincinnati, Ohio, and spelled Klondike the same way (there were numerous variations), but that isn't assurance of origin. Gum vending machines tended to be later.

Hillsboro THE HILLSBORO, 1897. Located in southern Ohio one major town and less than twenty miles south of Greenfield, the Hillsboro Wooden Ware Company in Hillsboro, Ohio, may have been the inspiration for naming the Waddell Wooden Ware Works ("The Four Ws") a few years later. Both firms made wooden trade stimulators, with the Hillsboro machines conventional coin drops. You had to be pretty lucky to hit a 2 or 3 on THE HILLSBORO, with almost all coins going in 1.

Bell (WRIGLEY'S) DEWEY (PIN MACHINE), 1899. Jonas D. Bell and Company, Chicago, affiliated with Wrigley's Gum Company early on with their WRIGLEY'S SLOT MACHINE, a single reel gum trade stimulator. Next was a conventional penny drop with enormous sparkle from a color lithograph of Admiral George Dewey, Hero of the Spanish-American War, on the pinfield. Placed by thousands as a free premium for dealers when they ordered eight boxes (160 five-cent packages) of gum.

"Tiered Backdrop," unknown manufacturer, circa 1900. Discovery of hitherto unknown and unidentified "Mystery Machines" is very exciting. It happens all the time. Coin release knob at the right bottom of this "MM" drops the coin after a visible payout. Backdrop is tiered so coins slide down by levels. Closest thing in patent art is an undiscovered latticed pinfield machine credited to cigar maker Frederick W. Bishop of Los Angeles, California, in 1893, except no knob.

Hamilton DAISY NO BLANK, 1907. No blanks coin drop made by the Hamilton Manufacturing Company in Hamilton, Ohio, as the ultimate minimalist cigar machine. Case is a single cast iron casting, with a coin release wire handle at the right bottom. The merchant saw what the customer won, and then pushed the lever down to drop the coin in the cash box. Earned the nickname "Bread Loaf Top" because of the shape of the cabinet.

A. J. Fisher (ORIGINAL) PREMIUM, 1910. A. J. Fisher and Company of Pittsburgh, Pennsylvania, was a Mills Novelty distributor that made their own games. PREMIUM converted the wooden pyramid format to a "G" gum vender in metal. "1¢" covers the coin chute at top, with the name PREMIUM cast into the top panel of the cash box. Design below lock is distinctive. Left and right award panels are missing. Fisher sold the machine to Mills Novelty, thus ORIGINAL PREMIUM.

Comstock PERFECTION WHEEL, 1898. Spinning out a new product. Comstock Novelty Works in Fort Wayne, Indiana, took their THE PERFECTION and added a pointer wheel at the top to create tall PERFECTION WHEEL. The pointer wheel says "Multiply your odds. Win free cigars." Most wheel segments are 1, but there are 2s and a 3. So get 3 below and 3 above you get a total of 9. Highly desirable "PERFECTION" coin chute carries "Jan. 22nd 1900" patent date.

Zeno YUCCA, 1908. Zeno Manufacturing Company in Chicago tried all sorts of marketing ideas in their final days as an independent company. They had long been known for their articulated wall mount Zeno gum venders, picking up a lot of competition in the early 1900s. Zeno had Mills Novelty design special vending machines in 1908, and also tried countertop coin drops to promote their gums. YUCCA is a major Zeno brand. Wrigley's Gum bought them out soon thereafter.

Miller Novelty LITTLE DREAM PLAY BASEBALL, 1907. Created by Miller Novelty Company of Chicago, LITTLE DREAM was sold to the Mills Novelty and initiated the Golden Age of wooden gum awarding coin drops with "G" payouts. Miller made the gum model LITTLE DREAM in 1904 and the follow on LITTLE DREAM PLAY BASEBALL scoring in runs in 1907, at which point Mills Novelty took over to continue making the gum machine. Caille Brothers started it in 1910. Widely copied.

Sloan THE LEADER, 1910. Major coin machine dealers, distributors and operators around the country came to the conclusion they could make a lot of money if they made their own machines. It was difficult when it came to payout slots, but as easy as lumber, nails, and stain if they made counter "G" trade stimulators. Sloan Novelty Company, Philadelphia, Pennsylvania, was a leader of the pack. And that's what they called their game. Widely copied, fairly common.

Hamilton DAISY, 1910. Runaway success of the Hamilton DAISY NO BLANK "Bread Loaf Top" led to oversold sales, and the need to speed up production. The game was redesigned to replace the fancy round topped casting with a slab sided one, nickel plated. The award panel on the front boasts "No Blanks," offering "One or more cigars for every nickel played." Wrigley's Gum picked it up and gave one away for a ten box order. It earned the nickname "Diamond Top."

Bradford LITTLE GEM, 1913. Bradford Novelty Company of Providence, Rhode Island, a Caille Brothers distributor, issued catalogs of Caille machines with their own proprietary names. Unique to the catalogs were simple wooden box coin drops called LITTLE GEM FORTUNE TELLER and LITTLE GEM, differing by the award panels (missing here) and the strips across the bottom. LITTLE GEM offers 1, 5, 10, and 25 wins; the other spells FORTUNETELLER. Knob at right discharges coin.

Mills Novelty THE PREMIUM, 1916. When Mills Novelty Company Acquired the A. J. Fisher (ORIGINAL) PREMIUM, they proceeded to make the game under the same name. Text "THE PREMIUM GUM MACHINE" is on the cast panel above the cash box. Mills made changes. Top coin chute now has an open front, and the award frames and front lock panel have some artistic embellishments not in the original. Mills soon changed the castings and continued to make the game as SILENT SALESMAN.

Hamilton DAISY, 1910. An attempt was made to promote the Hamilton DAISY "Diamond Top" as a cigar brand counter board, with samples made to gather orders. But none have been found to date. In all likelihood they were made, and were largely discarded in the 1930s when cigarettes took over most tobacco sales. But the search is on. DAISY coin drops show up periodically at shows, malls and auctions, and one of these days branded cigar versions will surface.

Cawood Novelty PLAY BALL, 1913. A complete surprise to trade stimulator and baseball collectors, PLAY BALL was thought to be a 1930s game based on its cabinet and graphics. With no maker name or manufacturer panel, it remained a "Mystery Machine" for years. Then your author made a fortuitous vintage paper buy which included a 1913 catalog of the Cawood Novelty Company of Danville, Illinois. And there was PLAY BALL, and another version called PANAMA CANAL.

Knight OUR LEADER, 1922. The diamond center pinfield of the Sloan Novelty Company of Philadelphia THE LEADER became a basic for simple wooden coin drops. The difference was in how the format was presented. A unique presentation is the Knight Novelty Company LEADER made in Clifton, Massachusetts. The award panels and pinfield pattern are pre-printed on a single overall sheet. This enables the game makers to position the pins correctly time after time. Rare.

Banner Specialty LEADER, 1922. Banner Specialty Company in Philadelphia, Pennsylvania, seemingly inherited the LEADER mantle from the earlier Sloan Novelty Company in Philadelphia. Banner Specialty was both a Mills and Jennings distributor, with extensive routes of their own. The firm survived into the present era, mostly in music. Banner Specialty LEADER is one of the most commonly found coin drops. The Banner red ribbon makes this a particularly prime piece.

Boyce Coin WEE GEE, 1925. Much experimentation was undertaken in counter games in the '20s using new materials and production techniques. Boyce Coin Machine Amusement Corporation of Tuckahoe, New York, burst upon the scene in 1925 with a line of wall and countertop amusement machines for stores, restaurants, and clubs, not mentioning their obvious speakeasy use. On WEE GEE you set a pointer to a question, and get a yes or no answer depending on where the coin lands.

Jackson CAN'T LOSE, 1930. The Jackson Company games seem to have been created for midwest routes in Indiana, Ohio, Michigan, Illinois, and Wisconsin where they are most often found. Format and play action is virtually the same for all of these games. CAN'T LOSE refers to the fact that a no blanks machine, with 1, 2, and 3 wi A higher payout version called HAVE SO FUN has four "5" winners, one "10" and "0," making the blanks outnumber the winners.

OUR LEADER, unknown manufacturer, 1922. So many versions of the diamond pinfield LEADER and OUR LEADER coin drops were made by distributors and operators, we will probably never be able to identify all of them. The wooden cabinet is basic, as is the playfield pattern. Usually a knob or dial is on the front center or right side to take the coin off the pinfield after the award has been paid. This OUR LEADER has a push rod in the front of the right side panel.

Jackson NEVER LOSE, 1928. The Jackson Company of Newcastle, Indiana, is a bridge between the trade stimulator past and counter game future. Taking the old coin drop idea, they created taller, more visible games in wooden cabinets with lower cash boxes. Large print and colored decals proclaim the NEVER LOSE game name. Cash box door is in the back, with a very old looking key which mislead collectors for years. These games are often found in midwest antique malls.

Northwest Coin CHURCHILL DOWNS, 1 Northwest Coin Machine Company (also Northwestern Coin Machine Company) i Chicago started with a single coin CHU HILL DOWNS in 1932. Spinner picks the winner and graphics show horses in a h race. In early 1933 that changed to a straight coin drop with horses racing pa grandstand. You pick Man-O-War, Burg King, or Twenty Grand to win based on which of the three top coin chutes you p Same basic name, new game.

[Pie]rce WIN-A-PACK, 1933. One of the more [e]nduring "Mystery Machines." WIN-A-PACK [w]as first made in wood by General Novelty [M]anufacturing Company in Chicago early in [1]933. They were out of business in ninety [d]ays. Next it shows up in this cast aluminum [c]abinet with Deco trim and display room for [f]ree packs. Pierce Tool and Manufacturing [C]ompany of Chicago habitually picked up [d]efunct makers. And this cabinet looks [Pi]erce. Maybe not conclusive, but certainly [c]onducive.

Exhibit Supply THE BOUNCER, 1936. You theoretically aim your penny in the right direction with the rotating coin chute at the top of THE BOUNCER. Forget it! The moment that coin drops to bounce off the steel spring below you've lost control. Clever game. Also camouflaged. Fortune teller panel at upper right tells you what you want to hear. But it's the panel at left that tells you your payout score. For some reason neither of the Exhibit coin drops were ever advertised.

Atlas Games CIVILIAN DEFENSE, 1942. World War II was one of the most vitriolic wars in history, with the propaganda particularly heinous and brutal. The realization that cities, and civilians, would come under aerial bombardment was recognized on both sides of the war from the Axis to the United Nations. CIVILIAN DEFENSE by Atlas Games of Cleveland, Ohio, depicts an American city under German air attack. Scary. Shoot down a German bomber and score.

[Ex]hibit Supply DOUBLE DICE, 1936. Exhibit [Su]pply Company of Chicago was well known [for] its dice and dial games. What they are [no]t known for are coin drops, yet they made [so]. DOUBLE DICE combines features from [oth]er game formats. Put penny in chute, [pu]sh plunger, and the coin drops over the [wh]eel to spin it, and then works its way [do]wn the pinfield to land in a die hole. [Ma]tch wheel and die numbers to double the [gam]e and you get the jackpot. Rare. One [kn]own.

Five Boro DUCK SOUP, 1937. Not traditional, but a coin drop nevertheless. Few of these water machines have survived, probably because their bottles or bowls are broken easily. One of a number of makers was the Five Boro Machine Manufacturing Company of Brooklyn, New York. The idea is to drop pennies into the glass and scoring cups below the coin chute. Optical illusions and the reaction of a penny dropping in water makes them slide all over the place. Hard to do.

Baker Novelty BOMB HIT, 1942. The war action took many forms in coin machines, but for counter games bombing was preferred. Of course, you were directing coins to targets, and dropping them from above. On the Baker Novelty Company BOMB HIT made in Chicago, baseline scores are Very Poor, Poor, Fair, Good, and Direct Hit in the center. If you were to project the values of coin machines into the future, the wartime themed games are the best bet as investments.

BOUNCERINO, unknown manufacturer, 1952. Once the postwar interstate highway program promoted by president Dwight David Eisenhower got underway in the '50s, and the shopping mall replaced the neighborhood stores, counter games became as extinct as dinosaurs. Small makers and locations still used some, with many unidentifiable without maker names. "Mystery Machine" BOUNCERINO has 50¢ and $5 holes, with discouraging blocking pins over both of them.

Runyan Sales KEEP EM BOMBING, 1942. If we were the good guys and they were the bad guys (depending what side you were on), KEEP EM BOMBING. Runyan Sales Company of Newark, New Jersey, made one of the most revengeful games, showing crude caricatures of Japanese soldiers under a bomb filled sky with "Smack A Jap" side copy. Remarkably, the passage of time has eased the message, and Japanese collectors adore these games. Five coins and a bomb release. Bombs away!

Frantz BULLS EYE, 1954. The post-WWII coin drop got a lot smaller and cozier. J. F. Frantz Manufacturing Company of Chicago acquired the Baker Novelty games in the early 1950s, and then sired some of their own, starting with BULLS EYE. In BULLS EYE you aim the chute at the bullseye below and drop the coin, only to have the pinfield bounce it all over to throw it off course. Great stuff for neighborhood bars, bowling alleys, and billiard halls, all good '50s locations.

TARGET

The target trade stimulator and counter game is a variation of the coin drop format, equally demonstrating Galton Board characteristics. Where in coin drop machines the coin is fed from the top in order to get the full effect of its careening capability, bouncing around from pin to pin to ultimately land in a scoring trough at the bottom, the target game feeds its play counter from the side, usually right, to impart speed and direction to the moving coin. This act alone implies some level of control over the progress of the coin, theoretically making the target a skill game. As in the Galton Board experiments, numerous coins fed into a target game over a long period of time will cluster in a characteristic pattern in the scoring troughs below, unique to the pinfield design, enabling the game maker to closely approximate the results and determine the high and low scoring areas.

Much as the coin drops developed, the target games were generally assigned to trivial pursuits, providing inexpensive awards for play. In a typical no blanks target game the left and right center (and often even the center) scoring troughs would be marked "G", meaning a stick of gum for the penny played, with much higher awards to the far left and right, such as 5, 10, and 25. While this supposedly

Latimer GAME O' SKILL, 1895. GAME O' SKILL production continued for a number of years, making the cigar machine almost as popular in the west as the FAIREST WHEEL was in the midwest and east. Award paper and glass graphics changed with new production runs. The game advantage is that both player and counterman can see the score, blank or win, between two panes of glass. Win is paid off, and then the coin dropped into the cash box by the knob on the side.

meant multiple sticks of gum, the tally was usually paid off in cash or equivalent trade, although rarely as these side scores are very hard to make.

Due to their supposed skill characteristics, and the fact that a coin is kicked across the playfield by a finger flipping component (generally called a shooter), target games do not need the deep playfield favored for coin drops as much of the action takes place as the coin flips back and forth across the top pins until the initially imparted momentum wears down, at which point the coin begins to trickle down between the pins.

It is this action, and the suggestion of skill play, that gave the game its original name. The creator was James D. Latimer of Latimer and Company located in the hotbed of slot machine invention in San Francisco, California, in the 1890s. His first GAME O'SKILL was made in July 1893, and quickly patented. It was followed by the similar LITTLE HELPER in 1896, with both machines in wooden cabinets. What made the Latimer games different from other coin drops was the right side shooter that flipped the coin across the playfield once the index finger of the right hand whacked the coin resting in a channel facing the playfield. The Mills

Latimer GAME O' SKILL, 1893. Rarely can a game format be attributed to a single origin, something that can be done for targets. Defining element is the coin shooter at upper right. Coin in chute, whack it with your index finger at the shooter below to bounce over the rubber pegs to score, six blanks, two "1," two "2," and one "3." Latimer and Company of San Francisco, California, applied for the patent in July 1893 as a toy to get around anti-gambling patent laws.

LITTLE HELPER, unknown manufacturer, 1908. One of the great charms of really old trade stimulators is the trashy colored paper stock often used for award panels. LITTLE HELPER is one of numerous Latimer type target machines of unknown origin. Format was widely copied, with few makers putting names on their products. LITTLE HELPER has a deeper cash box and a bigger rubber peg field. Prizes are higher 5, 10, and 25 range, or "G" for gum, an indicator of circa 1908.

Novelty Company in Chicago bought the rights to the game (as they did for many other San Francisco machines created in the 1890s and early 1900s through a local agent) and produced their own decorated cast iron cabinet GAME O'SKILL in 1902. That didn't prevent others from copying the game, with the White Manufacturing Company of Chicago making their own cast iron GAME O'SKILL in 1902, followed by a wide variety of wooden models similar to the Latimer original made by smaller producers. Condon and Company in far away Vinalhaven, Maine, made their GAME O'SKILL in October 1903, and a dead ringer for the original Latimer game was made by Earl A. Robinson Novelties in Providence, Rhode Island, in 1909 as THE NEW PIANO GAME, taking its name from the nickelodeon pianos then popular in saloons and dance halls.

The Latimer format returned in the mid-1920s when a number of eastern makers created new games around the old idea. Novix Specialties in New York City made a rash of the games in wood starting in 1927 as INDOOR BASEBALL, KOIN-KICK FOOTBALL and INDOOR AVIATOR among others while Coin Sales Corp., also of New York City, made MILLARD'S MINIATURE BASEBALL. These short coin flip games soon lost favor as more sophisticated counter games reached the market.

It was the Mills Novelty Company that expanded the format, and gave the target machines their generic name. Stretching the game upward, and keeping the coin flipping device at the side, Mills introduced their taller cast iron cabinet TARGET PRACTICE in 1918, just as coin machine production got underway again right after World War I. The target name came from a casting mounted at the left top of the playfield with three circular targets tantalizingly close directly across from the shooter, seemingly ready to catch a penny flipped hard enough to fly across the top into one of the slots. The problem is that the openings are precisely matched to the diameter of the coin, and the shot must be perfect to go into any of the targets for an award. Failing that, the coin trickles down to the scoring troughs below. The Mills TARGET PRACTICE came back in 1920 in iron once again, and was converted to a more decorative aluminum case in 1922 where it remained a part of the Mills line in continually upgraded cabinets well into the 1930s.

The copying of the Mills TARGET PRACTICE was nothing short of an epidemic. By the middle 1920s the Mills target machine was best selling counter game in the country, and a dozen or more manufacturers leapt onto the bandwagon. Industry Novelty Company in Chicago had it by 1918, with aluminum versions of the TARGET and an O. D. Jennings and Company variant called FAVORITE in the Jennings line into the 1930s. The Silver King Novelty Company of Indianapolis, Indiana, had the game in their own castings by 1918, with Royal Novelty Company, also of Indianapolis, bringing out their model in 1921, with both makers producing their versions into the 1930s. Banner Specialty Company in Philadelphia made the game in wood as BANNER TARGET PRACTICE in 1924, followed by a number of other smaller makers. Some games had strange names, such as KITZMILLER'S AUTOMATIC SALESMAN made by Matheson Novelty and Manufacturing Company of Los Angeles, in the mid-1920s. Blatant aluminum cabinet reproductions of the Mills machine called TARGET

A. J. Fisher LEGAL, 1908. It's the same game as the enlarged LITTLE HELPER version of the Latimer GAME O' SKILL, only in nickel plated cast iron. Award schedule is in gum, same as LITTLE HELPER. Once again, no maker name, so this machine remained an orphan for years. Closer inspection indicates many similarities to the cast iron cabinet for the A. J. Fisher and Company (ORIGINAL) PREMIUM made in Pittsburgh, Pennsylvania. Until proven wrong, this will do.

GAME O' SKILL, unknown manufacturer, 1912. It was inevitable that the Latimer GAME O' SKILL format would be converted to a pinfield to get away from the rubber cushioned pegs. Once the Sloan Novelty THE LEADER coin drop format was established in Philadelphia, Pennsylvania, GAME O' SKILL diamond center pinfields were soon to follow. Maker of this pinfield version is unknown. Cigar being promoted is "47" of the Saginaw Cigar Company, Saginaw, Michigan.

or TARGET PRACTICE were made by the Ad-Lee Company, Buckley Manufacturing Company, National Coin Machine Company, Pace Manufacturing Company, Reliable Coin Machine Exchange, Specialty Coin Machine Builders, and Specialty Manufacturing Company, all in Chicago, in the 1927-1932 period. The Barr Novelty Company in Shamokin, Pennsylvania, also produced a DOUBLE TARGET model in 1924. By 1930 the TARGET PRACTICE machine had been produced in more models, and in greater numbers, than any counter game before or since. They are extremely difficult to tell apart as maker's names are rarely applied, and the cabinet castings often differ only in small details.

But this was only the beginning. The target game format was adaptable to many other applications, such as sports themes. The Exhibit Supply Company in Chicago took a leap of faith with their wooden cabinet PLAY BALL of March 1926, quickly produced in an elaborate aluminum cabinet by August of that year. Full size baseball and football target games were made by a number of manufacturers, with the games generally rare and valuable. Among the most colorful are the FOOTBALL PRACTICE, BEER TARGET, and HAPPY DAYS targets made by Great States Manufacturing Company in Kansas City, Missouri, with the latter two shaped like steins of beer.

The target format was also adaptable to smaller and less expensive models, with cabinets made of castings and sheet metal, and often a combination of both. Manufacturers and marketing organizations such as the Calvert Manufacturing Company, of Baltimore; Blue Bird Sales Corporation and Blue Bird Products Company, of Kansas City; D. Robbins and Company of Brooklyn, and a number of others, produced these smaller models between the mid-1920s and late 1930s, by which time the fad for targets had run its course to be replaced by more complicated machines. One of the last target games to be produced in some volume was TID-BIT by The Munves Corporation, of New York City, in 1939, itself a rarity.

Mills Novelty TARGET PRACTICE, 1918. Mills Novelty made GAME O' SKILL in 1902 in a cast iron cabinet. They added their wooden pinfield LITTLE DREAM in 1907. In 1918 they combined both games plus SILENT SALESMAN into TARGET PRACTICE, a cast iron pinfield game with a coin shooter. It gave the genre its class name: target machines. It was a revolutionary success and led to a dozen or so copycat makers. "MNCO" in bottom of left casting indicates Mills Novelty Company.

Mills Novelty STAR TARGET PRACTICE, 1919. The cash box on TARGET PRACTICE was too small for long collections, so Mills Novelty beefed up the bottom to increase its size to open up the collections schedule. Sticker on pinfield at lower left is for Buckley Manufacturing Company. Pat Buckley was said to have run the largest TARGET PRACTICE route in the world. He put his name on all his route machines, and may also have reproduced them as "MNCO" is missing on this machine.

Silver King THE TARGET, 1922. Silver King Novelty Company in Indianapolis, Indiana, was an early target game devotee. They started out with the games in 1918, the same year Mills Novelty did. THE TARGET of 1922 shows obvious improvements in cabinetry and shooter, with the Silver King name all over the front casting. Case is cast iron, nickel plated. A marvelous specialty collection can be assembled using only TARGET PRACTICE machines of many makers.

National Coin NATIONAL TARGET PRACTICE, 1926. One thing that coin machine entrepreneur Pat Buckley of Chicago liked to do was form companies, and run each of them in competition with the business and each other. Buckley re-engineered the Mills TARGET PRACTICE for his own routes, and produced them under a number of firm names. National Coin Machine Company produced NATIONAL TARGET PRACTICE and added the "NATIONAL" name in the lower left casting panel.

Exhibit Supply PLAY BALL, 1926. From the many target games made in the '20s you'd think that the design was in the public domain. It wasn't, but everyone copied what they could and tried for a patent. Exhibit Supply Company of Chicago got one on PLAY BALL for its unique baseball scoring features. If you made a hit and lit the bulb you got your penny back. Most were made without the light. PLAY BALL GUM VENDER has a different front with a gum ball window.

Matheson ORIGINAL AUTOMATIC VENDER, 1925. The automatic vending part is that you get a gum ball out of an enormous container in the back that holds 1600, vended into a side cup below, with every play. Automatically. Semi-automatic, really, because you have to push in the plunger above the cup. Matheson Novelty and Manufacturing Company of Los Angeles, California, made the game and got it patented. Another variation is KITZMILLER'S AUTOMATIC SALESMAN.

TARGET PRACTICE, unknown manufacturer, 1927. Another Pat Buckley TARGET PRACTICE, but by which firm? Buckley TARGET PRACTICE games come in two aluminum cabinet styles, the "Deer Antler" design of the NATIONAL TARGET PRACTICE and a plain deckled front as shown. Possible makers are National Coin Machine Company, Service Coin Machine Company, Reliable Coin Machine Exchange, or Buckley Manufacturing Company, all Buckley firms and all located in Chicago.

Mills Novelty NEW TARGET PRACTICE, 1 Mills Novelty Company in Chicago was great believer in reaffirming market posi So they got their TARGET PRACTICE gam patented with improvements in the shoot and coin catchers, and brought out an entirely new version in an aluminum cab Where plain diamonds adorned the fror the old model, we now have olympiads taking a stretch with the discus. It gave t old TARGET PRACTICE game a new leas life.

Jennings THE TARGET, 1926. Just about every candy, cigar, and checkout counter in North America had a target game by the late 1920s, with most of the major coin machine manufacturers pumping out their own versions. O. D. Jennings and Company of Chicago entered the arena at the end of 1926 to fill out their line and discourage Mills sales. The cabinet castings are distinctive, introducing Native American graphics to Jennings machines, an enduring theme.

Jennings FAVORITE (PEANUTS), 1926. O. D. Jennings and Company upped the ante with their FAVORITE line of commodity dispensers using the target game format. Essentially an aluminum cabinet version of the wooden Matheson ORIGINAL AUTOMATIC VENDER of the year before, the Jennings version retains the enormous storage area in back of the machine, with a product display window. Headers varied: plain (for gum balls), CANDY and PEANUTS, the latter the rarest.

Coin Sales MILLARD'S MINIATURE BASEBALL, 1927. The GAME O' SKILL format lasted almost as long as the enduring WINNER DICE game, with new interpretations still being produced until the end of the 1920s. MILLARD'S MINIATURE BASEBALL by Coin Sales Corporation of New York City can easily be mistaken for the much earlier game. Early models had four bumpers and people still racked up the Home Run. So a center playfield wedge was substituted to make it tougher to score.

Ace TARGET PRACTICE, 1927. Enter everyone else. The TARGET PRACTICE business was so good, others jumped in. Ace Manufacturing Company in Chicago first made TARGET with a double-deck plain front, called "All Aluminum Target." They quickly came back with a designed front to compete against Mills and Jennings called TARGET PRACTICE. Side lugs permit wall mounting. A version called AUTO-TARGET vended a gum ball when you pulled a release rod.

Blue Bird Sales BASE BALL TARGET, 1926. Blue Bird Sales Corporation in Kansas City, Missouri, was a sheet metal shop, so their products have a flat sided look with colorful baked on enamel. The result is, for want of a better word, tacky. These are trash classic machines, and very twenties. The baking process cracked the finish, which Blue Bird called "our own special crackle paint." Standard color was blue. This is the baseball model in yellow.

Hiawatha Amusement TARGET PRACTICE, 1927. The Hiawatha Amusement Company of Los Angeles, California, demonstrates that imitation is the sincerest form of flattery. Mills Novelty was continually being honored by those who copied their machines, and cost them market share in the process. We will probably never know how many makers copied TARGET PRACTICE in the 1920s. Lucky for collectors, the Hiawatha TARGET PRACTICE has the company name in the front casting.

D. Robbins AUTOMATIC BASEBALL, 1928. Sheet metal-ly and crackly, there is no mistaking the origin of this game. Except it has the D. Robbins and Company name on it, a major Brooklyn, New York, distributor. The copy "Distributed exclusively by" appears before the D. Robbins name. AUTOMATIC BASEBALL was made by Blue Bird Sales Corporation of Kansas City, Missouri, exclusively for D. Robbins and Company. The crackle paint finish is unmistakable.

D. Robbins PENNY CONFECTION VENDER, 1929. This machine may have been destined to be called "Boxing Skill," the name on its prominent pinfield paper. But National Amusement Machine Company in the Bronx, New York, was making a Latimer GAME O' SKILL format game called BOXING SKILL at the same time. It is possible that D. Robbins and Company, 'cross town in Brooklyn, backed off and changed the name to the one on its cash box casting: PENNY CONFECTION VENDER.

Calvert SPOOK HOUSE, 1930. Target games that are Calvert and Calvert alone have the name "Calvert. Mfg. Co. Baltimo MD" across the lower front casting. Anothe clear indicator is the large bas relief Ameri can Indian head, seemingly straight off the U. S. "nickel" 5¢ piece. SPOOK HOUSE h one of the better Calvert printed pinfields with ghosts and goblins. The cabinet is a target version of the original Calvert INDI SHOOTER gun game with two gum ball windows.

Jennings FAVORITE (BALL-GUM), 1928. After a year or so of active use, O. D. Jennings and Company cleaned up the FAVORITE and settled in to the calm life of gum ball dispensing. Some of the castings were cleaned up and strengthened, such as the display window, and a new BALL-GUM header replaced all others. The experiment with diversity had ended. While the 5/2/1/0/ 0/0/0/0/1/2/5 scoring line suggests blanks, every play delivered a gum ball if you hit the dispensing rod.

Calvert ALL ABOARD, 1929. Similarities between Calvert Manufacturing Company of Baltimore, Maryland, and Blue Bird Sales Corporation of Kansas City, Missouri, are many. Both made a series of cheap and trashy target games, and both made games exclusively for D. Robbins and Company in Brooklyn. Calvert ALL ABOARD has the distinctive Indian head graphic element common to all Calvert targets, with a printed railroad pinfield. Robbins also sold it as their own.

Calvert GEM CONFECTION VENDER, 19 GEM CONFECTION VENDER is a gum b version of GEM. The D. Robbins and Company model does not have the castir data on the front. The Calvert version do It also has the depiction of an American Indian with a bow and arrow in the uppe left casting. But not this one. This is the v rare "Dunk The Darky" version in which African-Americans are the target of the coins. Politically incorrect, but the way thi were.

Atlas Indicator BASE BALL, 1931. The target format fits a baseball simulation well as you are batting a coin. After making a large arcade floor machine called BASEBALL, Atlas Indicator Works of Chicago miniaturized to create BASE BALL in February 1931. The pull-back bat to kick the coin was weak and unprotected, so a second version with the bat shield as shown came out the next month. Both versions are found. Sports themed games are increasing in value.

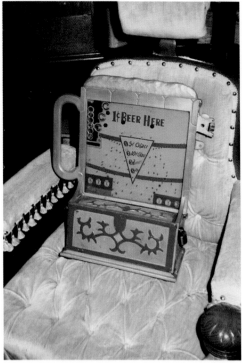

Great States BEER TARGET, 1933. Great States added more pinfield excitement to the target game format than all of the previous game makers combined in the first forty years of its existence. With Repeal the new law of the land, Great States made the target a beer mug to create a whole new game. The casting includes foam at the top of the mug. BEER TARGET came in plain and jackpot models. The Buckley "Deer Antler" front has been copied.

Great States HAPPY DAYS, 1933. With "Happy Days Are Here Again" the theme song of the winning Democratic party in the 1932 presidential election, and FDR taking over the national helm and ending Prohibition, Great States took the beer mug comparison a step further with HAPPY DAYS. The game name was added to cash box top casting, with beer barrels on the front. This is one of the greatest counter games ever made. You can practically taste this thing. Prime and rare.

Great States FOOT BALL PRACTICE, 1932. The ultimate sports expression in a target game. Great States Manufacturing Company in Kansas City, Missouri, was a new counter game maker and FOOT BALL PRACTICE (named on the award paper and not on the pinfield) is their first game. They put their name on top. One coin catcher upper left, two added scoring holes in the field. This isn't a soft core gum vender (the gum displayed doesn't vend) but a hard core gambling machine.

Skipper Sales SKIPPER, 1942. End of the line for the target format. SKIPPER is a wooden non-critical materials wartime game that avoids the obvious and doesn't go military or whack the enemy. Even the shooter is made of wood. Skipper Sales company was located in Philadelphia, Pennsylvania, a hotbed of operators and routes, and probably market enough to take most of the SKIPPER production. The cash box is compartmentalized to keep track of payouts. Rarely seen.

SINGLE REEL

Race game, roulette, wheel, pointer, dice, coin drop, target trade stimulator, and counter game formats bring their indicators (be they balls, pointers, or whatever else) to the scoring graphics to show the awards won, if any. The more advanced multiple reel card machine and the revolving dial mechanisms changed all that by reversing the process. They bring the winning symbols and high score graphics to the indicators, be they mechanical stops, stationary pointers, or viewing windows. The symbols and scores are mechanically moved by the reels, or dials, to bring them into scoring and viewing position.

It is axiomatic that the five reel card machines were the first to accomplish this reversal of fortunes at the end of the 1880s by displaying poker and spot symbols on moving reels that were stopped in front of a show window. If you consider coin machines alone, that is true. But if you look further back into the history of mechanical game wheels, two machines stand out as providing the same form of show on a single reel. They were developed by William Dennings, a Civil War veteran in residence at the National Military Home in Dayton, Ohio, as part of the maintenance staff. Dennings liked to diddle around with

Paupa and Hochriem THE ELK, 1904. One of the most sophisticated trade stimulators ever made, with a lot of internal machinery to dispense the proper token when a winner is hit, depending on which of five coin chutes you play. Paupa and Hochriem of Chicago first made THE ELK "Model 7" with a center push lever, as shown in its patent. They quickly changed it to a right side lever, but not before Watling copied and made it as THE MOOSE in 1905. Rare.

mechanical gaming devices, and created his first GAME WHEEL in February 1882, followed by his GAME WHEEL NO. 2 in August 1885. Both machines have a single reel, or thin drum, encased in a countertop cabinet that has numbers on the outside rim. By spinning the wheel, the numbers flash past a viewing window at the top until the wheel stops, displaying a final number, win or lose. These are saloon pieces through and through, and led to the later multiple reel machines, and a simpler single reel format that first appeared in 1897.

The trade stimulator adapter of the single reel idea was Jonas D. Bell and Company of Chicago, Illinois. Bell created two boxy wooden countertop machines with paper graphics called the NICKELSCOPE and PENNYSCOPE in January 1897, produced the nickel model the next year as HOW IS YOUR LUCK, and sold the concept to the Wrigley Gum Company in 1899 as the (WRIGLEY) TRY YOUR LUCK, later to become known as the WRIGLEY'S SLOT MACHINE when it appeared in a gift catalog. The machine was given away to Wrigley gum dealers at no charge for multiple case orders of gum. The idea was to set the game up at

Bell WRIGLEY'S SLOT MACHINE, 1899. Created as a giveaway, and ultimately as a throwaway, the Jonas D. Bell and Company WRIGLEY'S SLOT MACHINE was an outgrowth of their earlier PENNYSCOPE and NICKELSCOPE single reel models of 1897. Made as a Wrigley Gum promotional piece, the game was given free with an order for ten boxes of gum. It is pictured in old Wrigley's premium catalogs. Three award cards came with the machine to convert to gum, candy or cigars at will.

the cash register counter and sell a stick with every play, with the player perhaps winning more in trade. The game was inexpensive, light weight, and suffered greatly at the hands of its players due to its delicate nature. Few have survived as a result. That fact alone all but ended the single reel format for trade stimulators.

It was Paupa and Hochriem of Chicago that countered this situation by entering the trade stimulator market five years later with a highly sophisticated, heavy, and exceedingly durable single reel machine that automatically paid out trade tokens in side cups if winning symbols were hit. Paupa and Hochriem was one of the pioneer automatic payout slot machine makers of the 1890s and brought this experience, the creativity of Gus Hochriem, and the production knowledge of Joseph Paupa to bear on a field that had all but run the course with its initial race, roulette, wheel, and pointer machines and was in need of new thinking. The Paupa and Hochriem five-way ELK of February 1904, their Model 7, provided a new solution to the opportunity. It was a major advancement in trade stimulator play as it took the merchant out of the equation and eliminated any disputes over the show. The winning players who played their coin in one of the five slots that corresponded with a winning reel symbol got tokens in values from a nickel to a quarter which they could then exchange in trade. The more enlightened merchants had dedicated tokens made for their establishment, which meant that only their tokens could be exchanged for cash or merchandise, with tokens won on these machines at other establishments

Mills Novelty SPECIAL, 1906. Mills Novelty made a number of improvements on the Paupa and Hochriem ELK. The elk on the front of the Mills casting is larger and more detailed, with the Mills "Owl" logo on the right side of the case. In the SPECIAL the machine name and five coin chutes have become an integral part of the upper casting to avoid slippage into the "wrong" holes. The view window has been opened up to show near misses, an arrow pointing to the winning line.

Paupa and Hochriem ELK, 1905. The definitive ELK as made by Paupa and Hochriem and also produced by Mills Novelty Company on an agreement basis, with Mills ultimately buying out P & H in 1907. Reel viewing window is large enough to see the winning line and no more. P & H version has lettering "Fortune Teller" in base, while the Mills version says "Mills Novelty Co." in the same place. Symbols are playing card spots. Award card is for "Free Cigars or Drinks." Valuable.

having no value in their shop. These trade tokens alone are important trade stimulator collectibles.

The P & H ELK was a landmark machine as it created a whole new genre of trade devices, and was immediately copied by a number of other makers after its modified 1905 model hit the market. Its sophistication, and manufacturing requirements, eliminated the possibility of small shop production. As a result, only the major machine makers offered the machine, either as original production or as private label made by Paupa and Hochriem. The Mills Novelty Company, having a close personal relationship with Joseph Paupa, was producing the ELK by August 1905. Caille Brothers Company in Detroit also brought out their ELK in 1905, but renamed their machine when fellow B.P.O.E. Elk's Club members informed brother Art Caille that it was unseemly to produce a gambling machine with their lodge name. So in June its name was changed to the tongue twisting but highly descriptive IMPROVED AUTOMATIC CHECK PAYING CARD MACHINE, and advertised as such. The Watling Manufacturing Company of Chicago was making the machine as the IMPROVED ELK (they usually added the word "improved" to any machine they copied to distinguish their version, with improvements rarely made) and the Cowper Manufacturing Company, also of Chicago, had their model THE ELK by 1907.

The various ELK models had barely reached the market when Paupa and Hochriem came back with a smaller and more highly decorated nickel plated cast iron cabinet Model 8 PILOT in February 1906, followed by their Model 9 IMPROVED ELK in March. That started a whole new string of copycat machines, with Mills Novelty adding their PILOT in 1906, followed by the EAGLE in June 1907 and IMPROVED SPECIAL, a larger ELK, in 1909. They also broke away from the limitations of a trade stimulator and created a six-way automatic payout slot machine in the same, but a larger, format called CHECK BOY, itself soon to be copied by Caille and Watling.

BALL version by 1910, actually a modified THE TIGER, and Watling had their own single reel BASEBALL by 1915, as did Silver King Novelty Company in Indianapolis, Indiana, followed by the rebuilt 1918 BASEBALL of the Industry Novelty Company in Chicago, using older machines made by others.

The format had also leapt the Atlantic Ocean to show up in France. Cie Caille in Paris, the French branch of Caille Brothers Company, was selling three- and five-way versions of their LE TIGRE in 1905, adding the same variations for their LA COMETE in 1906. The leading French machine maker Pierre-Abel Nau in Paris was making their

Paupa and Hochriem PILOT, 1906. Smaller and more compact that the ELK, P & H "Model 8" PILOT is a six-way machine for store and soda fountain locations. It has one of the most beautiful and detailed cast iron cabinets ever made for a coin machine. Small details, such as the reel viewing window, are built around a sailing theme. Play, push, spin the reel, and if you win the amount flagged at the top, a bell rings and the proper check is dispensed. Still being found.

Sundwall THE EAGLE, 1907. Early example of revamping, the Sundwall Company in Seattle, Washington, converted the 5¢ P & H PILOT to their own 25¢ version called THE EAGLE for their extensive routes in the Pacific Northwest and the Yukon gold fields. New bronze castings replacing cast iron have THE EAGLE on top, with the lower nameplate area saying "The Sundwall Co. Seattle, Wash." To give collectors hope, this machine was found in Vancouver, British Columbia, in 1993.

Caille Brothers added their SPECIAL, short for the full name SPECIAL AUTOMATIC CHECK PAYING CARD MACHINE, in 1906, and initiated their TIGER line based on the IMPROVED ELK in 1907, leading to a proprietary single reel line of both domestic and export models. Watling proceeded with their own SPECIAL ELK in 1906 and PILOT in 1907 while Cowper dropped out of contention. It was about this time that Paupa and Hochriem sold out their single reel machines and rights to the Mills Novelty Company, who embarked on a series of their own single reel machines which included L'AEROPLAN and UMPIRE, the latter a baseball game format. Caille Brothers had a BASE-

own versions, including LES PETITE CHEVAUX in 1905, LES 3 COULEURS in 1910, followed by L'ELAN (ELK), SPECIAL and L'AIGLE (PILOT or EAGLE). The durability, intricate workings, and cast iron beauty of all of these machines, American and French, have made them among the most desirable trade stimulator collectibles, with correspondingly high antique values.

It was the durability and mechanical capability of these machines that led to another phase of their development. Last produced on both sides of the Atlantic before World War I, the machines survived well past World War II in revamped versions that upgraded them to more modern

A similar situation occurred in Australia after World War II. Prior to the war there had not been much of a counter game or slot machine industry "Down Under," but the war changed that. American air bases, with their accompanying slot machines (used to pay the bills of the officer clubs), brought the machines in. When the Americans pulled out the machines were left, first leading to a revamping industry, and soon, to an indigenous Australian coin machine industry. Charles Shelley Pty. Ltd. of Sydney came up with a unique idea, and bought up all of the American and French single reel token dispensers they could find in Australia, and finally, in postwar France. Modified reel mechanisms and new cast aluminum cabinets led to the SHELSPESCHEL of 1948 and for some years thereafter. Some of these machines are still operated in Australia and Tasmania, making the single reel machines the most enduring trade stimulators of all time.

d'Abel Nau LES PETITES CHEVAUX, 1905. Paupa and Hochriem single reel cast iron token machines were an immediate international hit, with the leading French producer of payout slots and trade stimulators quickly embracing the line. First production by Pierre-Abel Nau of Paris is LES PETITES CHEVAUX, translated as "The Pony," replacing the elk with a horse head. Three-way machine for large French franc. Jockeys are shown in their racing colors. Great marquee. Rare and valuable.

features. The revamping began during their active production period, with the Sundwall Company in Seattle, Washington, creating their own EAGLE machines before 1910 with their own front castings replacing those on P & H and Mills PILOT models. Even the Mills Novelty CHECK BOY payout slot became a trade stimulator in 1917 when F. W. Mills of Chicago converted the cast iron cabinet machines to six-way wooden cabinet trade token dispensers with new front castings as the PREMIUM TRADER. Then, after years of being off the market (although remaining available in the used machine market at very low cost), the P & H, Mills, Caille, Watling, and other single reel machines came back as a five baseball coin head themed counter game. Taking the old machines, modifying the reel mechanism and covering them with new and colorful aluminum cabinets, Daycom Inc. of Dayton, Ohio, introduced the REEL-O-BALL in 1932. Success led to the dedicated Reel-O-Ball Company of 1933, and finally the Yendes Manufacturing and Sales Company, a division of the jukebox operating Yendes Service, Inc, of Dayton, of 1934, selling the game into the mid-1930s.

Mills Novelty UMPIRE, 1910. Baseball was enormously popular in the years before World War I. Mills Novelty took the SPECIAL cabinet and created new castings and colorful baseball graphics for the symbols, award cards, and viewing window, not to mention terrific cabinet castings. A catcher is ready to receive a ball, with the bat on the baseline, "Owl" logo to the right. IMPROVED UMPIRE of 1913 put a baseball below the bat and raised THE UMPIRE name.

Mills Novelty L'AEROPLAN, 1910. Aviation was in its infancy when the single reel machines were in their heyday, so it was inevitable that the two would merge. Mills Novelty built a five-way French export version called L'AEROPLAN, misspelling the name by omitting the "E" from the French word aeroplane. Very detailed castings of early aircraft at a time when the French were the leaders in aviation. Area at base says "Mills Novelty Co. Chicago U.S.A." Two known.

Caille IMPROVED BASE-BALL, 1911. The baseball idea was infectious. Caille Brothers Company of Detroit called their version BASEBALL, with initial models using TIGER cabinets that screw a baseball casting over the face of the tiger. The cabinet casting became dedicated with IMPROVED BASE-BALL of 1911, adding "Play Ball" to the front. Cabinet castings, coin chutes, and marquee differs greatly from Mills, but symbols, reel, and window graphics are virtually the same.

Silver King BASEBALL, 1914. Silver King Novelty Company in Indianapolis, Indiana unabashedly adopted the machines of others—most often Mills Novelty and Caille Brothers—and made them their own. BASEBALL is without question the Caille Brothers IMPROVED BASE-BALL, although there are improvements. The front baseball carries "Silver King Novelty Co. Indianapo Ind." in its casting. Symbols, view window, and marquee graphics are completely new and easier to follow.

Caille NEW SPECIAL TIGER, 1911. Once Caille Brothers Company got away from the ELK name as a result of B.P.O.E. complaints, they established their proprietary TIGER line in 1907, with a tiger on the front. TIGER went through a series of evolutions, concluding with the NEW SPECIAL TIGER in 1911. The play handle was moved to the right side. It is the only single reeler made in this style. Strangely, Caille retained the old style coin head. One known. Valuable.

Caille THE COMET, 1911. Caille Brothers had a branch in Paris, France, called Cie Caille. Cie Caille both made machines in Europe, and imported them from Detroit. Ideas criss-crossed the ocean both ways. Cie Caille sold an export single reel called LE COMETE in both three-way and five-way versions. Caille Brothers in Detroit made a five-way version called THE COMET, the same name in translation. LA COMETE machines are found in France, THE COMET in North America.

F. W. Mills PREMIUM TRADER, 1917. Large machine by trade stimulator standards, th PREMIUM TRADER is a modification of the six-way automatic payout CHECK BOY machine made by Mills Novelty Company itself based on Paupa and Hochriem formats. F. W. Mills Manufacturing Compo Chicago, Illinois, was run by Frank Mills, brother of H. S. Mills of Mills Novelty. He bought up old CHECK BOY slots and converted them to wood cabinet token dispenser trade stimulators.

Daycom REEL-O-BALL, 1932. A Dayton, Ohio, operator named Yendes got the brilliant idea of buying up old single reel Paupa & Hochriem, Mills, Caille and other ELK machines and revamping them into new baseball games. Daycom Incorporated in Dayton converted the machines. They were rare. In the early 1990s a number of brand new, mint condition REEL-O-BALL machines were found in their original crates in New Mexico. Now there are more. Their value went up. Not down.

Shelly SHELSPESHEL, 1948. The conversions of single reel trade stimulators made in the United States and France between 1904 and 1917 didn't end with the Yendes REEL-O-BALL of the mid-1930s. It happened again after WWII. Charles Shelley Pty., Ltd. of Sydney, Australia, bought up hundreds of French machines just before the war. They were held in storage when they were illegal. After the war, and legality, they were converted to the modern looking SHELSPESHEL.

Yendes Service REEL-O-BALL, 1934. There are four makers of REEL-O-BALL with different corporate names. Daycom Inc. was first, then Yendes Manufacturing, next the Reel-O-Ball Company in 1933, and finally Yendes Service, Inc. in 1934, all in Dayton, Ohio. As the mechanisms varied based on the sources, American and French, so did the REEL-O-BALL castings over the years they were made. The machines are so integrated it is almost hard to believe they are old mechanisms.

BABY BELL

Rather than a specific format that describes a class of trade stimulators and counter games, the classification of baby bell, much like that of card machines, is a collective. It is based on the symbols used rather than the mechanism of play. In broad terms, baby bell denotes gambling.

When the first three reel automatic payout slot machine was created by Charlie Fey of San Francisco around 1905, its symbols included horseshoes, stars, card spots, and a bell. With the exception of the bell, these were symbols that had been used on earlier gambling wheels of the late 1890s and early 1900s. Gambling machines were yet to find their own indigenous symbols, but the Fey machine started them in that direction. The German born Fey named his slot machine the LIBERTY BELL in honor of his adopted land, and the cracked bell symbol was quickly picked up by other makers as the personification of a payout slot machine. The final homage to Fey's invention was extended when all three reel format automatic payout slot machines received the class name of Bell machines.

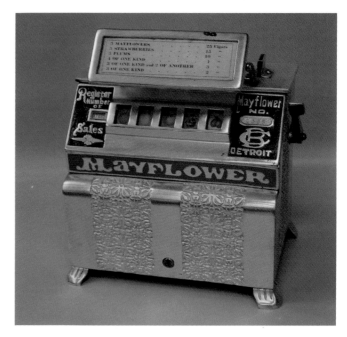

Caille MAYFLOWER, 1910. All metal card spot machine first made by Puritan Machine Company of Detroit in 1905, Caille Brothers acquired the line in 1910 and immediately put their "CB" logo on the upper right front. MAYFLOWER keeps track of plays in a counter at upper left. Caille produced the machine in four models; STYLE A with card reels, STYLE B with fruit symbols as shown, STYLE C with numbers 0 to 9 in three colors, and STYLE 4 numbers 1 to 13 in four colors.

Caille MERCHANT, 1908. Looking older than its years, the Caille Brothers Company MERCHANT is a cabinet throwback to the older UPRIGHT PERFECTION card machines of the past, and specifically the Caille BANKER of 1906. The difference is three reels and fruit symbols, something that only came after the LIBERTY BELL and OPERATOR BELL automatic payout slot machines of 1907 and afterward. A stick of gum with every play. Typical Caille, the front casting is magnificent.

Restrictions on gambling, and a growing body of law against slot machines in the early 1900s, led the Bell machine down a distinctive path. The big machine makers, Caille Brothers in Detroit and Mills Novelty Company in Chicago, promoted their Bell machines as gum venders to get around the law. To further this camouflage, they both developed reel symbols that expressed gum flavors. Caille Brothers included apple, strawberry, and pineapple among their symbols, while Mills Novelty created their own line of Bell fruit gum. These graphics were solidified in 1910 when the Mills Novelty Company copyrighted their new Bell machine fruit gum symbols, made up of lemon, cherry, orange, plum, bell, and a brand name Bell-Fruit-Gum logotype. The Mills fruit symbols became the standard of the industry, with other makers changing the black and white Bell-Fruit-Gum graphics to their own name, such as Pace and Silver King, while Watling used the word Mints. Smaller manufacturers that didn't have the room for their name, or just wanted to remain generic, substituted the word Bar. Most of these symbols are still in use to this day, with the fruit and bar symbols appearing on slot machines all over the world.

When a way was sought to provide merchants with counter games that offered over-the-counter pay in merchandise or cash in the mid-1920s, the baby bell was created. Smaller than its progenitors, without their massive mechanisms and payout systems, the baby bells looked for the world like miniature slot machines. The fruit symbols complete the comparison. The result was a wide variety of games and playing formats that used fruit symbols. The naive giveaway of a few cigars and small amounts of merchandise offered by the pre-WWI trade stimulators gave way to the larger payouts and "blanks" play of the 1920s and '30s, with losing reel shows just taking the player's money. After growing complaints, some of the baby bells added gum vending so the player at least got a gum ball for the effort.

Baby bell control was very rigid and handled in the same manner across the country. The operator usually gave the merchant the instructions for play, with the trade publication *Automatic Age* explaining them in detail in their February 1933 issue in an article by J. D. Roberts, a game operator. In part, here's what he had to say:

"Most Baby Vendors, when leaving the factories, are set to pay out an average of thirty per cent. In other words, for every dollar the customers put into the machines, they only get thirty cents in return. Most anyone will stick a few coins into a new or different machine just to see how it works, but if

Silver King BALL GUM VENDER, 1926. Wherever Caille Brothers was, Silver King Novelty Company of Indianapolis, Indiana, wasn't far behind. The Silver King BALL GUM VENDER of 1926 is virtually the same machine as the Caille PENNY BALL GUM VENDER missing the elegant Caille castings and marquee, with cabinet elements differently sized and placed. In effect, a plain pipe rack version. Machine takes penny, nickel, dime, with center window confirming coin played for payoff.

he is continuously disappointed by receiving little or nothing in return the novelty soon wears off and he ceases to play. That is the one prime reason why machines 'go dead' on locations, so it is up to the operators to overcome this by manipulating the pay-off or reward feature. The answer is to give the customers a better 'break' and you will find the little machines far more popular and an everlasting source of income.

First, open the back of the machine. Now slowly revolve the reel next to the gum compartment and note that the two symbols pass the indicator under the glass for every 'click' of the mechanism. This means that only every-other-symbol on the reel will ever have an opportunity to stop on the line between the indicators. It is then readily understood that there are two sets of symbols on every reel and all that is necessary to change from one to another is to slide the paper bands on which the symbols are printed up or down on the reel for a distance of one picture to effect a change.

You will note that there are five cherries on the reel next to the gum compartment and if you will count those that stop under the indicator when operating the reel by hand you will find that there are only two that do so. By prying up the edge of

Caille PENNY BALL GUM VENDER, 1926. The coming of the automobile in the 1920s made North America mobile and opened up thousands of new locations. New game ideas filled the gap, with the fruit symbol baby bell one of the most compelling. You feel like you are gambling, but at low risk for a penny. Caille Brothers got the ball rolling with a bellwether game called PENNY BALL GUM VENDER that gives you a ball of gum every play, if you push the rod. Widely copied.

the reel at the punch mark with a knife or screw driver the paper strip will slide freely on the reel. Slip the paper up or down, either way, for the distance of one picture, line them up in front, then clamp the edge of the reel down again and you have changed the pay-off from an average of thirty per cent to that of fifty, and you will note that now three cherries will come up under the indicator instead of two.

Additional changes may be made by changing the number three reel as well and extra paper strips which will increase the pay-off beyond fifty per cent may be had, but it is suggested you write the manufacturer or consult your local jobber before attempting to make any changes beyond that just recommended on the number one reel or you may experience some difficulty.

To keep Baby Vendors popular, operators should change the percentage of pay-out frequently. By operating a few days at a very high pay-out you will find many players will be attracted to them. There is no reason for such equipment to become unpopular unless operators make them so by taking more than their just percentage.

A word as to locating these machines may be of help to some. Many schemes have been tried out, but in the long run it has been learned that for most locations the 'penny-a-play' idea works best. In other words, the operator receives a cent for each coin in the machine regardless of denomination. This is easily figured and you will find it easier to place a machine in a new location on this basis. The merchant gets all the 'white money' and the operator all the pennies, but he also receives a penny for each additional coin or 1¢ for every nickel, 1¢ for every dime, and 1¢ for every quarter. This makes it unnecessary for merchants to keep a record of pay-out which of course meets with his approval.

Merchants should always be instructed not to pay out on penny plays. This is only a nuisance and will cause merchants to turn against the little machines as quickly as anything else. Pennies should be considered as played for the ball gum, nothing else."

It's easy to see that there was a lot more to operating these counter games than meets the eye. Just for the heck of it you might want to check the reels on your baby bells to see if they are two cherry or the modified three cherry display machines. If they aren't three cherry machines you might want to change them as it is just as much fun to get a winning show on a collectible today as it was in a store in the 1930s.

Mills Novelty PURITAN BELL, 1926. The mad rush for baby bell machines brought older games out for upgrading. Mills Novelty took their cast iron MILLS PURITAN still in production in 1925, left and above, and recast the cabinet in aluminum with wider reels and a single large window to create the fruit symbol PURITAN BELL, machine at left. It is commonly known as the "Aluminum Puritan." It was part of the late '20s production and material improvements of the Mills line.

Silver King IMPROVED PENNY BELL, 1926. Silver King took an additional step in moving the baby bell format closer to the payout slot machine by switching the BALL GUM VENDER mechanism around to put the fruit reels on the right side and the gum window and vending on the left, creating the IMPROVED PENNY BELL. The full front aluminum casting further enhances the comparison. Original models have award card in center, moved left to avoid conflict with the coin chute.

Caille FORTUNE BALL GUM VENDER, 1927. Quick upgrading of the Caille PENNY BALL GUM VENDER of 1926, the new "improved model" FORTUNE BALL GUM VENDER of 1927 adds full front castings to the oak cabinet, with typical Caille filigree and flourishes. Caille called it a fortune teller, ball gum vender, and miniature bell, "3 machines for the price of one." Award card tells fortunes with numbers in them to indicate payoffs. Renamed JUNIOR BELL in 1928.

Monarch PENNY GUM VENDER, 1927. Monarch Manufacturing and Sales Company, Indianapolis, Indiana, was a wholly owned subsidiary of Silver King Novelty Company handling competing routes, and machines. They may have been used to sell older Silver King machines. PENNY GUM VENDER has a new front, different coin chute and original center positioning of the Silver King IMPROVED PENNY BELL marquee. Monarch Sales, another subsidiary, sold a modified version as MONARCH.

Superior Confection FORTUNE BALL GUM VENDER, 1927. Superior Confection Company of Columbus, Ohio, was another satellite Caille Brothers counter game marketer. When the Caille FORTUNE BALL GUM VENDER with its new cast cabinet came out in January 1927, Superior Confection soon followed with their own in June. The differences are minor, but enough to distinguish one machine from the other. Front castings and marquee are simpler, with Deco lines and painted trim.

Engel IMP, 1927. The baby bell seems to have been invented in the state of Michigan. First by Caille Brothers, followed by the Engel Manufacturing Company of New Holland. The town is interesting. It no longer appears on maps, but seems to have been a suburb engulfed by the city of Benton Harbor. IMP looks like a miniature bell machine, something the other baby bells don't. Award card is coded 11-2-27/3-M, indicating 3,000 cards printed on November 2, 1927. Rarely seen.

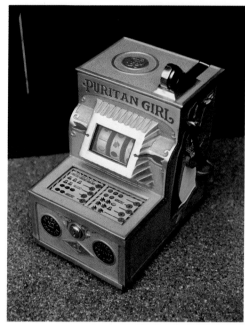

Burnham and Mills BABY VENDER, 1927. Quick to seize opportunity, Burnham and Mills of Chicago modified the Caille design for their own enormous midwest route needs, and also sold their BABY VENDER on the open market. Groetchen Tool made the machines for them. B. & M. Products Company, an alternate name, was a joint venture of F. W. Mills Manufacturing Company and Burnham Gum Company. It was a runaway success, and put Groetchen Tool in the counter game business.

Jennings PURITAN GIRL, 1928. The direct competition between Mills Novelty and O. D. Jennings and Company was heating up, never to stop, in the late 1920s when Jennings produced cover machines for the Mills counter game line. PURITAN GIRL is the Jennings equivalent of the Mills PURITAN BELL "Aluminum Puritan," and is actually a better looking machine. To stake their claim to the game, Jennings put the patent number 1,562,771 for their exclusive coin head on the front.

Midwest JACK POT PURITAN BABY BELL, 1929. Midwest Novelty (Manufacturing) Company of Chicago was the beginning of a major coin machine tradition. It was founded by Ray Moloney, who later reform his business ventures as the Bally Manufacturing Company. In the 1920s Moloney so coin machines made by other manufactur through Midwest, with J. M. Sanders in Chicago making some of his baby bell machines. PURITAN BABY BELL came in plain and JACK POT models.

Keystone Novelty KEYSTONE PURITAN BELL, 1928. One of the stranger looking counter games, and an early baby bell. Keystone Novelty and Sales Company was located in Indianapolis, Indiana. It uses the Puritan name as it has a coin divider cash box that spins off every fourth coin for the operator. First reel has fortunes, so you get fruit plus fortune. The cabinet is aluminum, but it looks like something that might have been made in cast iron.

Superior Confection THREE WAY SELECTIVE MINT VENDER, 1928. Ultimate development of the Caille Brothers PENNY BALL GUM VENDER format, and one of the most attractive counter games ever made. Superior Confection put a modified version of their FORTUNE BALL GUM VENDER on top with a new coin chute and window, adding a three-column selective roll mint vender at the bottom. Play a nickel, get your fortune, and pull the proper handle out to select the mint flavor.

Midwest THE ACE, 1929. Midwest Novel represented a number of counter game manufacturers and put their name on th products. THE ACE is a distinctive baby b seemingly sold only by Midwest Novelty clearly made by Lark Distributing Compo of Inglewood, California. Lark Distributi made THE CHRYSLER, an advanced mo of the LARK, for Midwest Two lions appe on the front of THE CHRYSLER. The diffe ence in THE ACE is the award card, and loss of the lions.

Field Paper Products KEYSTONE PURITAN BELL, 1929. Field Paper Products Company, Peoria, Illinois, acquired the Keystone Novelty KEYSTONE PURITAN BELL barely a year after it came out. Field ran the line out in a dozen models. BABY JACK-POT SKILL VENDER has three reel stop skill buttons in the small "Try Ur Skill" panel at top, gum window at left, coin window at right, and jackpot lower center. Same machine with cover for the jackpot is called 2-IN-1 JACK-POT VENDER.

Mills Novelty THE BELL BOY, 1931. A substantial counter game built like a payout slot machine by the largest producer of mechanical slots, the Mills Novelty Company of Chicago. Interesting play on words for a baby bell, using "BELL" in the machine name. A gambling version with three set dials and number reels is called NUMBERS. Match the numbers to win. THE BELL BOY does not seem to have been particularly successful, so the machines are not seen very often.

Chicago Mint PURITAN CONFECTION VENDER, 1931. Largest Puritan style baby bell, with the Chicago Mint Company name in the top casting. You would think they made it, but they didn't. Production was by J. M. Sanders for Chicago Mint, with award cards often saying "Distributed by K. and S. Sales Co., 4325 Ravenswood Ave., Chicago, Ill.," the same address for Chicago Mint. Also for Garden City Novelty and Pierce Tool and Manufacturing. Lower left panel vends mint rolls.

Giles and Simpkins HOLLYWOOD, 1930. Hooray for Hollywood, and the entire west coast coin machine industry. There was a hotbed of small counter game makers in California, including Lark Distributing in Inglewood, Fey in San Francisco, Mills Sales in Oakland, and others including Giles and Simpkins, Inc., also in Inglewood. Giles and Simpkins made HOLLYWOOD, almost a PARK look alike with improvements and a flashy name, suggesting a connection to Park.

Lion LION PURITAN BABY VENDOR, 1931. Lion Manufacturing Company of Chicago was another Moloney venture, leading to Bally. The Lion LION PURITAN BABY VENDOR was one of two counter games they sold, with other Lion named machines sold by Midwest Novelty, confusing to collectors. It would be nice to believe that the lion heads at upper left and lower right identify this as a Lion machine, and they do. But Buckley also put them on thousands of their own machines.

Pace DANDY VENDER, 1931. Not many baby bell machines have a distinctive appearance, and it is easy to confuse the products of one maker or marketer with another. But not with the Pace Manufacturing Company machines. They have their own distinctive castings and graphics. The colorful gum ball panels front is a flag for Pace production. The other area of confusion is the use of the DANDY VENDER name for dozens of different machines. You can't have it all your way.

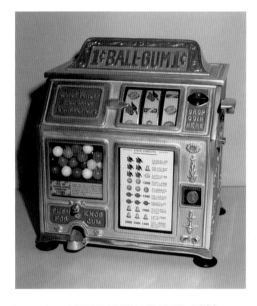

Groetchen NEW DANDY VENDER, 1932. One of the few distinctive looking baby bell machines. Maker is Groetchen Tool and Manufacturing Company of Chicago, who entered the business after making B. & M. BABY VENDER machines under contract. Initial 1930 production had a paper panel at upper left not unlike the Burnham and Mills machine. NEW DANDY VENDER of 1932 cleaned up the castings, with later production closing the upper left window with name text in the casting.

Daval DAVAL GUM VENDER, 1932. The partnership of Dave and Al at Douglas Machine Company went so well, they renamed their firm Daval Manufacturing Company, with DAVAL GUM VENDER their first product. Production was enormous, with casting patterns changed before the year was out. The Daval model looks very much like the Douglis version, except for the bottom of the lower front. The Daval has single art elements on each side of the gum ball frame; Douglis has two bars.

Chicago Mint JACK POT PURITAN CONFECTION VENDER, 1933. A comparison of the front castings, window frame, jackpot feature, side coin chute over the play hand side castings, and detail of the playing card Jack are so clearly the work of the J. M. Sanders Manufacturing Company it seems obvious who made the Chicago Mint Company baby bells. In March 1934 Chicago Mint officially changed their corporate name to Pierce Tool and Manufacturing Company.

Douglis Machine DAVAL GUM VENDER, 1932. The beginning of a big name in the counter game business. DAVAL GUM VENDER was first made by the Douglis Machine Company of Chicago, a firm run by Alexander S. Douglis. Production was extensive, still reflected in the fact this is one of the early 1930s counter game most frequently found by collectors in shops, malls, and sales. Douglis took in partner Dave Helenbein, and named the game after Dave and Al: Daval.

Sanders BABY JACK POT VENDER, 1933. J. M. Sanders Manufacturing Company invented the simple Puritan baby bell, producing them for many firms. They had a proprietary line, with BABY JACK POT VENDER noted for its artistic knave graphics on each side of the reel window. These castings are unique and do not appear to have been made by other producers. Top casting says "Puritan Baby Vendor." Jack pot fills automatically, is unlocked and paid off on three bars.

Pierce WHIRLWIND, 1933. One of the great flops in counter games, the Pierce Tool and Manufacturing Company WHIRLWIND came and went in a matter of months. First introduced in triple dial version in February 1933 based on the Jennings LITTLE DUKE slot machine, it switched to reels in a new cabinet in June. The jackpot was unlocked on the side for a three "Ball Gum" bar winner. Design was said to be based on architecture at the 1933 Chicago World' Fair.

Pace DANDY VENDER, 1935. The Pace DANDY VENDER remained in production from 1931 into 1936. The only apparent change starting in 1935 was in the cabinet castings. The original model has a gum ball front and "Dandy Vender" in the top casting, and a side ribbon banner with the dates 1776-1932 to denote the initial model year. The 1935 model added "Pace Mfg. Company" and "Chicago, Ill." to the top casting to go along with the machine name and a round decal spot on the side.

Daval ULTRA MODERN DAVAL GUM VENDER JACKPOT, 1933. Daval Manufacturing Company and DAVAL GUM VENDER were both on hand at the defining moment of the Golden Age of counter games, five years between 1932 and 1937 when more games were introduced than any five year period before or since. Continual refinement and add-ons kept the DAVAL GUM VENDER a viable product, although the later models without gum windows at the bottom, or with jackpots, are rarely seen.

Dixie Manufacturing DIXIE BELL, 1934. Inventor Floyd R. Marsh skirted the law pretty close with his DIXIE BELL, making it in his Dixie Manufacturing Company, Portland, Oregon. It is a skill game with stop buttons below the reels. He was taken to court over the machine and lost, so not many were produced. In the long run it didn't matter. In 1977, at the age of eighty, Marsh won a million dollars in a Canadian lotto and had a helluva good time with the money.

Pace JAK-POT DANDY VENDER, 1935. Another addition to the 1935 run of the Pace DANDY VENDER was a jackpot model, which Pace called a JAK-POT just as they did for their payout slot machines. This feature was an attachment with a lock on the side that could be opened when a player hit three bars for the payoff. A strange omission is the award card, suggesting that the awards were such common knowledge that a reminder of what paid what was unnecessary.

Buckley ALWIN, 1937. All sorts of things from years past combined by Buckley Manufacturing Company of Chicago. The cabinet came from the 1933 Buckley PILGRIM VENDOR and the jackpot is a counter game generic. What is different is the stamped steel name panel riveted to the top, four reels, and a mottled red finish. Reel one sets odds, with 2, 3, and 4 fruit strips. Awards are three of a kind plus odds. Jackpot is three "Fortune" bars and "Jack Pot" on odds reel.

Bally BABY RESERVE, 1938. A pinball maker that dabbled in arcade machines and slots, Bally Manufacturing Company of Chicago was also significant in counter games. They were short run producers as few are found. Typical of Bally, BABY RESERVE is less than attractive. Yet it was a powerhouse for operators. Fruit, cigarette, and number reel strips and instructions come with the game as shown, although the paper is often lost. Bally fruit includes apples. Swivel base.

Groetchen ZEPHYR, 1937. By 1937 the counter game business had practically settled into Daval and Groetchen with a few stringers on. One of the reasons was new mechanisms, with the two leaders continually upgrading their equipment. ZEPHYR is a gum vender, holding up to 300 gum balls in the compartment behind the window at left. The merchant could refill the chamber without unlocking. Two sets of reels came with the game for interchangeability; fruit and cigarettes.

Groetchen IMP, 1940. How could you make a coin-op game smaller? That flap at lower left also vends gum balls. There is barely enough room left for a mechanism after the cash box and gum ball storage. IMP comes in number, cigarette, and fruit reels, with a black cat. Award card flap flips over the top to make it look like a radio if the fuzz drops around to check for gambling. After a short but populous run, the game came back after WWII as IMP, and in 1949 as ATOM.

Daval CUB, 1940. Wherever Groetchen went, Daval was sure to follow. They pulled off the same miniaturization stunt with CUB ten months after IMP and the race was on. Daval extended the usefulness of their basic game by providing room for five reels, with two blocked off on the fruit symbol CUB. The game also came out as a five reel card game called ACE. All of this machinery in a palmful. Incredible. All three—IMP, CUB and ACE— sold like hotcakes.

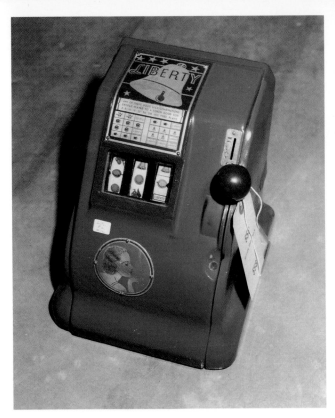

Groetchen LIBERTY, 1940. Both Groetchen and Daval had a standard line of counter games in addition to their small machines. The Groetchen baby bell basic was the token payout LIBERTY, run off in over a dozen coin, ball gum, "discreet" (holding the token behind a window), specialty reel, non-coin, and other models. Its mechanism put the Groetchen pull knob to the far right of the cabinet. Later production added the lady front. Cigarette counterpart is MERCURY.

Western Products TOT, 1940. Western Products Company of Chicago, previously Western Equipment and Supply Company before they went through one of numerous bankruptcies, came out of left field in 1939 and 1940 with a long line of counter games. TOT was created to compete with IMP and CUB, coming in plain and gum ball models, with token payout fruit reels or cigarette. Billed as "the world's smallest token payout counter machine." It was their last counter game product.

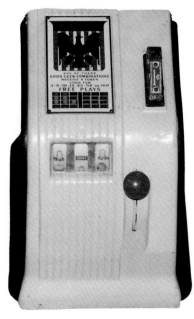
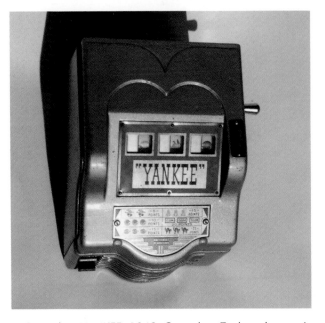

Daval AMERICAN EAGLE, 1940. Wherever Groetchen went, Daval continued to follow. This time they announced their patriotic named token payout AMERICAN EAGLE basic baby bell within thirty days of Groetchen. The games of the two manufacturers look very much alike, although the Daval mechanism places the pull handle to the right center of the cabinet. It's like vintage car spotting; you've got to know what the grille looks like. Many models, including gum. Cigarette version is MARVEL.

Groetchen YANKEE, 1949. Groetchen Tool made a major product switch in 1941, using the miniature IMP mechanism for a full line of small counter games that have the front grilles of 1941 model cars. Baby bell YANKEE is fruit symbol, KLIX has numbers, WINGS has cigarette reels with POK-O-REEL five card reels. WWII ended production quickly. Three of the games came back after the war, including POK-O-REEL, KLIX, and YANKEE, the latter in both fruit and cigarette reels.

CIGARETTE

It was America's wars that led to the hard to break cigarette habit of the present era, going back to their very beginnings. When the Spanish Conquistadors came to the Americas at the beginning of the 16th century they discovered tobacco. They didn't actually discover the fragrant leaf—the American Indians had done that centuries before, and made smoking a ritualized part of their society. The peace pipe of legend brought the Europeans face to face with social smoking, yet that was only the surface expression of a major phenomenon. The discovery of carved stone Indian pipes in burial sites from central through North America reveal that tobacco smoking was a widespread personal habit for hundreds and perhaps thousands of years, also suggesting wide ranging trade routes connecting most if not all of the American Indian nations.

When Cortez toppled the Aztec nation centered at Mexico City his troops brought a local habit back to Spain. The Aztecs rolled their tobacco into tubes made of corn

Lion PURITAN BABY BELL, 1931. Created as a baby bell, the cigarette award upgrading that reached across all counter game lines impacted on the Lion Manufacturing Company PURITAN BABY BELL produced for them by another maker, most likely the Buckley Manufacturing Company based on the casting details. Early cigarette revamp reels have basic brands established in early thirties advertising, including some not so well known, such as Clown, center, reel 1.

Monarch Manufacturing PENNY GUM VENDER, 1931. Cigarette machines were not a product of the twenties. Monarch Manufacturing and Sales Company of Indianapolis, Indiana, enjoyed sales with their PENNY GUM VENDER baby bell, but not as a cigarette machine. That came later. When cigarette machines began to become ubiquitous as counter games, older machines were upgraded with new reels and award cards to convert them to cigarette rewards. Ball gum marquee wasn't needed.

husks, and chain smoked just as enthusiastically as any modern smoker. There are no records of the rates of lung cancer or heart problems as a result of Indian smoking for the simple reason that nobody knew what those illnesses were, and few people lived that long anyway.

Tobacco cultivation and smoking began in Spain after the Mexican conquest, with early forms of cigarettes included. It was picked up by the last surviving Moors in Spain who in turn took the plant and its habits to Turkey and the Middle East, where ritualized smoking took on a whole new look. By the early 19th century cigarette smoking was a southern European treat, with hand rolled Turkish cigarettes the exotic champions of the art. For years the connection to Turkish tobacco (for example, Fatima cigarettes) was proof of an excellent product. By the 1840s hand rolled cigarette production had moved to France where the French aristocracy took up the habit. It soon spread throughout Europe.

Strangely, the art of smoking, originally American, came back to the United States from Europe in the form of the expensive hand rolled French and Turkish products.

Eastern dandies took up the habit, regarded as affecting and effete by the rougher American element west of the eastern cities. That is, until the Civil War. The Union invasion of the southland, and exposure to the more relaxed southern form of hospitality, which included tobacco, made the Army a hotbed of plug chewers, pipe and cigar smokers, and snuff sniffers. An American hand rolled cigarette industry was also created coming out of the war, giving the central southern states a national product, albeit a limited one.

It was in this climate of limited opportunity that James Buchanan Duke of North Carolina, called "Buck" by his family, decided to go into the cigarette business in 1881 at the age of twenty-four. Working at his father's W. B. Duke & Sons family tobacco brokering business, young Buck started out with hand rolled "cigar-reetes" (as they were then called) and soon faced the fact that production was hampered by all of the hand work. It was about this time that he made a connection with another North Carolinian named James Bonsack, who had invented a primitive mechanical cigarette rolling machine, feeding in paper and cut tobacco and producing a finished product. With products in hand (although it took a lot of development time to fully commercialize the machine and finally step up production) Buck Duke applied his special talents. He wasn't

Groetchen SILENT DANDY VENDER, 1933. Just another innocuous "Ball Gum" counter game? Possibly. It is extremely difficult well over sixty years after the fact to clearly identify each and every early 1930s "Ball Gum" machine. But there are clues. The plain gum ball front and screw-on award panel are distinctive to the improved SILENT DANDY VENDER baby bell by Groetchen Tool Company in May 1933. Cigarette version has its own dedicated award panel and reels.

Superior Confection Company CIGARETTE BALL GUM VENDER, 1931. This could well be the first dedicated cigarette reel counter game. Daval claimed it for PENNY PACK, but this is older. It is part of the Superior Confection Company BALL GUM VENDER family, a line that started in 1929. Modified models, such as the BASEBALL BALL GUM VENDER of 1930, have dedicated front castings. CIGARETTE BALL GUM VENDER has new design above "Gum." Marquee is original Superior Confection.

an engineer or a management man, but he was an incredible promoter. Creating flashy packaging, the early beginnings of brand names (his Duke of Durham, Criss Cross, and Sweet Caporals cigarettes were major national brands for half a century), giveaways and package coupons that could be traded in for colored pictures of showgirls, he made cigarettes more than a smoke. They were the open door to an exciting new world.

Even more excitement was created when Duke convinced rolling and sealing machine designer Bonsack that he could also invent a machine to vend cigarettes. Bonsack created the first cigarette vending machine at the end of the 1880s, which Duke then placed at stations on the new New York City electric elevated railway. Sales took an immediate jump. With success at hand, Duke formed the American Tobacco Company in 1890, with a commanding 40% of the cigarette market. By 1905, American Tobacco Company had a whopping 88% of the market until the government stepped in and busted its trust in 1911 with a Supreme Court decision. Soon afterward the Big Four American tobacco producers were sharing an ever growing market. Buck Duke did more than provide a new product, and a new way to sell them; he created a national American habit that will take a lot longer to eradicate than it did to establish.

Cigarettes remained an eastern habit for many years, chidingly said to be favored by riverboat gamblers and others willing to smoke the short white paper "pimp sticks" and "coffin nails" (with the descriptions uncannily close to the truth as later medical research would prove)—that is, until World War I. The A. E. F. came back from France as a "cigger-rett" smoking mass of men, with the habit roaring into the Roaring Twenties as an ever expanding mark of sophistication and personal pleasure. By the end of the 1920s cigarette advertising was among the largest in the nation, along with cars and radios, and women were being directly targeted as the next market expansion. It was about this time that cigarettes were initially being offered as awards for counter game play, with cigarette reels beginning to show up in the early 1930s. By the mid-1930s virtually every form of counter game had a cigarette variant, with cigarette package symbols on reels, dials, and other forms of game displays. National brands such a Lucky Strike, Fatima, Camels, Phillip Morris, Chesterfields, Wings, and others were illustrated on the reels, adding to their promotion.

Cigarette counter games came back after World War II, but only briefly. Other forms of advertising would prove to be more compelling, with radio, TV, mass print media, and billboards getting the lion's share. It would be years before cigarette advertising would be curtailed, and removed from television, with the habit only growing after the heavy promotion and growing marketing success of the World War II years. Cigarettes, regarded as a relaxant and soother of nerves for those in the military (not to forget a massive subsidy from the government, and a significant lack of wartime restrictions on cigarette production), gained greatly in the war years, making it that much more difficult to go after the product in the postwar years. But the tide is inevitable. Laws are now being passed to outlaw cigarette vending machines, and soon sales will be curtailed at the counter, either by increasingly higher prices or by law, or both.

Pace HOL-E-SMOKES, 1933. The original model of the Pace Manufacturing HOL-E-SMOKES made in Chicago uses the gum ball design cabinet of the Pace DANDY VENDER of 1931. It has all the earmarks of a rush job. The front casting substitutes "One Cent Play. Three of a kind wins one package cigarettes any 15 cent brand." in place of the "Your fortune" award card. A new HOL-E-SMOKES header is added, but in this example the ball gum front casting was never painted.

Silver King LITTLE PRINCE, 1933. WHIRLWIND was made by Pierce Tool of Chicago and Field Manufacturing Company of Peoria, Illinois, lastly by Silver King Novelty Company of Indianapolis, Indiana. Silver King changed the name to LITTLE PRINCE. Based on the improved Pierce model with reels. While they look a lot alike they have subtle differences. Pierce at left, Field center, and Silver King right, with advertising in its casting. Two known.

The disappearance of counter games in the 1950s was the first major loss of exposure to cigarette promotion after the war. Soon, and not so far from now, cigarette smoking will be a vestige of the past, at which time the cigarette reeled counter games will be a historical reminder of what was once the most pervasive habit of North America. In the future a collection of 1930s cigarette machines and their colorful branded packages (including "Lucky Strike Green") will be interesting. The time to get these machines is now, before they disappear the same way as smoking.

National Coin PENNY KING, 1934. Superficially, PENNY KING looks like most other aluminum baby bell and cigarette reel counter games of the early 1930s. Then you realize that it is different and more sophisticated. The big difference is a unique mechanism and the A.B.T. coin slide, at upper right. This is Charlie Jameson and his National Coin Machine Exchange of Toledo, Ohio, getting away from dice games and setting up for the future. Landmark game.

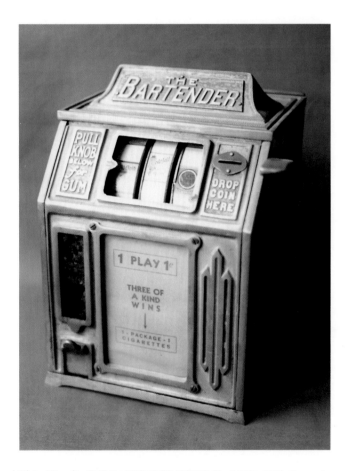

Jennings LITTLE MERCHANT, 1934. After some duller than dull card machines, O. D. Jennings and Company of Chicago wised up and went for some color. LITTLE MERCHANT is exceedingly clever. Reels are cigarette or numbers. You set up the win by dialing the three reels at the bottom with the knurled wheel sets. Then you play. The handle pull locks your lower selection and spins the top reels. Match top and bottom and you win. But it was too complicated, so it's rare.

Ohio Novelty THE BARTENDER (STYLE B), 1935. With Repeal, many new watering hole business starts meant the establishment of scores of tavern supply houses. Ohio Specialty Company in Cincinnati, Ohio, contracted with Groetchen Tool to produce a line of bar oriented counter games. Groetchen TAVERN came in beer (STYLE A) and cigarette (STYLE B) reels. Dedicated Ohio Novelty version THE BARTENDER followed with beer (STYLE A) and cigarette (STYLE B) reels.

Groetchen PENNY SMOKE, 1935. The unmistakable Groetchen Tool plain front, with the screw-on award panel and tall ball gum vender. Groetchen Tool made their own version of the Ohio Novelty cigarette reel THE BARTENDER (STYLE B) as PENNY SMOKE, differing by award panel on front and a new header. Described as "The famous Groetchen-built Penny Cigarette Vender" based on prior experience with the Ohio Novelty games. Awards average 45% of gross, based on 15¢ retail per pack.

Buckley CENT-A-PACK, 1935. With their enormous capacity for PURITAN BELL machines, Buckley Manufacturing Company of Chicago joined the rush for cigarette games near the end of 1935 by riveting a proprietary name CENT-A-PACK panel over the existing integral header, adding cigarette reels and a new award front panel. *Voila! Zigaretten!* Aluminum games often look drab and dirty. A quick wipe with vinegar and a water rinse make them sparkle with new life.

Superior Confection CIGARETTE SALESMA MINT VENDER, 1936. The indigenous Superior Confection counter machines are large and durable enough to allow modif tion to create new forms. CIGARETTE was stretched to a three column mint vender, side knob dispensing just like the larger s machine front venders. In keeping with th huskier machines, the simple coin chute v replaced with a difficult to slug coin slide upper right. Also gum ball model.

Pace HOL-E-SMOKES, 1935. Once the initial rush was over, and when the time came to change match plates for future castings, Pace Manufacturing Company upgraded their three reel HOL-E-SMOKES cabinet to its own look. Except by that time they were sharing the newer stars-and-stripes cabinet design with other Pace games, including SUDS and the redesigned DANDY VENDER, and five reel HIT ME, ARMY 21 GAME, and CARDINAL. HOL-E-SMOKES header clearly identifies the cigarette game.

Superior Confection CIGARETTE GUM VENDER, 1935. After aping Caille, Watling, and Mills Novelty products, Superior Confection started making their own machines in their Columbus, Ohio, factory. They made a series of unique designs produced by no one else. CIGARETTE in plain and gum models is sleek and modern for 1935. It also has twenty-stop reels like a payout slot, which makes it run like a well tuned clock. Gumball compartment is enormous. Cabinet paints are distinctive.

National Coin SMOKES, 1936. National Coin Machine Exchange of Toledo, Ohio was a pioneer in upgrading the lowly co chute reel counter game to a durable co slide counter machine, starting with their PENNY KING in 1934. SMOKES carries the theme, including divided cash boxes merchant and operator, the original and strong National Coin mechanism, and a elegant deckle and polished aluminum cabinet with a dedicated header. Three o kind wins a pack.

aval SPIN-A-PACK, 1936. Cigarette achine lash-ups were all the rage as a tidal ave of smoking and drinking gripped the nited States after the years of government pression of bad habits through Prohibition. er and cigarettes also skyrocketed sales of illed American cheese sandwiches and ndwich makers. The country was obviously ing to hell. But it was tasty on the way. abinet similar to the latest version DAVAL JM VENDER with 1¢ in a redesigned Gold dal area.

Bally BABY, 1936. The beginning of the super miniaturized counter game, and Bally Manufacturing Company of Chicago did it first. BABY is a cutie! First two, then three, different reel sets and award cards for cigarette, free games, and numbers came with BABY (Bally was noted for including a thick envelope of interchangeable reels, award cards and merchant instructions with their games) but usually got lost long ago. A rare find has them all. Seven inches high.

Garden City GEM, 1936. Garden City Novelty Company of Chicago sold a series of small and inexpensive counter games, some also marketed by their parent firm Pierce Tool. Their exclusive line started with the cigarette reel GEM in 1936, a modern looking game that set the pattern for a number to follow. Curved edges give the machine an automotive Chrysler "Airflow" look. Cabinet casting is generic, with varied model nameplates at top, award card center, gum balls below.

val CENTA-SMOKE, 1936. The only erence between the Daval SPIN-A-PACK d CENTA-SMOKE is the change in the der. It looks like one may have been the ck conversion test model for the other. But ch one? There's even a third: WIN-A- OKE. CENTA-SMOKE turned out to be the prietary production model, but that may e been a forced decision as Buckley nufacturing also made a SPIN-A-PACK arette version of their basic baby bell.

Jennings STAR PENNY PLAY and STAR VENDER, 1936. O. D. Jennings and Company of Chicago went for counter games in a big way in 1936 with CLUB VENDER, a token pay cigarette machine almost as large as a full size slot. They took the same mechanism and put it in a smaller cabinet without the token dispenser as STAR PENNY PLAY, adding a gum vending model called STAR (PENNY PLAY) VENDER. In an attempt to make them look classy with a bronze front, they ended up looking grim.

Groetchen SMOKE HOUSE, 1936. When Groetchen Tool brought out their new line of large five reel counter models in the summer of 1936 it included card, dice, number, and horse reel games. But no cigarette model. So they went back in time to quickly fill the hole. Five reel SMOKE HOUSE of August 1936 looks like the Deco cabinet 21 VENDER of 1934, with SMOKE HOUSE nameplate over the hold button area, spinning the five card reels of ZIG ZAG. Instant revamp. Didn't last long. Rare.

Buckley DELUXE CENT-A-PACK, 1937. Buckley Manufacturing Company CENT-A-PACK of 1935 uses the PURITAN cabinet w a new nameplate, plus cigarette reels and paper. DELUXE CENT-A-PACK of 1937 do the same thing with its companion PILGRIM cabinet. First reel sets up the odds while re 2, 3, and 4 indicate win. Buckley picked K cigarettes as their top winner. Maybe Pat Buckley smoked them. Three Kool plus 20 odds is 100 packs. Game comparison sho differences. One known.

Groetchen ZEPHYR, 1937. Starting in 1937 Groetchen Tool had two or more new cigarette games annually until U. S. involvement in WWII, then at least one every year after the war until 1949. Also in 1937 it was a lot easier to accomplish as the effort didn't require dedicated machines, but rather the substitution of reels and award cards. ZEPHYR gives you interchangeable fruit and cigarette reels and award cards, run as you wish. It came back in 1938 as a fortune teller.

Western Equipment MATCH-EM, 1937. Western Equipment and Supply Company of Chicago, Illinois, tried, tried, and tried to become a major manufacturer, but kept missing. It blew them into bankruptcy and reorganization in 1939 as Western Products, Inc., but not before they made a strong appearance with MATCH-EM. They made one of the most modern cabinets ever put on a game, but they left their name off so these machines were unidentified for years. Old ads settled that.

Garden City PRINCE, 1937. MATCH-EM PRINCE look like they came from the sam company. In a way they did, but not exac PRINCE is the last offering of both Garde City Novelty Company and Pierce Tool of Chicago. Again, no name plates so they were "Mystery Machines." Pierce and Garden City flyers from 1937 resolved th Western Equipment was the producer, bu PRINCE wasn't successful enough to kee three firms alive. 1936 Jennings CLUB VENDER in background.

Groetchen GINGER, 1937. Frequently found machine, and a great collection starter as it has good looks, durability, snappy action and Lucky Strike "Green" cigarette packs on the reels. GINGER is also a token payout which adds to the fun. Be sure you get tokens if you buy one as they are hard to replace. Cigarette reels are in penny and nickel play, plus extra fruit reels in penny and fortune reels in nickel. The languid smoking lady is a sensational trashy graphic.

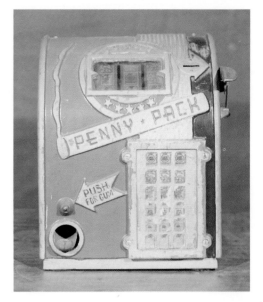

ennings GRANDSTAND, 1937. The mechasm for the token vending Jennings CLUB ENDER ended up in the two non-token ayout STAR PENNY PLAY machines of 936, and came back in 1937 in the garette nickel play token pay GRANDTAND. The mechanical change is the ddition of a one-piece jam proof coin chute ructure at the top with a slot that is an provement over the circular coin chute of e earlier model. A 1¢ version came out in 938 called PENNY CLUB.

Groetchen SPARKS DELUX, 1938. Adding token payouts to counter games made them gambling machines of the same order as automatic payout slots. The difference was called "discreet," with payouts discreetly on the side of the machine and no blazing banners on the front talking about the big payoffs. Cigarette reel Groetchen SPARKS DELUX models come in 1¢ or 5¢ play, with the game also found with beer and horse reels. Gum ball available for every play.

Daval 1940 PENNY PACK, 1939. The most recognizable Daval machine, and one of the best known counter games ever made, PENNY PACK didn't pioneer cabinet design with its modern curved top cabinet in 1938. But it is the game most people remember. Red, white, and blue flag effect and oversize lighted cigarette casting with its upward wafting smoke have a lot to do with that. 1940 DIVIDER PENNY PACK has deeper cabinet cash box. ABCO Novelty rebuilts sold into the 1950s.

rris LUCKY, 1938. Best known for their eptional MASTER peanut, gum ball, and k vending machines, Norris Manufactur- Company in Columbus, Ohio, made ir pitch for counter game sales in 1935 n a five reel card machine called REEL USEMENT VENDER. The game was esigned into a three reel format for the 38 introduction of LUCKY with cigarette ds, plus gum balls on the reels. Three ky Strike—for LUCKY—pays a token. You ays get a gum ball.

Bally WAMPUM, 1938. The profusion of Bally counter games was such that the firm was pumping out four or five new machines a year, with over a dozen each in 1937 and 1938 alone. That came to a screeching halt by the end of 1938, with the cigarette token payout game WAMPUM the last of the prolific runs. It uses the Bally teeny, tiny, mechanism and adds a big gum ball compartment and the token vender. That's a lot in a little box. Modernistic cabinet. Rarely seen.

Daval TALLY, 1938. Daval did a lot of experimentation with cabinet design and game features. TALLY brings a lot of that together. The original numbers reel version (giving the game its name) has a tally meter at the top that inches up a digit every five plays to increase the jackpot. Match the numbers and win it. In the cigarette version shown you get a pack for three of a kind, and get the tally payoff if you get three Spuds. Spuds? Maybe Dave or Al smoked those.

145

Groetchen MERCURY, 1939. Counterpart the Groetchen LIBERTY, the token vender cigarette MERCURY and fruit and other symbol LIBERTY are essentially the same machine with different reels, new machine names and cabinet colors. Groetchen said the game has "windstream design." DELU MERCURY of 1940 adds a universal multi coinage chute and a mechanical clock, replacing the vacuum pump timer of the original MERCURY models. Also a ball gu vender model.

Sanders TOKETTE, 1939. J. M. Sanders of Chicago never had a broad line, but they kept pace with industry developments. Three of a kind win a pack, but three of the top cigarette packages wins a token, automatically dispensed on the side. Not particularly attractive, but hard to find. Original version has TOKETTE name in the casting below the window, later replaced by a paper award panel with a coin build-up jackpot below. Finish looks wrinkled, but it's smooth.

Daval JIFFY, 1940. Miniaturized penny only cigarette game with a full ball grip handle. Uncommon game that often goes unidentified as it carries no maker name on the award card or cabinet. Daval promoted it as having a completely new mechanism, paving the way for the forthcoming miniature ACE and CUB machines later in the year, with JIFFY coming out at the same time as the Groetchen IMP. Somewhat later it came out as a coin head reel game called HEADS OR TAILS.

Daval COMET, 1940. By 1940 Daval had changed their corporate name to The Daval Company, Inc., and two of the big numbers introduced in January of that year were the token payout cigarette reel EX-RAY and JIFFY machines. Omitted from any mention, or advertising, was the visible captive non-payout token version of X-RAY called COMET. It takes the gum ball window off front center and discreetly hides it in a thin line at upper left. Three coin escalator on both games.

Daval MARVEL VISIBILITY, 1940. Daval counterpart to Groetchen MERCURY, wit both competitive games having "M" nar an easy way to remember the cigarette machines. A clue to the manufacturer is where the handle pull sits; inboard on Daval, and flush right on Groetchen. MARVEL is a token "payout," awarding token when three Spuds are hit. MARVE VISIBILITY, shown, keeps token visible be glass inside machine until paid off, to b covered until hit again.

Daval MARVEL BALL GUM VENDER, 1940. Appearance change in the MARVEL BALL GUM VENDER is a lifted coin chute, on all later production of MARVEL machines, and the addition of a large gum ball window front center. This was a MARVEL innovation. Gum balls got dirty or chipped on previous venders, so these are fakes. They are actually colored marbles in a sealed window made to look like gum balls. Daval fruit counterpart is AMERICAN EAGLE, matching all MARVEL models.

Groetchen SPARKS CHAMPION, 1940. It may have the SPARKS name, but SPARKS CHAMPION of late 1940 is an entirely different machine than the SPARKS games that preceded it in 1938. It is a colorful and wonderful looking golden machine. Thin gum ball window on the left side, gold award window on the right. Basic model is cigarette, with many other reel configurations. Two token payout; trade token on the side for three-of-a-kind and gold award in front for three GA symbols.

Sanders LUCKY PACK, 1941. The distinctive J. M. Sanders Manufacturing Company look that took a dozen years to crystallize. But who knew? There isn't a visible mark on any of these games, or other Sanders games that look like it, such as LITTLE POKER FACE NO. 2, to give you any idea who made them. The interior coding for this machine is JMS4. J. M. Sanders! LUCKY PACK is five reels, derived from a poker game. Three of a kind, a pack; four 4 packs and five 5 packs.

Groetchen IMP, 1940. The keystone machine in a counter game collection because of its uniqueness and size. IMP machines aren't rare; they border on the common. Some models are rarer than others and in the long run they'll be worth more. The most common models are the pre-WWII cigarette reels as shown, followed by fruit with the black cat. Post-WWII IMP machines have cigarette reels, most common, and fruits. Be sure award card flap is there. Many were lost.

Sanders ZIP, 1941. J. M. Sanders made his small five reel LUCKY PACK cabinet even smaller for a three reel cigarette machine. Nothing fancy. Three of a kind for a pack of smokes. Your author did war work in a small shop in the Austin district on the west side of Chicago before going into the Air Force, working on the punch presses that made Sanders parts. The old guy that ran the place could hardly wait for the war to end to get back to work for Jack Sanders.

Groetchen WINGS, 1941. When Groetchen Tool made their major product design switch in 1941, putting the IMP mechanism into a full line of small counter games, WINGS was the cigarette model, YANKEE fruit symbol, KLIX numbers, and POK-O-REEL a five card machine. Pearl Harbor ended production in a hurry. The only one of the four that didn't come back after the war was WINGS, with YANKEE available in both fruit and cigarette reels. WINGS is the least likely to be found.

Groetchen ATOM, 1949. It is difficult to try and explain how often the word atom was used soon after the end of WWII. The atomic bombs that dropped on Japan in August 1945 had changed the world (while they saved millions of Japanese and American lives), and soon advertising, promotion, products, plans, and bathing suits (before the bikini atomic tests!) were named Atom, or were Atomic. The trademark office must have gone nuts. Groetchen renamed IMP the ATOM in 1949.

Daval FREE-PLAY, 1946. Some collectors are surprised that any new counter game ideas were produced after World War II. The surprise is that quite a number were, even by Daval and Groetchen. But they didn't last long. FREE-PLAY was a new award approach that had not been tried in counter games. It is based on the free play awards of a pinball game, in which the wins are racked up in a counter to be played or paid (they called it "cancelled") off by the merchant.

Comet NON-COIN COMET, 1950. When Daval closed shop at the end of 1948, they sold out their machine inventory, parts, and rights to Comet Industries, Inc. in Chicago. Comet Industries brought the Daval games back in 1949, including the cigarette MARVEL in coin and non-coin models. In 1950 they changed the machine names and color schemes and added a machined aluminum play handle to make them all look new, with MARVEL becoming the company namesake machine: COMET.

SPECIALTY REEL

Colors, flags, numbers, and other unique graphics are specialty reel symbols that go back as far as the early 1900s, appearing on additional versions of multiple reel card and single reel machines to mask their gambling characteristics, or provide an interesting alternative to poker or playing card spot play. A number of the cast iron pedestal five reel card machines made by Mills Novelty, Caille Brothers, Bernard Sicking and others substituted fortunes and pictures of future mates on their reels, although these are generally regarded as arcade machines because of their placement in amusement centers, and because no chance element or awards went with the play. It would be years before non-awarding counter games would appear, with their play and score features the full reward for dropping a coin.

It was in the mid 1920s on into the 1930s when specialty reel counter games were popularized, with the new symbols tied to the masking of gambling characteristics, the identification of specific manufacturers, unique game features that depended on the symbols to provide the play display, or even current events. The primary substrate for a specialty reel machine was usually a baby bell, with substitution of the innocuous specialty reel graphics for the obvious gambling usage of a counter game when it was associated with fruit symbols. The substitution most often was that of numbers, with many machines offered in fruit and number versions, and in later years, an additional cigarette version. Most baby bell machines had a specialty reel version at one point or another. In a few cases the roles were reversed. When the PURITAN trade stimulator was introduced in May 1904 by the Puritan Machine Company Ltd. of Detroit it had three colored reels. Its CHECK PAY PURITAN version of 1905, made to compete with the Paupa and Hochriem ELK and later PILOT models, had number on color reels which became the PURITAN standard. Mills Novelty in Chicago had a number reeled PURITAN by the end of 1904, with Caille Brothers taking over the Puritan firm in 1905 to have their own number reel PURITAN and CHECK PAY PURITAN machines. It wasn't until 1926 that the reeling was reversed, with Mills Novelty putting fruit reels on their new "Aluminum PURITAN" as the PURITAN BELL. Caille Brothers followed with their own PURITAN BELL the same year, followed by O. D. Jennings and Company in Chicago with their PURITAN GIRL available in both number and fruit reel models in 1928.

Kelley THE KELLEY, 1903. Second most populous machine in old store photographs. Began as the Kelley Cigar Company, with Waddell THE BICYCLE machines found with Kelley Manufacturing Company names. Introduced their own machine in 1903, a modified COUNTER PERFECTION with number reels and a Zeno stick gum dispenser. Intermixed reel spin (three numbered forward, two blank back) for awards. Contacts in grocery goods put machines all over the country.

While these few examples show evidence of a reversal from specialty to fruit reels, the major thrust was just the opposite, with most specialty reel machines evolving the other way around. It started early. The Leo Canda Company had number reel POLICY and FIGARO machines in 1893 and January 1897 respectively, with the latter the pattern for trade machines to come. The five reel THE KELLEY, marketed by the Kelley Manufacturing Company of Chicago in 1903, had originally been created as a card machine, with the reels switched to numbers along the lines of the FIGARO as THE KELLEY was a grocery and drug store penny trade stimulator that had to avoid any direct connection with gambling.

The real reel change came in the late 1920s when the baby bells were also marketed as other variations, changing the reels and sometimes the actual cabinets. The PURITAN BABY VENDER marketed by the Lion Manufacturing Company in Chicago in 1928 had a "Good Luck" fortune reel version with black cats, horseshoes, hearts, and stars as their replacement symbols. When the Buckley Manufacturing Company of Chicago embarked on their extensive line of baby bells in 1931 they included both number and "Black Cat" fortune reel machines in the fruit symboled line. Major counter game producers, including Bally, Jennings, Mills, and Pace plus many of the smaller game makers, provided fruit, cigarette, and number reels on different models of their machines, offering distributors and their merchant customers a variety of options for the same game.

Caille REGISTER, 1906. This machine looks so much like a National cash register of the period it is easy to see how the police would have missed it on a check out saloon or cigar counter. Great subterfuge. REGISTER is a hard core three number reel cigar machine. Total reels for awards: 17 total gives two cigars; 27 total gives 12 with other awards for 19, 21, 23, 25. Every seventh coin goes in operator cash box, the rest into the merchant drawer. One known.

Kelley THE IMPROVED KELLEY, 1905. Collectors get all excited when a new machine, or a new version of an existing machine, shows up. When an example of THE IMPROVED KELLEY showed up in Florida in 1994, with the word "Improved" in the marvelous front casting, it was big news. But there must be more. A Wheeling, Illinois, Pace Auctions listing from 1980 includes one. Collectors should look at their front castings. Only the three numbered reels contribute to score.

The most involving use of specialty reels was with machines that required specific symbols to further their game play. One of the first examples of this latter day requirement was for the baseball games that were revamped from existing Baby Bell and FORTUNE BALL GUM VENDER models, with the Superior Confection Company BASE BALL AMUSEMENT of 1929 such a modification. Numerous others followed. Greater use was made of specialty reels for machines that were created to use them, rather than that of converted games. The three reel JOCKEY CLUB of 1933, made by the A. B. C. Coin Machine Company in Chicago, had a numbered horse on its first reel, a matching number on the second reel (if you were lucky enough to get it) and the award odds on the third reel, creating a whole new form of counter game out of the baby bell format. Groetchen Tool of Chicago made a similar game called HIGH TENSION in 1936, although it added a lot of tension by spinning five reels. Reel one sets the award odds; reel two indicates position such as win, place, or show, and the next three reels have alphabetical letters. If they spell the three letter name of a horse on the award card, you win. Buckley made a four reel version called HORSES in 1937 that set the odds on reel one, and used reels 2, 3, and 4 to spell the name. Their GOLDEN HORSES version

Cowper PURITAN, 1906. Original coin separator trade stimulator. PURITAN throws every fifth (or seventh, depending on setting) coin into the operator box, the rest into merchant cash box they can access with their own key. Counter upper left keeps record of plays to keep everybody honest. Created in 1904 by Puritan Machine Company of Detroit, Michigan, and also sold by others. Rare Cowper Manufacturing Company of Chicago version is on swivel base to confirm payouts.

in a gold painted cabinet was one of the flashiest counter games of the 1930s. They also made a position and odds version called MUTUEL HORSES. Groetchen made one called HIGH STAKES, with horse race reel games also made by a substantial number of other smaller producers.

One of the most used specialty reel formats was that of the 21 game, in which two numbers were given to the player, with the option for two more hits, after which the fifth house reel is revealed. Daval, Groetchen, Mills, Jennings, and many more of the smaller makers produced similar games. Then there were the truly unique specialty reel games that relied on the symbols. The Daval TIT-TAT-TOE of 1936 shows nine windows, with X and O symbols on the reels. Awards were made for anywhere from three to nine correct lineups, straight across, up and down, and criss cross. The Daval REEL DICE of 1936 uses dice symbols, and the very clever REEL SPOT sets up the award on the first reel, leaving the player to pick out the spot on any of the available three covered reels. It is a mechanical version of the three shell game, no spot, no win, odds 2:1. Many other clever game formats were worked out created around symbols that had meaning only for that specific game.

Among the most interesting specialty reel machines are those keyed to human activities or current events. When Prohibition was repealed by president Roosevelt in 1933, effective in stages stretching into 1934, a rash of beer reel counter games followed. Buckley was first with their beer reel PURITAN BELL of 1933, followed by their beer reeled

Mills Novelty IMPROVED PURITAN, 1907. Cast iron PURITAN number reel machines with coin separator cash boxes made and sold by others—Puritan Machine, Caille, Cowper, Detroit Coin—paid homage to the original maker. But not Mills Novelty of Chicago. They put their name on it in 1904 as MILLS PURITAN, ignored the patent and reengineered the machine in 1907 as IMPROVED MILLS PURITAN. Large coin head. Stamped steel parts. Mirror at top so barkeep can see the windows.

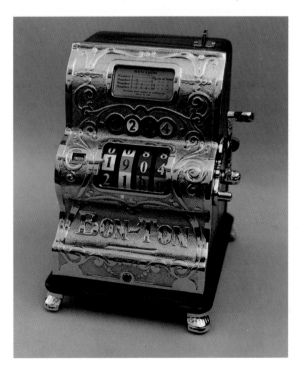

Caille BON-TON, 1907. Plays like a PURITAN, with four number reels and the consecutive play counter at far left. BON-TON by Caille Brothers Company of Detroit is a trade machine, with awards ranging from 2-1/2¢ to 30¢ in merchandise (or cash as the merchant may elect) in penny play, 5¢ to $1 in nickel. Numbers must be consecutive 1, 1 and 2, 1 and 2 and 3, or all four to win tokens in the proper amount. Early models did not have play counter, mechanisms differ.

IMPROVED PURITAN VENDER in 1934. Pace Manufacturing Company came out with their SUDS in 1935, as did Groetchen Tool with their TAVERN games in a variety of styles, also made private label for some distributors. By 1936 the beer reel fad was over, to be replaced by other current themes. Daval made a three reel number reel game called CLEARING HOUSE that was keyed to the daily newspaper U. S. Treasury report, taking the last three digits (the last dollar, and two digit cents figures) of the figure with the merchant posting it next to the machine. If the player hit the right three digits on the right day, winner. Due to the labor intensive requirements of posting the winning number sequences by the merchants, similar market and numbered keyed machines soon disappeared. They were often replaced by sports symbol games, with the Groetchen Tool LIBERTY SPORTS PARADE of 1940 a classic example. The final special reel format was inspired by World War II, with military reels showing up on a number of original machines in 1941 and 1942, with numerous war reeled revamps in the later war years.

Over the years hundreds of specialty reel applications were tried, with the games among those most desired because of their unique play and display features.

Wall HUSTLER, 1927. One of those great looking games and super names that add to the fun of counter game collecting. Wall Novelty Company of Huntington Beach, California, came out of left field with their HUSTLER in 1927. It combines a Fey look with the direction machines were taking in Chicago and the east. A commodious compartment holds gum balls. Reels are numbers, but playing cards tax stamp in award panel suggests that California regarded it as a card machine.

Mills Novelty PURITAN, 1926. Windows too small, and cabinet still cast iron. So Mills Novelty took their IMPROVED MILLS PURITAN and further "improved" it as part of the late '20s upgrading of the Mills line. The big step is an aluminum cabinet with wider reels and a single large window. Reissued as PURITAN BELL in fruit and PURITAN in number reels. It is called the "Aluminum Puritan" to differentiate it from the older machine still in use.

Keeney and Sons BABY VENDER, 1927. So many companies were involved with the BABY VENDER it is hard to keep track of the variations. One that is a nuts on dead ringer for the Burnham and Mills version of the modified the Caille design is the BABY VENDER by Keeney and Sons, Inc. of Chicago, Illinois. A major distributor and mail order coin machine outlet, Keeney and Sons made, and put their name, on a lot of machines in the late 1920s. Fortune reel strips and card.

Midwest LION PURITAN BABY VENDOR, 1930. The false safety net of a lion's head on a cast aluminum cabinet to identify a Lion Manufacturing Company machine, as sold by Midwest Novelty Manufacturing Company of Chicago. Not so fast, stranger. Starting with these castings, Buckley Manufacturing Company of Chicago kept them on their games for another half dozen years. Fortune symbols with a black cat. Structure at top is a special anti-slug penny only coin acceptor.

A.B.C. Coin JOCKEY CLUB, 1933. The beginning of novelty reels and clever scoring. JOCKEY CLUB of A.B.C. Coin Machine Company, Inc. of Chicago started the trend toward horse race reel games. Reel one, at left, stops to show a horse and its number. If reel two, center, stops on a different number you're out. But if it is the same as reel one you are a winner, and reel three suddenly gets very important because it sets the payout odds. Most are 2:1, but it goes up to 20.

Dean Novelty PENNY DRAW, 1934. When these card spot games first showed up they generated excitement. Who made them, where, and when? Fey? Some were called PENNY DRAW, others in a sleeker cabinet PENNY ANTE. The more found the more company names. Finally one had the name Penny Ante Amusement, Riverside, California, 1934. It turns out A. J. Stephens in Kansas City copied the game, and got sued over it. They made PENNY DRAW for Dean Novelty Company, Tulsa, Oklahoma.

Midwest PURITAN BABY BELL, 1933. With Bally Manufacturing Company of Chicago established for pinball games, Ray Moloney's Midwest Novelty (Manufacturing) Company of Chicago still kept selling coin machines. PURITAN BABY BELL is a perfect substrate for a variety of playing plans. Fortune reel strips with horseshoes, four-leaf clovers, and mystical signs is standard. What is new here is the "Free Beer" award card with payouts in 10¢ beers after Repeal.

Wonder Novelty SQUARESPIN, 1933. Here is an absolutely wonderful machine that has a lot of color, and distinctive mechanical workings under glass so you can see everything that happens. And it is not a "Mystery machine." The maker's name is plastered all over it: Wonder Novelty (Company). Front Deco style casting says "Original mechanical SQUARESPIN." So what's missing? What, where, when? Everything but the name. One mysterious example known. Wonder Novelty indeed.

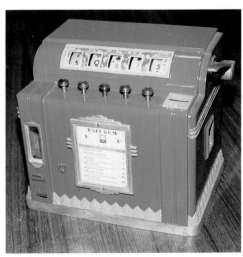

Rock-Ola HOLD AND DRAW, 1934. Typical Rock-Ola, more promotion greeted this game than any other at the time it was announced in September 1934. Five sets of reel strips come with the game to "...allow you to use a new and different appeal each month for five months!" according to its advertising. You get card spots as shown, numbers, baby bell fruit, cigarette, and dice and award cards to match. Quite a few are found. It must have been a reasonable success.

Stephens MAGIC BEER BARREL, 1934. The A. J. Stephens and Company MAGIC BEER BARREL shouts Prohibition is over, Repeal is in. The Kansas City, Missouri, maker created one of the trashiest and socially descriptive counter games ever made, and no one else has it. It is pure, unadulterated A. J. Stephens. The premier piece has pretzel reels on top of the barrel as shown, and vends pretzels from the column. CIGARETTE KEG has cigarette reels.

Mills Novelty DIAL, 1935. Counter games got bigger just as payout slot machines got smaller to be less obvious. Mills Novelty Company of Chicago bridges both machines with DIAL. Mechanism is the popular Mills Q. T. payout slot with a similar set of moon and star reels and 2 to 20 pay award card. Get a win and the Douglas DC-3 airplane starts to notch its way around the globe in free plays. Stay to play the plane around, or get paid off to reset the aircraft to zero.

Groetchen THE FORTUNE TELLER, 1935. The Groetchen Tool plain vanilla front with its screw-on award panel and tall ball gum vender on the left side, decor on the right, shows up again. The fortune telling reels a particularly charming in THE FORTUNE TELLER as they are simple symbols in color with numbers in them to permit reel scorin just like a fruit or number reel game. Fortu is read on the reels. Award card just carrie a symbol and color code for scoring. One known.

Groetchen TURF FLASH, 1935. You have to practically be a handicapper to understand this game. And it's big. For starters, it has thirty coin slots. Thirty! They say ten players can play at a time, but that makes for a crowd around a counter game. You bet on any of ten horses for win, play, or show on front. It takes another penny above the handle to start the race. Three reels with horse's names, win, place, or show positions. An attendant has to pay off, so it flopped.

Groetchen TAVERN (STYLE A), 1935. Groetchen TAVERN came in beer reels (STYLE A) and cigarette (STYLE B). The games were the same with the exception of the award card—which can look a bit screwy, as TAVERN STYLE A puts Lucky Strike cigarettes at the top of the card as the big winner. Below that are beer mugs, beer steins, beer glasses, and beer bottles. There is just about no other way to say or show beer except by the barrel. We'll leave that to A. J. Stephens.

Pace SUDS, 1935. Edwin W. Pace of the Pa Manufacturing Company of Chicago knew his booze; he was a drinker. Beer wasn't beer. To him it was SUDS, with the best ba reel strips in the business. They even show growler, the tin container that kids used to take to the saloon to get a bucket of beer Dad. SUDS has the stars-and-stripes cabir used by other Pace games, including HOL SMOKES, DANDY VENDER, five reel CARE NAL, ARMY 21 GAME, and HIT ME. Two known.

ce NEW DEAL JAK POT, 1935. There is thing kinder or gentler about NEW DEAL K-POT. No gum balls every time or no anks play. This is a hard core gambling achine that is about as greedy as a counter me can be. Gimme the money! Reels are combination of card spots, symbols, and mbers, all of which fit in to a scoring plan sorts while game play is really directed at tting the five of a kind needed to make the erchant unlock that jackpot. One known.

Daval REEL 21, 1936. It's blackjack, with numbers instead of cards. Coin in and push the handle to spin the "Deal," reels one and two. The other three reels spin at the same time, only they are shuttered and you don't know their value. You have two draw reels to enhance your hand, the object to get 21, or at least beat the house. Push draw three and it pops open. If you go as far as four, same story. Under 21 you push the house button to see what you have to beat to win.

Daval REEL DICE, 1936. Clever, clever, clever. Coin in, down handle, spin. Reel one sets the point, with any pair paying 2. You do better with some points than others, including craps, with 12 scoring big. Then the "Roll" occurs in reels two and three and you instantly see if you won or lost. Mostly lost. As a dicer on reels REEL DICE plays the field, plays a point, plays the natural, and pays off on the odds. Gum ball every play to take something home to the kids.

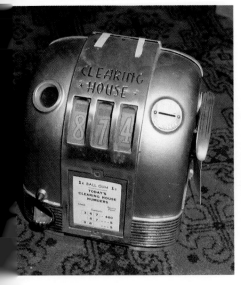

al CLEARING HOUSE, 1936. CLEARING USE is based on the last three digits of U. S. Treasury balance for the day as orted in the newspapers. If on the first ay in April 1936 the daily balance was 942,226,118.74 (2 billion, 942 million, thousand, 118 dollars and 74 cents, ing by today's figures) the number for day was 874. The merchant was ected to keep the daily total next to the chine. Match all three, the first or last to win.

Daval RACES, 1936. The "Humpback" look was distinctive to the Daval line in 1936, and a lot of game ideas were packed into its confines. In RACES there are three positions on each reel, or nine bits of data to interpret. Reel one sets the payout odds; reel two names the horses for each position, win, place or show; reel three names the winning horse. If it shows up on reel two, wherever it rests, win, place, or show, it pays the facing odds on reel one. Breathless game.

Western Equipment REEL RACES, 1936. Western Equipment and Supply Company in Chicago thought the Daval "Humpback" games looked so good, they copied them using similar but modified game ideas and proceeded to undercut prices. Reels click to a 1, 2, 3, and 4 stop. The first two reels must match numbers simulating horses, and if they do you are a winner. Reel three tells you where your horse came in, win, place, or show. Reel four shows payout odds for all three positions.

Groetchen HIGH STAKES, 1936. Where Daval went for "Humpback" aluminum cabinets, Groetchen Tool went for cabinetry. Churchill Cabinet in Chicago made their cabinets, leaving their name inside, often leading game finders astray. HIGH STAKES Reel one sets odds win, place or show. Reel two shows the position your horse crosses the line. Reels three, four, and five give you a chance to build the horse up from matching parts and colors to make it the winner.

Groetchen HIGH TENSION, 1936. Reaching back, HIGH TENSION is a horse race themed game using the two plus three reel format of the Groetchen 21 VENDER of 1934, only this time all the reels are visible after the spin. Reel one sets up the odds win, place or show. Reel two tells the position of the horse win, place, or show. All you need now is a horse. It's done by name. Ever heard of horses named Dan, Sam, Sue, Don, or Sun? Spell any with reels 3, 4, and 5 and you win.

Buckley GOLDEN HORSES, 1936. If HORS attracted attention, wouldn't GOLDEN HORSES attract more? Buckley seemed to think so, and they had just found a new durable gold paint to prove the point. GOLDEN HORSES was announced three months after HORSES as a "Gold type mo that remains perpetually new in appearance." Sixty years later it was still true. Sar game, except they increased lower payout odds from 2 to 4, and dropped high ones from 40 to 50.

Groetchen TWENTY ONE, 1936. Interesting to see how Daval and Groetchen Tool handled the same subject. TWENTY ONE is a blackjack game played with numbers. Coin in, handle down, all reels spin. The first three are shuttered. You see the "Deal," reels four and five (marked "2" and "1"). Use reels two and three (marked "4" and "3") for draws. If you are on or under 21 and have a chance to win, flip the "House" lever (marked "5") and get the news. Same, only backwards.

Buckley HORSES, 1936. Before Groetchen came out with their five reel HIGH TENSION spell name horse game in August, Buckley Manufacturing Company in Chicago did it with four reels in April. One less reel made for a simpler game, giving Groetchen a chance to add sophistication with their position reel. But Buckley was stuck with their smaller five reel PILGRIM cabinet. They masked for four reels, odds on reel 1, spelling Zev, Top, Sue, Dan, and Lil on reels 2, 3, and 4.

Buckley MUTUEL HORSES, 1937. If horse race game excitement was increased with fifth reel, Buckley could do it too. So they went back to the five reel format of the original PILGRIM cabinet, took out the thi reel masks, and added a mutuel feature reel number two. Reels 1, 3, 4 and 5 wer the same as before, with the same horse names ("Zev" became a common joke among horse racing fans). Reel two kicke with "Win" and two "Daily Double" pays, other places losers.

oetchen DIXIE SPELLING BEE, 1937.
oetchen Tool cabinets moved into 1937.
YAL FLUSH splits five reels in two to look
ten. After spin you can release up to five
tters to try and fill out a flush. DIXIE
MINOES also splits the reels, two dice on
h. Far right spins. 7 or 11 and you win.
a point and you have four draws to
ke it. An unadvertised version, as shown,
alled DIXIE SPELLING BEE with letter reels
pell Bingo, Keeno, or Dixie.

Bally THE NUGGET, 1937. No two Bally counter games look alike. They all look like they were created out of whole cloth. You can spot a Groetchen or a Daval by appearance, but you have to know your games to be able to spot a Bally. Fortunately for game collectors, THE NUGGET was advertised as a coin-operated salesboard, and the push-down play arm looks like the BALLY BABY machine. Three number reels to match the payout numbers and awards on the face of the machine.

Garden City BABY JACK, 1937. Western Equipment and Supply Company of Chicago seems to have done better with their private label games than their own. BABY JACK was made for Garden City Novelty Company in the same tradition, meaning cabinet, as PRINCE. Differences abound. BABY JACK is a five reel blackjack game with its award panel up where you can see it. Five reels spin, one and two your hand, levers for draws on four and five, and the house hand in the middle.

al REEL SPOT, 1937. Daval went beyond
Humpback" styling in one year, sticking
tern Equipment with a dated look.
al's comeback was REEL SPOT with a
d top and flat sides. Essentially, it's the
walnut shell game, with three shells
e shuttered reels) and a dried pea (a
red spot behind one of the shutters).
all four reels with the coin drop, with
one at left setting the payout odds. Now
ush the right button for the pea.

Bally GOLD RUSH, 1937. Luckily, the Bally THE NUGGET was advertised. Unluckily, GOLD RUSH never was. This game long defied dating or identification. Inexpensively produced, with a cabinet of bent sheet metal, the manufacturing process could be '40s, '50s or even '60s. Then there is the side handle that plays like a slot machine, completely unlike the Bally THE NUGGET. Also, no names or identification. But mechanism and awards are much the same. It's a Bally.

Four Jacks CLUB JACK, 1937. Can you really believe the markings on CLUB JACK below the two reels: "No coin necessary. Strictly for amusement only. No charge made to play." So what are you playing for, or against? This is a unique non-coin return to the origins of trade stimulators, when the guys in the bar played against each other to see who would buy the drinks. The bartender wins. Four Jacks Corporation was located in Chicago, and made a line of non-coin games.

Daval TRACK REELS, 1938. More cabinet experimentation from Daval. This dial set boxy beauty sets up a three reel horse race, with the player picking the horse to win by the pointer. Once set, and the coin dropped and handle pushed, the first reel tells you if you won or lost. If you have a winner reel two tells you the position, and reel three sets payouts for each position. Taller, and devoid of the dial set, the cabinet came back in TALLY later in the year.

Mills Novelty WILD DEUCES, 1938. Another piece of counter cleverness from Mills Novelty Company of Chicago. This time simplicity itself; no holds or draws, or doubling odds. Plain stuff. While this is classified as a card machine, the reels are card spots rather than the cards themselves. WILD DEUCES plays no different than the old SUCCESS pedestal card machines of forty years earlier except that deuces are wild. Two pair awards 2, royal flush awards 50.

Bennett DOUGH BOY, 1940. Paul Bennett made their games smaller by moving the gum dispenser to the left, away from the cabinet stretching bottom. Fewer gum bal but neater machine. DOUGH BOY was a attempt to capitalize on the draft and the men in uniform, but they used a dated W expression. "G. I. Joe" hadn't been coinee yet, and wouldn't until the American invas of Italy. Four reels ending in 10 to 50 pay 10¢, number 2000 pays $25. Most reel symbols are a doughboy.

Mills Novelty KOUNTER KING, 1938. Looks simple, but it's a tough call. KOUNTER KING stores up free plays as you rack up points. Reel numbers are only 1, 2, or 3. Match reel one and reel two you get a 2:1 payoff, or exercise the option to open reel three. If they don't match, you lose. But if all three match it's 6:1, or the option for reel four at 20:1, on up to 60:1 for reel five. With only three numbers it's a cinch, right? It's a killer.

Bennett DEUCES WILD, 1938. Same game, twisted name, plays the same. Wooden DEUCES WILD made by Paul Bennett and Company in Chicago was for the same locations as the metal Mills Novelty WILD DEUCES, costing about half as much. Mechanism and cabinet were first made for a three reel cigarette game called LUCKY PACK. Compared to Mills, awards are higher. 2 for two pair, 3 for three of a kind, 12 for a flush, 15 for four of a kind. Mills paid 2, 2, 8, and 15.

Bennett SKILL TEST, 1940. Same game DOUGH BOY (sans the doughboy on re one) makes Bennett SKILL TEST an entire different game with a new award panel. even plays the same, but with different numbers. Numbers ending in 11, 22, 3 44, etcetera award 2 skill points. 99 an as last two digits award 5 skill points. 5 6666, 7777, etcetera award 50, and 00 awards 100 skill points. Bennett games difficult to identify as they do not carry t company name.

Daval X-RAY, 1940. The beer reel equivalent to the cigarette X-RAY by the newly renamed Daval Company, Inc. X-RAY is a token payout that pays off on three of a kind "for added amusement" as it says on the front. No award card. It was all word of mouth; you just knew what you'd get. The other front copy says "Your coin buys ball gum," dispensed at left with fake marbles in the window. Token payout as X-RAY and visible captive token as X-RAY VISIBILITY. Coin escalator.

Groetchen LIBERTY SPORTS PARADE, 1940. LIBERTY is the Groetchen substrate game for 1940, off in a lot of directions. These games are often found, usually with baby bell or cigarette reels. Rarer models are LIBERTY SPORTS PARADE in straight, "discreet," plain and ball gum models with sporting reels showing tennis rackets, footballs, boxing gloves and other items. The smoking lady on front of the cigarette models is replaced by a badge showing two guys boxing.

Daval "21," 1941. Using CUB and ACE as a starting point, Daval packed a lot of entertainment in a tiny package with "21," a blackjack game with all the features of much larger machines. The downward push handle spins all five reels, shuttering 3, 4, and 5. Reels 1 and 2 set up your hand, allowing up to two hits on reels 3 and 4. Make or stay under 21, and push house button 5 in the center and see how you did. Lousy mostly. This game is very hard to beat.

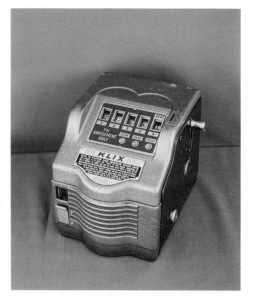

Daval HEADS OR TAILS, 1940. Big grip for a small game. Counterpart to the cigarette ODDY. Clever play, and unique to Daval. Your coin sets the play, and you insert it heads or tails. Your coin shows up in a window at lower right. You want all three reels to be heads or tails as you selected, with the center reel also setting payout odds number. The window carries the pasty looking marbles that were supposed to make your mouth water for a gum ball.

Groetchen SPARKS CHAMPION, 1940. While the basic model may be cigarette, SPARKS CHAMPION comes in a lot of other reel configurations, including numbers, beer, sports, and horses. The horse game is the most colorful, building the nag up from matching parts and colors across three reels in the manner of HIGH STAKES of 1936. You have a shot at two tokens; three GA symbols give you a gold award in front while matching a horse drops a payout token on the side.

Groetchen KLIX, 1941. The blackjack game in the Groetchen Tool quartet of IMP mechanism counter games, with YANKEE fruit symbol, WINGS cigarette, and POK-O-REEL five cards. KLIX plays the same as the Daval "21," except things are moved around. Push that handle down and all reels spin, numbers 3, 4, and 5 shuttered. Reels 1 and 2 are your hand, filled out by draws on reels 3 and 4. Reel 5 is the big surprise as the house hand. These small games are really fun to play.

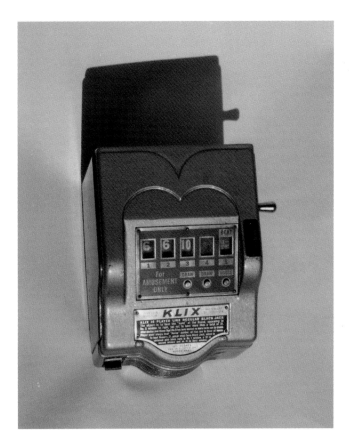

Groetchen KLIX, 1949. Home from the wars. Groetchen had so overproduced the four KLIX type counter games they were generally available throughout World War II if a distributor or operator was willing to pay the price. It wasn't until the end of the 1940s that the supply finally ran dry. So they came back. The cabinet differs, sprayed silver over new castings, with an added trim line at the top. This time around they were calling the game KLIX "21" BLACK JACK. Fun.

Comet KING, 1950. When Chicago based Daval sold out to Comet Industries Inc., also in Chicago, Comet kept pouring out the same old games with the same old names until 1950 when they pulled the switch and everything was renamed. ACE became KING, with coinage stepped up to penny, nickel, or dime. The latter was Korean War inflation at work at the smallest level. A lot more ACE games from the '40s are found than KING games from the '50s. Straight poker play.

NOVELTY AND SKILL

There are quite a number of trade stimulators and counter games that defy classification due to the novelty of their play features, or the requirement of skill to achieve the final result, something most often omitted from the early machines. Both features were sometimes combined in a single machine, with this diverse range of games loosely classified as novelty and skill machines.

Novelty games were seen early in the development of trade stimulators, and constitute a significant number of some of the more popular machines over the years. They are usually extensions of the basic race, roulette, wheel, pointer, dial, coin drop, and reel machines in one way or another, presenting these drive and display features in a different way. Few were completely original in concept. Among some of the more unique games utilizing a variation of the wheel format are the Brunhoff Manufacturing Company tops, including DIAMOND EYE and the smaller and more refined SPINNING TOP, both made in Cincinnati in 1899. The pointer format was used in a variety of ways, with creative examples including the enormous floor standing WONDER CLOCK made by the Yale Wonder Clock Company of Burlington, Vermont, in 1899, with a variety of subsequent models that followed. The Park Novelty Company RED BIRD made in Kalamazoo, Michigan, in 1903, was designed to advertise a specific brand of

Ganss GAME TABLE, 1889. Barroom amusement often meant a table game. Created in the 1880s and 1890s, few have survived. Non-coin Ganss, Clarke, and Dengler GAME TABLE made in East Saginaw, Michigan, keeps track of scores for various card games. Scoring drums for each player are set under glass at the table center, set by a handle at each playing position. The original version has a knee high shelf under the table for glasses. It kept the players buying beer while playing.

Weston SLOT MACHINE, 1892. Clockwork mechanism automatically activated when coin is dropped. Center lift rod tilts the dual bagatelles back to move four steel balls, two each, to the top. Pinfields are then lowered and the balls bounce down to settle in scoring channels below. Competition for the Clawson AUTOMATIC DICE SHAKER. Top scoring total of four requires four balls stacked two high in center channels. Weston Slot Machine Company was in Jamesville, New York.

cigar. When the dial is spun, with brass brads on its periphery, a small cork disk bounces all over the place behind the glass to land in the winning number (most often at the bottom, but sometimes slightly to the left or right) when the dial stops spinning. The flat circular cork becomes the pointer, although it could be said that the game was also a roulette clone in the vertical plane. One of the best known variations of the pointer format is the Twentieth Century Novelty Company SPIRAL made in Springfield, Ohio, in 1903 and for some years thereafter, becoming a basic country store trade stimulator that vied with THE BICYCLE for attention, although it is rarer. The dropped penny works its way down a rotating spiral in a tall glass walled box, with the spiral spinning a pointer dial at the bottom. When the coin reaches the bottom the indicator is pointed at the score. A dial variation is the floor standing 1901 STAR TRADE REGISTER made by the Star Trade Register Company of Montpelier, Vermont, in competition with the YALE WONDER CLOCK made in the same state.

Two rare novelty games utilizing the basic coin drop principles, including Galton Board results, are the Weston Slot Machine Company SLOT MACHINE, a form of double playfield bagatelle game similar to the Clawson AUTOMATIC DICE (SHAKER), made in Syracuse, New York, in 1892 and the FLIP FLOP, or LOOP-THE-LOOP, trade stimulator marketed by the Kelley Manufacturing Company of Chicago in 1901. The target format is represented by the Chicago based Pierce Tool and Manufacturing Company BUGG HOUSE of 1932, with five target shooters and a row of symbol pockets to be made across from each one. It takes five coins to play the complete game.

Clawson Slot Machine THE NEWARK RAINBOW, 1896. Set color on dial, drop nickel in at top, pull down handle. Hit your color and you get a value of 15 in merchandise for white, 10 for yellow, 4 for green, and 2 for red or black. Very early cast iron, size of THE FAIREST WHEEL. Only known example was assembled from parts found in a bushel basket by the author in the old factory building in 1978, based on original photos and advertising. There must be more.

A machine so rare it has never been found (yet!) is the G. F. Hochriem Manufacturer BOOSTER, a five reel counter game that has three dimensional horse castings mounted on the upright rotating reels, a combination of the race game and specialty reel formats. Hochriem made the game in Chicago in 1935. Finally, the single reel format is represented by the inexpensive and fairly common SPIN-IT made by the Shipman Manufacturing Company in Los Angeles starting in 1947. SPIN-IT spins a dial at the top of its nut vending mechanism that shows horses, numbers, or other symbols depending on the model, making it a variation of the single reel machine, although the reel drum spins on a horizontal axis.

Not all novelty games followed the basic drive formats, with some completely unique. The Yale Wonder Clock AUTOMATIC CASHIER AND DISCOUNT MACHINE of 1905 flashes colored lights to come to a stop and a score while flipping advertising cards inside the machine. This lighting format later showed up in the Moon & Avers FLICKER of 1933 made in the unlikely town of Baraboo, Wisconsin, and the later more exciting Bally LITE-A-PAX

Clawson Slot Machine LIVELY CIGAR SELLER No. 2, 1893. Pennies bounce all over bent wires behind glass tumbling down to land heads or tails at the bottom. Put in five pennies across the top, push down wire lever, and bing, bang, plop! Three heads up, 1 cigar; four up, 2 and five up, 3. Clawson Slot Machine Company, Newark, New Jersey. LIVELY CIGAR SELLER Nos. 1 and 2 have different wire patterns. June 1893 No. 1 has been reproduced; October 1893 No. 2 hasn't.

and PONIES counter games of 1938. The William C. Jones AUTOMATIC WIZARD CLOCK of 1903, first made in Niantic, Illinois, and its later Progressive Manufacturing Company WIZARD CLOCK of 1905, made in Pana, Illinois, being made by numerous other manufacturers after that, were completely unique, with an erratic token payout mechanism behind a mantle clock, paying one or a few more tokens per play. One of the most creative trade machines is the MISTIC DERBY of 1947, made by the Maitland Manufacturing Company of Chicago, in which the horse you play is lighted while a record plays a race, selected at random. If your horse comes in first, vocally, you win. The MISTIC DERBY is a combination of the race game format with the jukebox, therefore unique. Other interesting formats included punchboards, represented by the D. Robbins and Company EVERLASTING AUTOMATIC SALESBOARD of 1930, followed by numerous other versions which include the Chicago made Boduwil PROMOTION SALES REGISTER of 1931, B. A. Withey AUTOMATIC SALESBOARD of 1932, Mills Novelty TICKETTE of 1935 and the Daval AUTO-PUNCH and Groetchen PUNCHETTE games of 1936, among others.

Brunhoff SPINNING TOP, 1899. Somewhat more sophisticated and cleaned up, the final and patented version of SPINNING TOP is smaller than the earlier Brunhoff DIAMOND EYE trade stimulator. SPINNING TOP adds a coin head and anti-slugging window to confirm actual coin played if an award is indicated. Awards are paid based on the number indicated by the pointer at the top, with actual return indicated on award card below the coin head. Exciting machine.

Brunhoff DIAMOND EYE, 1899. There were a lot of ways to make a cigar wheel, and Brunhoff Manufacturing Company of Cincinnati, Ohio, tried a number. The most outlandish is DIAMOND EYE. The top carries the advertising for Diamond Eye cigars, with numbers on its periphery. It spins under a glass dome when the coin is dropped. Pointed arm indicator at top points to winning number to indicate cigars won. "Pat Apld For" on side.

Due to their novelty features, and often clever mechanisms, most novelty trade stimulators and counter games tend to be among the higher valued pieces, although the Shipman SPIN-IT and some of the other high volume post-WWII games disprove this theory by being exactly the opposite.

The skill and "whirlwind" games go back to the early days of trade stimulators, but received their real impetus in 1930 with the arrival of the Peo Manufacturing Company of Rochester, New York, LITTLE WHIRLWIND, an internal spiral game scored by small steel ball bearings. Peo followed up with a whole line of spiral games, with BARN YARD GOLF appearing in 1931. Unfortunately, so did the whole coin machine industry, with over a dozen copycat makers producing ripoffs of the Peo spiral games in a matter of months after the original appeared. A few makers produced them under license, D. Gottlieb and Company among them, while most just copied. Among the major makers of spiral games were Gottlieb, Keeney, Genco, and Pace, with many others right behind them. The format survived for years, with Daval making double playfield versions as late as 1947 called BEST HAND, MEXICAN BASEBALL, and OOMPH.

Mills Novelty THE MANILA, 1899. Much sophistication built into THE MANILA made by the Mills Novelty Company in Chicago. The machine was created by Paupa and Hochriem of Chicago in 1899, named after the Spanish-American War victory by Admiral George Dewey. By the end of the year Mills was making it. Shoots the nickels played, with target hits awarding from one to ten tokens. Side removed to show payout mechanism. Two known, this fine example was found in Tasmania in 1993.

Another distinctive collective format in this catchall classification are the combination novelty and skill games, made in much fewer variations due to their distinctiveness and clear association with specific manufacturers. An early example is the COIN TARGET (BANK) made by M. Siersdorfer and Company of Cincinnati in 1894. The small handgun at the front shoots the played coin across an open area encased by a glass cylinder at a far away target with coin size holes in its face. Shoot the coin into one of the holes (almost impossible!) and you win. Paupa and Hochriem in Chicago made a much larger version called THE MANILA in 1899, also sold by the Mills Novelty Company as THE MANILA. The machine has a series of taken payout tubes for winning shots.

Another unique format is the "Climber," a series of ramps that can lift a ball from the bottom to the top of a playfield through careful manipulation of the lifting knobs. Groetchen Tool of Chicago introduced the principle in a series of large arcade games at the end of the 1930s, and soon had a 1941 counter game version called PIKES PEAK. The idea came back after WWII in an inexpensive counter game called TILT-TEST made by Atlas Games in Cleveland, Ohio, in 1947, and was later picked up by Auto-Bell Novelty Company of Chicago who continued to make their own TILT-TEST throughout the 1950s.

A final, and one of the most significant novelty and skill game formats, is the "Kicker-Catcher" style. Originally pre-WWI European, with the German Jentzsch und Meerz BAJAZZO (CLOWN) of 1910 a classic of the style, the first of these games was made in the United States by Paupa and Hochriem in Chicago, adopting the German idea to the American game of baseball in their BASE BALL ball catching game of 1917. A ball is kicked into a pinfield by a "kicker," a variation of the target format, and then a movable "catcher" is manipulated across the playfield left-to-right to catch the ball for score, a free game, or a coin return. The Caille Brothers Company, and a number of other producers, were making their own versions by the mid-1920s, with the resulting CLOWN and JESTER games among the most valuable. Gottlieb made a similar game at the beginning of the 1930s, followed by the more successful Berger Manufacturing Company CATCH-N-MATCH of 1936. By 1941 Baker Manufacturing Company of Chicago was making a game called KICKER-CATCHER as one of the most popular counter games of the immediate American pre-WWII period. During the war years these were revamped, with numerous other manufacturers making similar games, such as KILL THE JAP of 1943. After the war, J. F. Frantz Manufacturing Company obtained the rights to the Baker KICKER-CATCHER games, and proceeded to make a broad line of variations into the 1970s, when they in turn were picked by Johnston Products, later to become Fun Industries, with the games still in production as of this writing. It is among the most enduring trade stimulator and counter game formats of all time.

Novelty, skill, and novelty/skill games have been produced in great numbers in great variety, with this synopsis touching on only a few of these unique games. If you go to the coin machine shows, and keep your eyes open, you will find many more, and some you probably can't live without. So get them.

Yale YALE WONDER CLOCK, 1899. YALE WONDER CLOCK by the Yale Wonder Clock Company of Burlington, Vermont, was made for stores and restaurants. Play a nickel and the lights flash in sequence. When they stop, the white light vends a token good for a 5¢ cigar, blue good for 2, green good for 5 and red for 10. Regina music box mechanism adds music to the play. Restorer Richard J. Gilbert of Oconomoc, Wisconsin, stands next to the machine to give it scale.

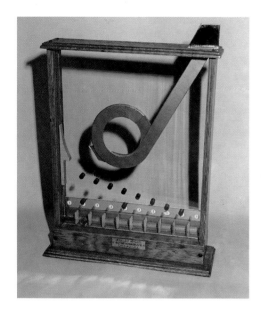

Kelly FLIP FLAP, 1901. FLIP FLAP (also called LOOP THE LOOP, but the original name is accurate) was made by Kelley Manufacturing Company, Chicago, Illinois, originally a cigar distributor. Coin goes down the colored cardboard chute from top right, whips up some speed as it makes a loop, and comes zipping out of the left end over the rubber bumpers to make a 1, 2, or 3 scoring trough to win a cigar or two or three. Difficult to control as speed is always the same.

Park Novelty RED BIRD, 1903. Almost a dead ringer for the Bennett and Company boxed cigar wheels made in Kalamazoo, Michigan, in 1900. Except for two things. Maker is Park Novelty Company, also of Kalamazoo, dating a few years later. The big exception is the play. A small cork disk is held in front of the pinned wheel behind glass, with catchers for each number like a roulette wheel, only vertical. When the wheel spins, the cork bounces, and lands in a hole.

Twentieth Century SPIRAL, 1903. Known to collectors as SPIRAL, this may or may not b the correct name. The problem is that no example, or corroborating contemporary advertising, has been found to reveal the actual name. Which is strange as numerou examples are known, suggesting a relative high population machine. Thomas A. Watt of Springfield, Ohio, applied for its patent on February 3, 1903. The Twentieth Centu Novelty Company of Springfield is the maker.

Mills Novelty LITTLE KNOCKER, 1902. Big and heavy, but it doesn't look it in a picture. The machine is two feet high and weighs almost fifty pounds. Five-way coin head in the manner of a payout slot machine. Pick your color red and black, white, yellow, blue, or green by putting your nickel in the right slot, then strike the plunger hard. The owl rides up a rod, and if it lands on your color the awards are from five cents to a dollar in trade. Only example found in 1991.

Loss (Novelty) SKILLIARD, 1903. Taking the Latimer GAME O'SKILL target game a step further. This is the back of the SKILLIARD, having a November 1903 trademark date in its casting. From this side it looks like a GAME O' SKILL in a cast iron base. Turn it around and you find a peg on a chain with a hole below each number. Put the peg in the hole, hit the coin with the shooter and land in the pegged number to win. Too difficult to win so it probably wasn't popular.

E. D. Parker SPIRAL, 1904. Difference between Twentieth Century and E. D. Par SPIRAL is the nameplates. Twentieth Cent has an older ivory aged celluloid namepl that makes the machine, with the newer version having a decal saying "Mfd. by E Parker Co. Springfield, O." The action is exciting. When you put the coin in the to slot and push in the button the coin runs down a twisted circular runway in the ce spinning both the spiral shaft and the po below, indicating 1, 2, or 3 times value

Yale AUTOMATIC CASHIER AND DISCOUNT MACHINE, 1905. Success of the YALE WONDER CLOCK of the Yale Wonder Clock Company of Burlington, Vermont, led to a line of even larger machines that flash advertising cards and lights and play a tune. Last in line was AUTOMATIC CASHIER AND DISCOUNT MACHINE with Regina No. 216 musical disk movement that plays a variety of tunes backed up by twelve bells. Trade check payout for stopping on colored lights. Ice cream parlor machine.

Dunn Brothers WRIGLEY PERFECTION "Straight Globe", 1906. With success of the "You Can't Lose" WRIGLEY DICE MACHINE given to shopkeepers free for stocking orders of gum, the Dunn Brothers WRIGLEY PERFECTION wasn't far behind. Glass walled mechanism shows the play action, and accumulated cash. Coin in, plunger down, and two steel balls pop. Eight colored holes; two each red, white, blue, and yellow. Always a cigar or gum package, but double if the two balls match colors.

Loheide WIZARD CLOCK ("2 column"), 1907. Once WIZARD CLOCK entered the mainstream of acceptance, half a dozen producers made the machines. Cigar maker Loheide Manufacturing Company in St. Louis, Missouri, was one of three local manufacturers or assemblers of the clocks that gave stores both a timekeeper and a trade stimulator. Nickel in and clock chimes once, twice, or three times while it issues corresponding value checks at the right side.

Dunn Brothers PERFECTION, 1905. Dunn Brothers of Anderson, Indiana, had a nice niche in their small countertop cigar and gum trade stimulators in addition to their store scales and cheese cutters. PERFECTION is a no blanks cigar machine. Original "Round Glass" has a glass globe, frequently broken when the two steel balls inside were popped. Later models substituted a capped globe held by posts. Eight holes; six "1" and two "2." Match numbers and win that amount.

Progressive Manufacturing AUTOMATIC TRADE CLOCK ("2 column"), 1906. WIZARD CLOCK, later generic name for the trade token issuing 8-day clock found on cigar counters and in grocery stores, has a diverse and convoluted history. Invented by two men: William C. Jones in Niantic, Illinois, in 1903 and James G. Huffman of THE FAIREST WHEEL fame in Decatur, Illinois in 1906. Huffman's Progressive Manufacturing Company of Pana, Illinois, made it for Albert Pick & Company.

Loheide WIZARD CLOCK ("4 column"), 1908. Loheide Manufacturing WIZARD CLOCK counter machines come in two models. "2-column" is smaller, less expensive, while "4 column" has more elegant trim and marble top. Ostensibly made for cigar counters, with virtually all United Cigar Stores having them on their cigar display counters, the WIZARD CLOCK is also the most common trade stimulator found on grocery and general store counters in old photographs. They were everywhere.

Pogue, Miller WIZARD CLOCK ("4 column"), 1910. Largest of the WIZARD CLOCK machines, the Pogue, Miller and Company of Richmond, Indiana, version has gilded trim and metal feet, nameplate at top edge. Otherwise similar to the Loheide clocks, although four inches wider. Lucky finds are made when these machines are found as clocks rather than coin machines. Many more remain to be found, with new maker's names anticipated. If Pogue made a "4-column," they probably made a "2."

Jentzsch and Meerz BAJAZZO (CLOWN), 1912. The German firm of Max Jentzsch und Meerz located in Leipzig, eastern Germany, was the most influential coin machine maker in Europe, and exported games to France and Great Britain in substantial numbers. Coin drop and catch BAJAZZO became their best known machine, first made in 1906 and repeated with changes for another twenty years or more. Called THE CLOWN in England, LE CLOWN BAJAZZO in France, and CLOWN in America.

Caille JESTER, 1925. Caille Brothers Company in Detroit made an American version of the German Jentzsch und Meerz BAJAZZO in 1924 as CLOWN. Smaller in size, it is readily identified by its top rounded corners cabinet, the only version made that way. Caille called it "An entirely new idea," which it was in North America, but not in Europe. A short run 1925 version called THE JESTER shows the CLOWN in blackface as illustrated. This rare example showed up in 1995.

National Institute A GOOD TURN, 1910. Use of trade stimulators and counter games as charity cash collectors goes back in time. A GOOD TURN is the oldest example known. Placed in stores and public places by the National Institute For The Blind. Coin chute says "Pennies Only" which spins the man with the pointer to your fortune. May be English as its penny chute is as wide as a British penny and A GOOD TURN doesn't turn on an American one cent piece. Two known.

A.B.T. PLAY BALL, 1924. British, French, and German pre-WWI wall machines, coming back in great numbers after the war, inspired direct copies or indigenous spin-off American counterparts. PLAY BALL is the latter, adding gum ball vending from a large container behind the top of the playfield to Americanize the concept. Four colorful baseball catchers are there to score first, second, third, or home hits with matching colored balls. First A.B.T. counter game.

Whiteside Specialty BASE BALL GAME OF SKILL, 1925. Combination target and ball shooter, BASE BALL GAME OF SKILL fills a tray of colored hard rubber balls at lower right when the coin slide is inserted. Insert balls in the container at top one at a time, and shoot with plunger. Score of six balls wins one to ten cigars. The higher the score the less you win. 25 and 21 pays one cigar, 15 pays three and 8 pays six. Whiteside Specialty Company was in Los Angeles, California.

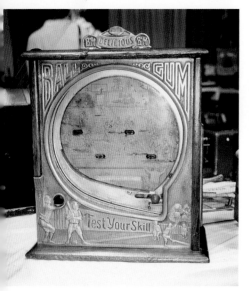

urnham and Mills PLAY BALL, 1927. edesign of the A.B.T. PLAY BALL of 1924, uch simplified to eliminate the graphic lutter. The Burnham and Mills version liminates the four catchers and matching olored balls and substitutes tab knockovers o score base hits and home runs. PLAY BALL credited to both Burnham and Mills and B M Products, which in reality are the same. neat graphic touch, B & M Products are eing sold in a playing field billboard.

Webb Novelty MIRROR OF FORTUNE, 1927. One of those continuing mysterious machines that never seems to acquire a maker's name. MIRROR OF FORTUNE has no name on it, although it does carry the patent number 1,640,453. But that's a dead end which turns out to be a coin detector patented by a lawyer. Then an example surfaced with a paper label that revealed the maker as Webb Novelty Company of Chicago. A.B.T. also made a version, smaller, with the same name.

R. and S. CHIP GOLF GAME, 1929. The end of the 1920s saw a plethora of strange and sometimes exotic cast aluminum and wooden cabinet counter games with physical scoring features. R. and S. Company of New York, New York, contributed CHIP GOLF GAME. Coin in at side, side lever down, and you get a steel ball at the bottom. Then you chip it with the flip handle at the bottom to make a hole in one, two or three across the top. Land in the middle bunker and the ball is lost.

ational Novelty BASE BALL, 1928. One of any derivatives of the Jentzsch und Meerz JAZZO coin catcher games, the National ovelty Manufacturing Company of Chicago SE BALL is based on the Paupa and chriem BASE BALL of 1917, a much ericanized version of the original German achine. Newer version was created by the rley Novelty Manufacturing Company of icago in 1923. Their contribution to the chine is large gum ball storage double inet at the right.

Erie Manufacturing THE EMCO (NERVE EXERCISE), 1928. This machine should be called "Nerve Wracker" because the game hooks you and drives you nuts because it is all but impossible to beat. Made by the Erie Manufacturing Company of Hartford, Connecticut, makers of the ERIE DIGGER arcade claw machine. Drop in a coin at right, lightly tap a plunger on the right side, and the coin is supposed to roll on edge along a bar to come back to you in the cup at left. Yeah. Sure.

Arcade Supply THE CLOWN, 1929. It looks European, but it's American. While these games may have been imported, the top aluminum marquee certainly isn't. Arcade Supply Company of Chicago put these on the market in 1929, just in time for the Great Depression. They are often confused with the Caille Brothers CLOWN, but they are larger machines with flat tops plus that THE CLOWN marquee that also says "Can you make him catch the ball." Rub tummy, pat head, catch ball.

Calvert INDIAN SHOOTER, 1929. Calvert Manufacturing Company of Baltimore, Maryland, got its start in coin machines with the INDIAN SHOOTER, a combination gun and target game. The difference is you shoot from the bottom. What is missing here is the grip pistol that fits in the center hole. The gun shoots the penny you play up the playfield to hit and knock out any of the four Indian heads in full feather regalia mounted on horses. Left and right are gum ball windows.

FLYING ACES, unknown manufacturer, 1932. One of the outstanding games of 1932 was STEEPLECHASE by Keeney and Sons and Exhibit Supply. Six colored marbles representing horses shuffle positions over a vertical playfield to land in random 1,2,3,4,5,6 order at the bottom. Widely copied, with FLYING ACES a somewhat more sophisticated version. Name is in logo style of FLYING ACES magazine about WWI flying "Aces" started in January 1928 and very popular in 1932.

Jentzsch und Meerz 1930 BAJAZZO, 1930. German game makers still had an export market in the '20s and early '30s before Adolph Hitler came into power in 1933. Leipzig, Germany, was called "The Chicago of Europe" because more coin machines were made there than anywhere else outside of Chicago. 1930 BAJAZZO was the third reincarnation of the famous coin catcher game. Brass trimmings of the past were now aluminum, with the catching clown as bright as ever.

Novix Specialties TRY-SKILL, 1929. Another one of those damnable nerve testing, nerve "wracking" machines. In TRY-SKILL you drop the penny in the top and try and balance its way down level by level to the bottom with the playing field tilt lever on the side to get it back into the cup without losing it in one of the many wide coin slots along the way. Novix Specialties of New York City was active in the 1927-1931 period, and then disappeared. Maddening game.

A.B.T. THE SIX HORSEMEN, 1932. One of the most exciting counter games ever made, it wasn't even made as a game. The enormously successful DUTCH POOL pinball games sold by A.B.T. in 1931 were quickly superseded by newer machines. Rather than lose market share, A.B.T. created THE SIX HORSEMEN as a replacement mechanism and playfield to fit in the cabinets. Numbered colored disks race around and come to rest in 1,2,3,4,5, and 6 positions. Fast! Not many sold.

International Mutoscope JUGGLING CLOWN, 1930. Another American version of the German BAJAZZO style CLOWN, only this time the International Mutoscope Reel Company, Inc. of New York, New York, created a much smaller counter game in a metal cabinet that includes the penny back feature and a gum ball vender. It was the first counter game fielded by the new International Mutoscope management, a market they were out of in a year.

Dean Novelty EVERREADY TRADE BOARD, 1930. Dean Novelty Company of Tulsa, Oklahoma, was a smaller game maker more primarily interested in operating than manufacturing. They made their games for their own routes, and made modest attempts at selling them nationally. EVERREADY TRADE BOARD is a punch board with trade simulator features. A nickel gives you a wheel spin for an award of 1, 2, or 3, the reward being the number of board punches you win.

Superior Confection AUTOMATIC CARD TABLE, 1930. Coffee shops, restaurants, and waiting rooms were potential placements for coin operated table games. Surviving numbers suggest that the idea was less than compelling as these games are rare. Half a dozen producers were involved in the genre, with the Superior Confection Company of Columbus, Ohio, one of the few companies to have a patent position in the art. Forty-eight cards in table are "dealt" and "drawn" by lever.

Gottlieb BASEBALL ROLL-ET, 1930. The Peo LITTLE WHIRLWIND was so widely copied it is likely that some of the games and makers will never be identified. D. Gottlieb and Company of Chicago was one of the copycats, simplifying the game with a side coin chute, reducing scoring pockets from six to four, with a ball flipping bat on front to replace the side flip lever to create a baseball version. Vertical alignment of the six scoring holes in the spiral is retained.

CHAMPION SPEED TESTER, unknown manufacturer, 1930. An astoundingly clever game with an unknown heritage. Dial casting says "Patents Pending," but no names or patent art have shown up to identify the game. Sports figures, particularly Babe Ruth, make this a killer game. But whose? Coin in and gum balls shoot out the spotter, which you are supposed to turn fast enough to eject them into the scoring holes. A possible maker, based on style, is B. Madorsky of Brooklyn.

Peo LITTLE WHIRLWIND, 1930. A sea change in counter game design. An operator turned manufacturer, Howard Peo of Rochester, New York, made LITTLE WHIRL-WIND as his second counter game, and locked the game up with protecting patents. Revolutionary spiral game hit the market like a bombshell, making the newly formed Peo Manufacturing Corporation a significant manufacturer and protector of the product as others all over the country copied the game.

Gottlieb ROLL-ET, 1930. In addition to the baseball model, the same playfield was used for a straight version called ROLL-ET. The cabinet is identical with the exception of the game name at the top, including the baseball mitts, bat, and ball in the four corners of the playfield casting. But ROLL-ET led to trouble. Peo threatened a lawsuit, bringing Gottlieb to the table and a licensing agreement. The action curtailed continued copycatting by others.

LARK, unknown manufacturer, 1931. INDOOR STRIKER was made and marketed by Pacific Amusement Company, Los Angeles; D. Robbins and Company in Brooklyn, New York; Indian Head Novelty Company of Los Angeles and Consolidated Automatic Machine Company of Los Angeles, California. These aluminum cabinet games all look alike, and may have been made by one producer. LARK looks different, with the INDOOR STRIKER name on the front unlike any other version. Remains unidentified.

Lilliput LILLIPUT GOLF, 1931. When a number of the 1931 aluminum front coin flip and kicker games showed up in the early 1990s they were thought to be considerably older. Photos suggested they were cast iron games. Reason (such as golf wasn't popular until the late 1920s), and old advertising, finally straightened out their origins and ages. LILLIPUT GOLF is made by the Lilliput Manufacturing Company of New York City, the company's only product. One known.

Gottlieb MINIATURE BASEBALL, 1931. Once the Gottlieb vs. Peo spiral copying situation was resolved D. Gottlieb and Company announced a license agreement with Peo, and promptly introduced MINIATURE BASEBALL with a colorful scoring panel at the top. The remainder of the game was the same as BASEBALL ROLL-ET, although the casting added the text "MINIATURE BASEBALL World Champion." Keeney and Sons made the same game as ELECTRIC BASEBALL with the casting revised accordingly.

B. Madorsky FOOTBALL, 1931. Another variation of the A.B.T. and B. and M. Products PLAY BALL kicker game, this time a metaphor for football with actual opposing team goal kicker figures substituted for the shooter. Maker is B. Madorsky of Brooklyn, New York. Two player, with plungers on each side to kick the ball. Kick a goal and it rings a bell. Early games had a smaller marquee and more complicated playfield. Battery operated. Perfect kick returns coin.

Pace WHIZ-BALL, 1931. Whatever Peo of Rochester, New York, did, Pace Manufacturing Company of Chicago did even more so. When Peo came out with LITTLE WHIRLWIND at the end of the summer in 1930, Pace followed up with a virtual duplicate in a cast aluminum cabinet called WHIZ-BALL in January 1931. You can hardly tell the games apart except for the base below and marquee at the top. They quickly followed up with two more of the games based on baseball and golf.

Parmly THE LUNATIC, 1931. Cleveland, Ohio, was never regarded as a major coin machine center, but over the years numerous counter game makers came and went, leaving their products behind as testimony to their involvement. In 1931 Parmly Engineering Company made a complicated flat tabletop game (which also came with legs) called WORLD SERIES BASE BALL. The cabinet casting was modified as a five green golf game called THE LUNATIC that scored in troughs similar to a spiral game. One known.

Goodbody THE WILDCAT, 1931. Distributor John Goodbody was located in Rochester, New York, the same town as Peo Manufacturing Corporation. He sold LITTLE WHIRLWIND, as well as a replacement front that converts the game to WHOOPEE BALL. Then he made his own version of the spiral game, different enough to avoid a lawsuit. An enlarged spiral is at the lower front with a large ball gum storage area at the top. A gum ball with every play and maybe your coin back.

Pace DEUCES WILD, 1931. Whatever Peo did, Pace was sure to follow. But now they were doing it legally. After the Gottlieb resolution and licensing agreement to make spiral games, Peo worked out similar licenses with Pace and Genco in Chicago and Royal Manufacturing Company in Syracuse, New York. The Pace DEUCES WILD marries the Peo PLAY POKER playfield with the Pace cast aluminum cabinet front and lower playfield scoring format. Plus the Pace "Duce Wild" marquee.

E. E. Junior PLAY BASKETBALL "3 Shots", 1931. The first E. E. Junior PLAY BASKETBALL offers three shots for a penny, but the ball wasn't placed well to shoot. Background shows a handoff. It came back in a new cabinet as a "2 Shots" model with the ball hole moved to the player's hand and a new backdrop showing the "JMCO" team on the bench. The cabinet was further improved for the final "3 Shots" model shown, giving you one gum ball and whatever else goes in the basket.

PLAY POKER, 1931. Peo Manufacturing Corporation was of an inventive turn of mind. A little over a year after LITTLE WHIRLWIND they turned the scoring hole arrangement half way around to put them six in a row horizontally to create PLAY POKER, theoretically obsoleting the vertical hole arrangement. Playing card theme is following in scoring troughs at the bottom. Peo mottled metal cabinet is distinctive and unique. Came with or without gum ball vender.

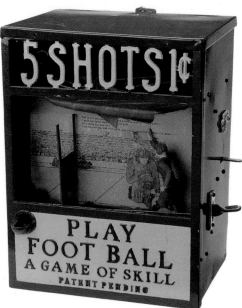

E. E. Junior PLAY FOOTBALL, 1931. E. E. Junior Manufacturing Company, Ltd., Los Angeles, California, made pinball games and ball gum venders, the latter with action figures that really shoot the ball. They became the rage of southern California. PLAY FOOTBALL is their second game. Insert penny, pull out lever, and five gum balls line up at the kicker's foot. Tap lever to kick them over the goal posts. You get at least one gum ball, plus any that go over the goal.

International Mutoscope TIP THE BELL HOP, 1931. Combination of the INDOOR STRIKER and JUGGLING CLOWN by International Mutoscope Reel Company, Inc. of New York City. Game and graphics are a product of their times, although little of what International Mutoscope did could ever be regarded as politically correct. Even then. They called this a fortune teller, but the scoring troughs set up a hand of cards as achieved by the five steel balls popped up the playfield.

Field VEST POCKET COUNTER TOP BASEBALL, 1931. Gottlieb, Pace, Genco et al may have reached licence agreements with Peo, while the Field Manufacturing Corporation of Peoria, Illinois, avoided the honor. They made things just differently enough to skirt real trouble. First they copied the Peo VEST POCKET BASKET BALL to make VEST POCKET BASEBALL. Then they changed that to add a half spiral to dodge direct comparison to make VEST POCKET COUNTER TOP BASEBALL.

T. H. Sloan THE POKERALL GAME, 1932. Small manufacturer counter game ideas often grew with bigger game makers. THE POKERALL GAME created by Theodore H. Sloan and his T. H. Sloan Manufacturing and Sales Company of Charlevoi, Pennsylvania, internally shuffles card spot disks in a row at the top when the handle is pushed down. Hold-and-draw buttons improve the hand. Fifty were tested in Pittsburgh, where they should be found. Only one known showed up in Canada.

Mills Novelty CATCH THE BALL, 1932. Mills Novelty Company version of the clown catcher with an aluminum front casting remindful of the International Mutoscope JUGGLING CLOWN. Mills version actually has earlier origins with application for a patent in 1926, based on the Paupa and Hochriem BASE BALL game of 1917. Play heightened by bouncing lively plastic balls made of "Plaskon." Catch the ball by twisting the front knob and get your penny back.

Marcus SCOTCH GOLF, 1931. The life of Peo spiral games was extended by placement fronts and playfields making new games out of old. Half spiral front fronted by Field Manufacturing became a revamp game format. M. M. Marcus and Company of Cleveland, Ohio, was one of a group of new front makers that permeated the field in 1931 and 1932 until newer games came along. Five balls. SCOTCH GOLF marquee missing. As an award game it pays on 12 or less, 27 or more.

Bally BALLY, 1932. Modified T. H. Sloan THE POKERALL GAME without the hold-and-draw feature. Bally Manufacturing Company in Chicago bought rights to the Theodore H. Sloan game and patent to create BALLY, a seven game machine that includes disk sets for poker, fruit symbols, dice, horse race, colors, numbers, and mystic fortune. BALLY was tested in Chicago, where it should be found. Only a few examples known, usually with 52 disk card set and missing others.

Field DING THE DINGER, 1932. Unique clever game that seems to have no antecedents or future development. First introduced by Field Manufacturing Corporation as INDOOR SPORTS (the name is still on the background graphics), "The Dinger" is played for a poker hand or numbers, with both in the backdrop. Drop in coin, turn knob at top to release the last hand played and then tap each of the five levers to lift weights on each rod up to a good hand. Dang, that's fun!

ean VAUDETTE, 1932. A collection of ounter games is sorely lacking in imagina-on if the smaller makers are not repre-nted. Dean Novelty Company in Tulsa, klahoma, was a significant contributor to versity. VAUDETTE is a classic Galton ard scoring game in the target class, cept the coin is flipped up and over in an c. Coins can be penny, nickel, or dime, th a dime game piling up some real oney in 1932 terms. Later NEW VAUDETTE s a visible jackpot.

Pierce Tool BUGG HOUSE, 1932. Multiply a target game five times and you have BUGG HOUSE, the first counter game by Pierce Tool and Manufacturing Company of Chicago, the survivor firm coming out of Chicago Mint Company. BUGG HOUSE has five shooters, one for each of five rows of symbols. You really have to whack the coins to get them to go to the far end of their row. Score 7500 and you get 150 points. To get 7500 you have to place a coin over every face on the playfield.

Peerless Products ODD PENNY MAGNET, 1933. Same game sold by two firms in the same geographic area suggests one made the game, and the other sold it as their own. Peerless Products Company was located in North Kansas City, Missouri, when A. J. Stephens was making games in the area. ODD PENNY MARKET is found with the Peerless Products name more often than the A. J. Stephens name, yet they are the same. The identifier is the paper. Garden City made it as LEAPING LENA.

ddick Engineering PAR-KET, 1932. A iation on the skill flip that simulates a eball game. First introduced in August 31 as SKE-BAL-ETE by Buddick Engineer-Company, Inc., of New York City. Like , Buddick was willing to sue anyone who ied their game. The game came back in 32 as PAR-KET. The early 1930s graphics errific, particularly the clothing and hair es. Balls come out chute at bottom right are flipped by the lever.

A. J. Stephens ODD PENNY MAGNET, 1932. A new name entered the counter game universe when A. J. Stephens and Company of Kansas City, Missouri, introduced ODD PENNY MAGNET in August 1932. Small and simple with a bent metal sheet steel cabinet, ODD PENNY MAGNET is a difficult game to beat. The coin drops down the center chute and hits a wire spring. The lucky bounce tosses it forward and into one of the fruit slots. A. J. Stephens became a significant game producer.

Hercules Novelty BUY AMERICAN, 1933. One way to beat the Great Depression was to keep the money at "home," in America. Hercules Novelty Company in Chicago made the BUY AMERICAN theme a game at the height of economic hard times. Nothing radical here, but a different form of the clown catcher game. Six balls for your penny or nickel. Catch five and you win five coins. Catch six and you get six more. Payout is automatic, front or back, but can be shut down for trade play.

Garden City PICK A PACK, 1933. Fairly big and flashy, PICK A PACK makes use of the older "Windmill" vending idea using the blades as pointers to pick the pack you win for your penny or nickel. Four-way coin head. Pick the pack by where you play your coin, push down the plunger, and watch those blades spin. Slowly. A penny on the right pack wins one pack, a nickel five. Coins to confirm payouts are seen in a window at the back. Pointer must stop between two tabs at the top.

Pierce Tool CIG-O-ROLL, 1934. Everything old is new again frequently occurs in counter games. Pierce Tool and Manufacturing Company provides grist for that verbal mill with CIG-O-ROLL, a variation of the Park Novelty RED BIRD of 1903 in which a ball substitutes for the cork disk. First you pick the number to win, and spin. It's not unlike a vertical roulette. Original model has cigarette pack and color wheel. Later version has card spot and gypsy reader as shown.

Royal Manufacturing BOUNCING BALL, 1933. Six balls for a penny with more twisting and turning than you can imagine. Imagine the playfield turning and you get the picture. Royal Manufacturing Company of Jersey City, New Jersey, was a division of the Royal Scale Company, and BOUNCING BALL is their only counter game. Simple construction. Lever at right gives a blast of air from a bellows to flip the balls. Lever at left turns the center playfield. Touchy tilt.

Pierce Tool GYPSY, 1934. Another variatio of the same theme, the Pierce Tool and Manufacturing Company GYPSY scores by ball that floats over the wheel behind glas and settles at the bottom to indicate the winner or loser. Pick the winning number the selector dial at left, then coin in and s with the handle. The big difference is the wheel with three 40:1 winning positions, which is a pretty big bang for the penny, nickel, dime, or quarter back in 1934.

A. J. Stephens LUCKY COIN TOSSER, 19 Another one of those almost impossible win penny toss games that A. J. Stephen Kansas City, Missouri, was so fond of producing. Drop the penny in the wide s at front to land on the tossing platform, turn the side knob to toss it forward. It s through the encapsulated air, often bang the top glass on the way, and never seer land on the "5" or "10" platforms at the end. A study in frustration.

Lu-Kat Novelty LUCAT, 1952. Never advertised, not known, until some showed up at a show. A truckload were found in an old bar in Grass Valley, California. Made by Lu-Kat Novelty Company, San Francisco, California, as a bar piece. Five were kept, the rest tossed in a local dump. Dime play, pull the tail at right to vend a ticket out its mouth. Match numbers to card and win. Button on left vends gum ball between paws. Brass mechanism. It looks like 1930s but was made in the 1950s. Want more? Find that dump.

Berger CATCH-N-MATCH, 1936. An interesting variation of the clown catch idea. CATCH-N-MATCH is a product of Berger Manufacturing Company of Chicago, and ended up as their only game. Five colored marbles with a brace of matching pockets in black, white, green, and red. You have to catch balls in their proper colored pockets. If you miss the balls come back, but getting the right color in the right pocket is the trick. Winners are a black and 1, 2, or all 3 colors.

PLACE THE COPPERS, unknown manufacturer, 1936. How could a game this graphically distinctive be unidentified? Worse, there is only one example known. PLACE THE COPPERS is penny play, the "coppers" referring to one cent pieces. Tough to beat. It takes eight cents—a copper in every street—to win a 15¢ pack of cigarettes. Two in the same stack breaks the run. Clue to the maker is the metal cabinet and corner designs, often used by A. J. Stephens and Company.

Groetchen PUNCHETTE, 1936. Build a better mousetrap and the world will beat a path to your door. But if it doesn't catch one, no path, closed door. Competitors Groetchen Tool and Daval in Chicago tried virtually everything, usually at the same time, including punch games. Daval made AUTO-PUNCH and Groetchen produced PUNCHETTE, "The mechanical punchboard." Payout up to $25 on a nickel when you got a chance to punch the "Mystery Tape." It got punched out.

Sands SKILL KATCH, 1936. More than a counter game, the Sands Manufacturing Company SKILL KATCH is an automatic payout variation of the clown catch idea. Never advertised, the only distributor that promoted the game was located in Dallas, Texas, so Texas production seems likely. Nickel in the coin slide at top, trickling down over the pinfield. Catch it and it dumps the pot below the catcher. The problem was the more common pots were usually empty anyway.

QUEEN TOP, unknown manufacturer, 1937. By the mid-1930s counter games were no longer an exclusive American province. British, German, and French machines were beginning to appear, soon to be squashed by the coming of World War II. Among the more attractive is an unidentified French game called QUEEN TOP. Its solid aluminum casting and triple fruit dial could have made the machine a great success in the United States. Match "Queen" and "Top" on the dials and win.

177

Mueller Specialties CHANGE MASTER, 1938. Change trays, often advertising cigars and usually made of glass, were on most cash register counters. Mueller Specialties of Wichita, Kansas, made theirs of porcelined cast iron. Merchant puts change in tray. Customer plays any coin at top, and lights flicker to a stop on two panels. Match them and get two times for two stars, three times for full moon, and five times for partial moon. The tray also says "Thank You" for a sale.

Star Manufacturing FLIP FLOP FLUZZEE, 1939. Counter games designed to disobey physical laws have long flipped or dropped coins over loops, springs, and into water to make them do abnormal things. In FLIP FLAP FLUZZEE made by Star Manufacturing and Sales Company of Kansas City, Missouri, you are supposed to drop a penny with a plunger push to make it bounce off a platform to flip into the glass tube and land flat on the platform below to win a pack of cigarettes.

Fortune Sales LUCKY ROLL, 1939. A diffic game to place in time as it looks older tha its years. LUCKY ROLL by Fortune Sales Company of Los Angeles, California, is a variation of the early 1930s spiral game. LUCKY ROLL shoots the coin, not balls, in the holes. Land in top holes A, B, C, or D and coin shows up at the bottom for an award. For years this was thought to be an M. Brodie Company, Long Beach, Califor machine as Brodie was its only advertiser.

W. H. Kelly PENNY JITTER BALL, 1938. Once electricity entered the scene, it didn't take much to make a different counter game. PENNY JITTER BALL is the product of W. H. Kelly Company, Tulsa, Oklahoma. Simple game. Slide a penny in and it pops three steel balls. If they all land in the same color holes of red, black, yellow, green, or white you win. But what? It doesn't say. A simple switch pops the base. This only known example showed up in an antique shop in 1992.

Munves Corporation TID-BIT, 1939. After long tenure as the leading arcade machi distributor in the United States, Mike Mur of New York City set up the Munves Corp ration to make and market original arca and counter games through contract buy and exclusive distributorships. TID-BIT lis a gum vender with a "new free play prin ciple," meaning you got your penny bac you flipped it into one of the four cups. S graphics are the 1939 New York World' Fair.

ills Novelty FLIP SKILL, 1939. The ultimate imbing game, with the player getting five alls for a penny, or nickel, depending on e dedicated coinage. If any game can be urely qualified as a skill game, this is it. alls enter at the bottom, and you get to alk them up the 32 inch tower by flipping em left to right and upward. Scoring hole at the top. Scoring balls go to center ndow. The more you play the better you et.

Erie Machine BOMBER, 1939. World War II worked its way into coin machines in a hurry. As inconceivable as any world war may seem, at the time people everywhere were deeply involved and everybody knew the reason why. It is not quite as comprehensible over half a century later. Erie Machine Company of Cleveland, Ohio, used gum balls above to bomb ships below in BOMBER. Land in scoring holes below and you get more gum balls for your money.

Scientific Machine TOTALIZER, 1940. Scientific Machine Corporation of Brooklyn, New York, was one of a number of large arcade machine makers that added counter games to their lines in the early 1940s, just before WWII included the United States. TOTALIZER is a simulated basketball game that gives you five balls and five scoring baskets to shoot at. The game gets its name from the fact that a scoring window at the top automatically totals your score as you play.

nco PUNCH A BALL, 1939. After years of ying, Genco, Inc. of Chicago finally went in their own directions in the late 1930s. NCH A BALL is a non-coin machine. You y the counter man a nickel a ball, so all cash goes to the location. Push the nger as many times as you paid for, and come BB size balls. Ruler lets the rchant count the total. Most are plain, n gold, blue, green, and red balls worth 50, 100, and 500 points respectively.

Western Products OOMPH, 1940. Western Products, Inc. of Chicago seemed to try everything, with few things working out for them. That makes their games few and far between. OOMPH is a modernized version of the old lung tester without the disadvantage of a blowing tube that everybody and his germ laden brother lip sucked. Put in a penny and OOMPH vends a straw. Stick it into the glass tube and blow to raise that Popeye look-alike indicator. Now that's different!

Howard Sales BRODI LIBERTY, 1940. About as small as a counter game can get. Actually, it's a 2-1/8" wide palm game, non-coin, made for travel play. Sold in Woolworth's and Kresges in the 1940s as a dime store toy. It was also advertised in *The Billboard* magazine by Howard Sales Company for professional use. Slickers skimmed the change out of marks on many a wartime train ride with these good looking mini-slots. All you have to know are the symbol odds.

Groetchen PIKES PEAK, 1941. The best known and most popular of the mountain climber games. PIKES PEAK came out in January 1941, and had barely a year of sales before wartime material restrictions ended production. They must have sold a lot as one year of production resulted in a high population survival rate for collectors. A great deal of fun. Your job is to lift a rolling ball up each level to the top scoring hole with the up and down knob on the front. Takes practice.

Standard Coin SKILL-A-RETTE, 1941. One of the more complicated cigarette games, SKILL-A-RETTE by Standard Coin Machine Company of Chicago combines a pinfield with a spinning wheel. Play the coin in the slide below, and get four balls for your penny. Shoot them up to the top and bounce them around the pinfield to land in the hole and spin the wheel. Or not. Baker Novelty Company bought out the inventory and rights to bring it back as their game in 1942.

Pioneer Coin SMILEY, 1946. From wartime snarls to postwar smiles, SMILEY characterizes the abrupt change from serious living the fun of the closing '40s when much of th world was suddenly released from the fear death and destruction. Maker is Pioneer Coin Machine Company of Chicago, a postwar start-up manufacturer. Five balls g into holes and behind scoring at lower left center. Original playfield has five pockets, later playfield is spiral as shown.

Groetchen ZOOM, 1941. While it looks like a throwback to the old spiral games with straight line scoring holes, ZOOM owes more to the British ALLWIN wall games made just before WWII. Graphics are wartime showing U. S. fighter planes trying to make hits on battle cruisers. It is more like the forerunner of the post-WWII Japanese Pachinko game. Score in a "Hit" hole, three out of the seven holes at the top, you get your coin back and three more balls to play. Big game.

Groetchen KILL THE JAP, 1943. World War II propaganda is said to have been the most vitriolic in wartime, surpassing that of WWI. To the Japanese the A-B-D Powers (American-British-Dutch) were hateful. To the Allies, particularly the Americans, the enemy was worthy only of death and destruction. Groetchen Tool Company of Chicago was one of the coin machine makers and revampers that used wartime hate themes. These games are rare and becoming valuable.

Standard Games WINDMILL, 1946. The wartime counter coin drops taught the co machine makers one thing at least; put graphics on the pinfield. Just when the games were disappearing they finally go good looking. After WWII they came bac with a rush. And ran into a brick wall of resistance. WINDMILL by Standard Gam Company of Columbus, Ohio, is wood, t and colorful. It's big, 17 inches high. A c catcher is at the bottom, but most player ignored it.

Standard Games WINDMILL JR., 1946. Standard Games Company of Columbus, Ohio, made WINDMILL in two sizes. WINDMILL JR. is smaller and narrower and has a different cast aluminum marquee. A lot of promotion was put behind WINDMILL, but nary an ad ever ran for WINDMILL JR. Playfield graphics are the same, except the walking ladies are closer to the bottom, the clouds aren't as full and the roses are much shorter on WINDMILL JR., suggesting they used the same boards cut off.

Central Manufacturing HI FLY, 1946. New Chicago game maker Central Manufacturing Company introduced HI FLY baseball skill game at the end of August 1946 just as serious interest in the first postwar World Series was getting underway. Central said HI FLY was tested 18 months, meaning it was created before the war was over. Coin in and push plunger at top to reset the game and play until three outs. Production runs changed cabinet and playfield colors.

Amusement Enterprises PITCHEM, 1946. Another postwar start-up game maker, Amusement Enterprises was located in Houston, Texas. They rushed into the industry with an automatic payout console, arcade machines, and a number of counter games. PITCHEM is a basketball themed game much like the competitive Baker Novelty KICKER AND CATCHER. The catcher is a basketball player rotated by the knob at left. Production runs changed cabinet and playfield colors.

Central Manufacturing HIT A HOMER, 1947. Practically a clone of the Central Manufacturing HI FLY of 1946, HIT A HOMER of 1947 has the same mechanism and scoring holes and pockets as the earlier game. The difference is a new name and much more alluring graphics using the same cabinet and playfield format. Good use is made of the generic baseball color: green. These games are sometimes difficult to date and identify as no names appear on the cabinet or award cards.

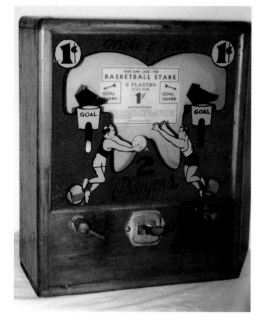

Associated Amusements BASKETBALL STARS, 1947. Pre-WWII games tended to base their names on the playing principles of a game. Wartime themes were about blowing up things or killing people. The prevalent theme after the war was sports, remaining that way ever since (not counting the intergalactic war and karate of video games). Associated Amusements, Inc., Boston, Massachusetts, made shooting a two player game with BASKETBALL STARS, also called "Mike and Jake."

Daval BEST HAND, 1947. The spiral games are back, in doubles. Daval Products Corporation, renamed after WWII, created a line of three double spiral games that make use of the same aluminum body castings. The games can be played singly, or by two coins for competitive play, with two separate cash boxes to keep track of the double plays. BEST HAND is set up as a card game, allowing two players to play poker against each other. MEXICAN BASEBALL scores hits and runs.

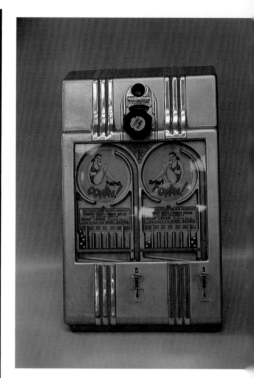

Daval OOMPH, 1947. A rarity, a counter "Girlie" game. It may have been a good idea for contemporary pinball, with named artists committing their pin-up thoughts to glass, but it was a dud idea for counter games. The lady in the playfield may have something to do with it, for she is hardly anyone's dream girl. The Daval Products double spiral games probably didn't do very well as few have been found, and they are still unidentified surprises to many collectors

Bonus Advertising ? JACK-POT BONUS ?, 1947. Upgraded version of the cash register trade stimulator, the Bonus Advertising System ? JACK-POT BONUS ? made in Los Angeles straps a jackpot on the back of a cash register facing the customer. Match the numbers and say the magic word (like the name of the store) and you demonstrate skill to earn the pot. Earlier JACK-POT BONUS ? model doesn't have the slick graphics of the 1947 model. Simple mechanism, coupled to the register.

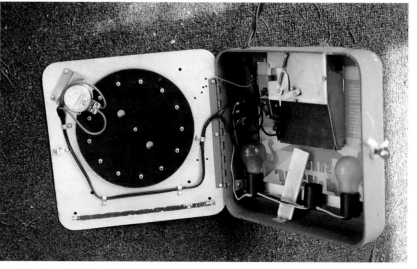

Marvel SLUGGER, 1948. Flip and catch ball sports themed counter games were a generic by the late 1940s, with half a dozen or more manufacturers making some form of the amusement device. Marvel Manufacturing Company of Chicago started out as a wartime pinball revamper, and graduated to the production of their own pin and counter games. Their baseball POP-UP counter game was introduced in 1946, came back as DIAMOND in 1947 and SLUGGER with new playfield pennants in 1948.

Binks WHIZ-BOWLER, 1954. After years of working in Chicago for Exhibit Supply, Keeney, and United as an electromechanical automatic payout console slot machine inventor, industry maven Mel Binks set up his own Binks Industries, Inc. in 1954 and made simple mechanical counter games. Too little and too late. He made ZIPPER and WHIZ-BOWLER, both ten ball penny flip games in vertical pinball formats. WHIZ-BOWLER scores in strikes, spares, high scores, and double scores.

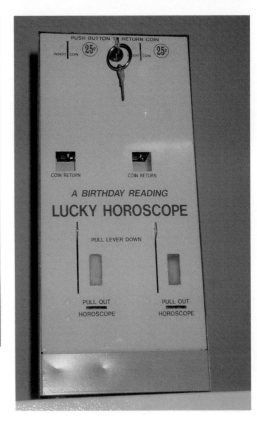

Marvel LUCKY HOROSCOPE, 1959. Plain and unassuming in appearance, LUCKY HOROSCOPE is actually a chance machine rather than simply a fortune teller. Original model has a large graphic at the top showing a Hindu fortune teller and horoscope tickets. Plays nickel, dime, and quarter. Fortune ticket is tab style, with winning numbers or stars inside for merchandise or money. Also by Auto-Bell Novelty. Game holds 1,000 tickets.

Auto-Bell VICTORY, 1951. Auto-Bell Novelty Company, Chicago, was one of the busiest counter game makers in the 1950s for the simple reason few of the games were being made, and they held a commanding position in a diminishing market. VICTORY is a basketball shooter played with five ping pong balls. Also made as HIT-A-HOMER in a baseball format, and TOUCHDOWN as a football game. Easy to change games as graphics are on the glass. Also sold by ACO Novelty.

J. F. Frantz POT OF GOLD, 1955. A steady stream of development leads from the Baker Novelty Company of Chicago KICKER AND CATCHER of 1941 through 1946, to the J. F. Frantz Manufacturing Company of Chicago acquisition of the game in the early 1950s and a complete redesign of the games leading to the POT OF GOLD version in 1955. These are enduring games and fun to play. Continued production at Fun Industries in East Moline, Illinois, into the 1990s.

UPI GOLD MINE, 1965. Proof that the counter game is alive and well. Well, maybe not so well. GOLD MINE is a circular mountain climber game with gold mine shafts along the way. You get five balls for a dime inserted in the side slide, and shoot them up the mountain with the lever in front. The object is to sink balls in the five mine shafts, and when you do the top lights up. Electronic, battery operated with a circuit board. Scoring balls show up along baseline.

BUILDING A COLLECTION

The beginning of a collection is knowledge. Or it ought to be. Usually things don't start out that way. Most budding trade stimulator and counter game collectors see a machine at an auction, in a shop or at an estate sale and surmise "That's kinda neat," and buy the thing. Once home, they wonder what the hell they have. If they get really interested, they take the next step and want to look things up.

Back to knowledge.

Knowledge is power. It helps you understand what you are looking at, gives you a buying advantage at a sale (if it's too high, drop out; if it's too low, grab it!) and adds to the joy of collecting. But knowledge is never an ill gotten gain. It takes hard work. You've got to be interested. You've got to dig. You've got to read reams of material. And you've got to stick with it. The search for knowledge makes you do things you thought you left behind in school. But once hooked on the search, you'll find it as exciting as collecting the machines themselves. And it's a lot cheaper.

Researching such seemingly insignificant devices as trade stimulators and counter games can be difficult. When you go back in time as far as a century, or more, the chances of getting some good first person interviews about any subject are somewhat limited, to say the least. You'd think your chances were really nil. But that isn't the case at all when it comes to researching these barroom amusement coin machines. True, there aren't a lot of people around that can provide a first person story about the card machines of the 1890s, but there are people who knew people, and knew the machines, well enough to tell you something about them. Many old operators that ran these games in the '30s and '40s are still around, and are often eager to talk about their experiences if you can track them down. Try looking them up in old telephone books in your library, get their current number, and make the connection. They might even have some old machines.

Then there is the contemporary media. It existed, and although the hard copies are difficult to find and are few and far between, there is still a lot to be offered by checking the microfilms of those saved by UMI (University Microfilms, Inc.) as part of their serials and newspapers program for libraries and researchers. It is hard to describe the thrill of pouring over microfilms of early 1900s publications to make the discovery of a machine in its pages. Very exciting! You should try it.

Collector Jack Freund of Springfield, Wisconsin, brings home some trade stimulator treasures. Jack's collector interests center on turn-of-the-century wooden cabinet trade stimulators used in saloons and stores in the 1889-1910 period, although his holdings include quite a number of later machines as well as cast iron vending machines.

Collector Alan Sax of Long Grove, Illinois, proudly puts his arm around his latest trade stimulator find, the revamped single reel Sundwall Company THE EAGLE with a cast bronze cabinet. Made in Seattle, Washington, in 1907, this example was a 1993 find in Vancouver, British Columbia.

I would be remiss if I didn't mention and credit the greatest resource of all, the United States Patent Office, and the inexpensive availability of issued patents from the 19th century to this day. As this book is being written they are selling for $3 a patent, and that's a bargain. The only catch is that out of the many, many millions of patents issued over the years you have to tell them exactly which ones you want. By patent number. And that takes a bit of digging. To get your own copies, send the patent numbers and $3 per patent to: U. S. Patent and Trademark Offices, Patent and Trademark Copy Sales, Box 9, Washington, D.C. 20231. And wait two or three weeks.

It took me over twenty years to assemble the patents for the trade stimulators and counter games illustrated and described in this volume as they are not listed in the patent records as such. You have to pour over every page of the weekly *U. S. Patent Gazette*, usually well over 500 pages per volume (52 volumes a year times 30 years makes 1,560 volumes and makes you cross-eyed), and pick out what looks right. I did it three times, finding things I missed each time around, and I'm still not sure I caught everything. But it seems to me that we have the most representative machines covered as I have yet to stumble across a trade stimulator or counter game with a patent date or patent number on it that is not represented in the listing of machines in this volume. That's not to say that collectors have found everything. You may be the person that makes discoveries that make history by finding the data and machines that have so far eluded all of us. Go for it!

There are also a number of good books on the subject. Not many to be sure, but enough to familiarize you with hundreds of the machines. No book to date has as many color pictures of trade stimulators and counter games as this one, but every bit of knowledge helps. To save you the time of trying to track them down, the major books covering these machines are listed in the resources section following.

There is also the original paper, such as advertising, mailing pieces, old engineering drawings, and the actual machine manuals. Patents aren't manuals by a long shot, but then again manuals were never printed for 99% of these relatively inexpensive machines and the patents will often do in a pinch as they sometimes have the engineering drawing (and in the case of electrical machines, schematics) you need to get a machine in working order.

A small sampling of the Bill Whelan collection in Daly City, California. Bill keeps his machines in display cases and on shelves for their protection, and to move them to museum exhibits and special showings in the San Francisco area.

Collector Tom Gustwiller of Ottawa, Ohio, explains the workings of THE OREGON made by the Portland Novelty Works in Portland, Oregon, in 1901. Gustwiller is the author of the excellent book *For Amusement Only* and has amassed one of the largest trade stimulator collections in the world.

Once you have read this chapter, researched a few machines, and read some of the available books, you are ready to go out into the countryside and start looking for collectible pieces. When you have mastered a few points, dating a trade stimulator or counter game isn't all that hard. If the cabinet is stained wood with or without a cast iron marquee (you can check that easily if you carry a refrigerator door magnet with you at all times) that holds an award card that says "Free Drinks" or "Free Cigars," you're probably looking at an early machine from the 1890-1900 period. If the cabinet is plated cast iron with the same marquee, it's probably from the 1900-1910 period. If the cabinet is stained wood without a marquee, with a paper or cast metal panel that says "No Blanks," it's probably 1905-1920. If the cabinet is cast aluminum, generally natural finish without a lot of paint, and has figures of people on it, it's probably 1922-1933. But the moment the cabinet starts having geometric designs or an obvious Art Deco treatment, you're looking at 1928-1936. By 1935 up until the American involvement in WWII in 1941 the cabinets looked quite modern, with the later machines having solid enamel colors or a "Hammerloid," or bumpy, finish. The post WWII machines of the '40s and '50s look virtually the same and are hard to tell apart from the machines of the 1939-1941 period. Paint is also an indicator. Pre-WWII paint chips, modern postwar paint doesn't.

The symbols on the reels can also be a clue. If the wheels and reels on an old wooden machine carry poker or playing card symbols on wide reels, the 1890-1910 period is a good choice. Fruit symbols started to be used in the middle 1920s, followed by cigarettes starting around 1930 and beer reels running briefly from 1933 to 1935. By the late 1930s and thereafter most symbols were fruits and cigarette packages. Even these can be a clue. A Lucky Strike green package dates the machine prior to 1943, with the white package postwar.

Good sources for finding machines are antique shops and malls, and particularly small estate sales and, of course, dealers and other collectors. Additional good sources are auction sales, often advertised in the antique trade media and in newspapers. But beware here. Auction ads are notoriously vague, even when the auctioneer has a smattering of knowledge about what is being sold. The problem is that hardly anyone but die-hard trade stimulator and counter game collectors can readily identify the machines. I know of many stories about payout slot machine collectors driving long distances to see an "old time slot machine," only to discover the machine was a fairly common counter game from the late 1930s. An auction held in Hesperia, Michigan, in the summer of 1996 excitedly promoted "three old slot machines" by telephone after they were discovered in the back of a barn. They turned out to be fairly common baby bell counter games from the 30s. But sometimes mention of a slot machine might really relate to rare old trade stimulators. You just never know. The safest way to know what you are doing when it comes to

Part of the Ken Durham collection in Washington, District of Columbia. Ken specializes in counter games of the 1930s through the 1950s, including vending. Ken is also the leading internet source for vintage coin machine information: http://Game Room Antiques.com.

an auction is to see a machine picture in an ad or flyer, or phone the auctioneer to get a full description of the piece before you commit to a long ride.

Another problem with auctions is that you rarely know the condition of the machines, even when you briefly get a chance to view them before the bidding starts. And that can be troublesome. No matter where you get your machines, the matter of condition can be critical as there are no spare parts. Most of the necessary parts for the large payout slot machines have been reproduced over the years, which is not true for trade stimulators and counter games. Serious collectors often like to have two machines of each example in their collection in order to cannibalize one to keep the other one running and looking good. That isn't always possible, and the older the machine the greater the problem.

The most obvious place to find old trade stimulators and counter games is where the games are, and that's at vintage coin machine shows. This is where you can see some actual trade stimulators and counter games if you've never seen them before (unless you are close to a collector). There are a number of national and regional vintage coin-op shows that are held around the country, and increasingly in Europe and the U. K. You can learn a lot about vintage coin machines at these shows without making an investment until you are ready. The shows themselves are worth the price of admission, whatever the price. You will see what's available, and quickly learn what isn't. You will also get a very clear idea of the pricing of the various forms of machines as well as what it means to restore, and over-restore.

The biggest show, and a mustn't miss for practically all coin machine collectors, is the Chicagoland Antique Advertising, Slot Machines and Jukebox Show held twice a year at the Pheasant Run Resort in St. Charles, Illinois. It's an event that is hard to classify, but you'll find antiques, so that makes it an antique show. Yet you are liable to find just about anything in the three packed exhibition halls. Be sure to hit them all. The bus service between them is free. Hop a ride and see everything, because the booth you miss just might have that something you have been searching for for years, only you didn't know it.

In the spring and fall people from all over the world gather at this show to buy, sell, and swap some of the most interesting things you can imagine. This includes classic bicycles, Merry-Go-Round horses, pinball games, and old cast iron street lights not to mention trade stimulators and counter games, jukeboxes, and slot machines. The almost 500 dealers that bring their wares to the ever-expanding Chicagoland Show provide an instant cross section of what is available, and their prices. The experts in the field are also at the show, including your writer who has THE ANSWER MAN Booth at location SD08 in the St. Charles Ballroom area. If you make it to the Chicagoland Show, drop over and say hello.

If you are interested in going, The Chicagoland Antique Advertising, Slot Machines, and Jukebox Show is held at the Pheasant Run Resort, St. Charles, Illinois. Forty minutes west of O'Hare Field, 65 miles west of Chicago on Route 64 North Avenue, 2-1/2 miles west of Route 59. Daily admission is $5. A dealer preview is held the Friday before the show, entrance fee $50. Exhibits are in three

Selections from the collection of Alex Warschaw of San Antonio, Texas. The Warschaw collection covers the full range of trade stimulators and counter games, from the "woodies" of the past to the one-of-a-kind and unidentified "Mystery Machines" of the '30s and later.

An interesting study in contrasts, and modern counter games. These are the J. F. Frantz "Kicker Catcher" games in the Alex Warschaw collection made in Chicago in the 1950s and '60s, and include, left to right, LITTLE LEAGUE, LONG SHOT, KICKER & CATCHER, POT OF GOLD, and BIG TOP.

buildings, with free buses running every ten minutes between all show buildings. For further show information and dates contact the promoter, Steve Gronowski, tel. 1-847-381-1234.

There is another advantage to going to the Chicagoland Show. The town of St. Charles, Illinois, is just five miles west on Route 64 and is noted for its many shops and three antique malls, open on Saturday and Sunday. You will also enjoy the charm of this Fox River Valley town and its restaurants.

But maybe you can't make it to Chicago. Never fear, there is probably a show in your area, as regional vintage coin machine shows are springing up all over the country. Among the biggest are the Philadelphia Game Room Show, held at the N. E. National Guard Armory in May, "The largest East Coast show of its kind," at Hackensack, New Jersey, in October, and others in Fort Worth, Texas, and Atlanta, Georgia. For dates and data about these shows call Bob Nelson at 1-316-263-1848. The west coast is covered by a growing California show called the Coin-Op Super Show, probably the second largest in the United States. It is held at the Pasadena Convention Center in the spring and fall, usually in March and September. For show dates and information call Rosanna Harris at 1-303-431-9266. If you go to any one or more of these shows you will quickly become proficient and knowledgeable in the field of collectible trade stimulators and counter games. These and other major shows around the country are listed in the resources section.

For all the easy ways to find machines, in shops, malls, auctions and at shows where vintage coin machine dealers display what they have to sell, it is the hard way that more often pays off in the greatest finds. That's beating the bushes. Sometimes it takes a sixth sense. That's when you have an open eye, ear, and mind about finding machines, and put them all to work at the same time to make a find. Many collectors have driven by junk yards, big trasher yards, antique shops, and other possible source locations only to get an eerie sensation at the back of their neck that makes them turn around and go back. When they did, they found a machine. That's the sixth sense at work. Sometimes the sixth sense fails. But when it works, the results can be legendary.

The classic tale of such a find, among mechanical bank collectors in any event, is the discovery of the first (and only known, to date!) McLoughlin GUESSING BANK barroom games of 1876. The machine is known to bank collectors as the "Banker Who Pays" based on its patent No. 191,065 issued to Edward J. McLoughlin of New York City on May 22, 1877. The patent calls the game a "Toy Money-Box," and bank collectors believed it for no other reason than that was what the patent said, and no examples of the machine had ever been found to say otherwise. The toy bank collectors adopted the machine as their own, not guessing that it was really saloon gambling piece that was only described as a toy to get around the patent anti-gambling regulations. They also had no idea who made the machines, or even if any were produced and sold, although it was illustrated in catalogs of the early 1880s.

It took almost a hundred years to come up with one. Mark Haber, one of the leading toy bank experts, made the find and told the story to F. H. Griffith, a toy bank collector and writer for *Hobbies* magazine. In a story in the March 1962 issue, Griffith quotes Haber's letters about his find.

"The discovery of these banks was entirely accidental and unusual, and the lead was furnished by an old picker who informed me that he thought he noticed something that might be a bank or a statuette being used as a door stop at a house in South Windham, Connecticut. His meager description of the object and the location left me no other alternative but to comb every street in South Windham until I finally espied the object. It was rusted and weatherbeaten, but unmistakably the bank patented by E. J. McLoughlin.

I had little trouble purchasing the bank from the occupants, Mr. and Mrs. George E. Sherman. Fur-

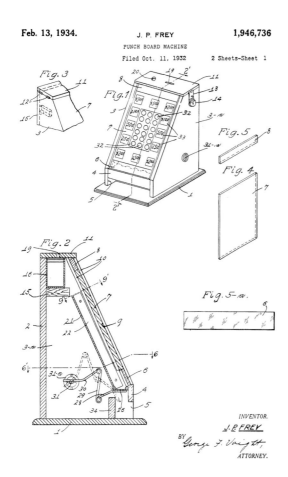

Research can be exciting, particularly finding new machines in patents. Patent No. 1,946,736 is for a counter game, circa 1932. The applicant is James P. Frey of Modesto, California. The game is unidentified and has never been found. There are hundreds more to be discovered in patents. You can get copies for $3 each from the government.

ther inquiry revealed that Mr. Sherman's grandfather had purchased the patent rights and had a number of these banks made up for distribution to jobbers who were to show these to the trade through their salesmen. To the best of his recollection, the orders for these banks were so meager as to make it an unprofitable venture into large production. Being possibly a gambling device in the hands of a child it did not seem to have any appeal.

Further inquiry on subsequent visits revealed that there were a few of the banks left in an old barrel in the original packings in the fine old stable on the estate. I managed to purchase two or three at each visit for sums of money plus some fine first editions which I always brought with me for Mr. Sherman as gifts. Mr. Sherman collected first editions and appreciated my thoughtfulness."

These are the only known examples of the GUESSING BANK yet to be found, and in the 1996 market the better examples are valued at around $5,500 each. It just goes to show that keeping your eyes and ears open can some-

Field research can be particularly exciting. Would you believe that the buildings of the Waddell Wooden Ware Works Company in Greenfield, Ohio, are still standing. The famous THE BICYCLE WHEEL and THE BICYCLE trade stimulators of 1896 through 1917 were made here. Relax. The stuff is gone.

times lead to success. Are there more finds like this awaiting discovery? Probably not as great as the GUESSING BANK find. But then again, who knows? There are many trade stimulators and counter games that are only known from their patents, catalog pages, or advertising, and it is likely that many of them will be found over the years. Such finds are still being made every year, either through an original and often lucky discovery, or through the dispersion of knowledge. In the latter case quite a number of collectors have been pleased to discover that machines in their collections are rare and sometimes the only known examples.

This book will probably unearth more.

Field research can also be rewarding. Discovery of the Clawson Machine Company in New Jersey by your author in 1978 led to the finding of dozens of AUTOMATIC DICE mechanisms from the 1890s. Collector/restorer Mike Gorski of Westlake, Ohio, manufactured reproductions using the original mechanisms. They are clearly marked as such and are now valuable collectibles in their own right as they work better than the vintage machines.

Very early cast iron. Case and machine parts are of iron, and casting details can be seen in the cabinet. Patent dates are often included in the outside casting surfaces. This is the precedent setting Amusement Machine Company STANDARD "Iron Card" pedestal card machine of 1891.

Pieced together bentwood cabinet shows its age. This is an early Leo Canda PERFECTION CARD "Round Top" circa 1898 with bronze working parts and cast iron trim. The nickel plating on the trim has disappeared with time.

The exceptional detail and alluring subjects of the decorative cast iron pieces from the 1900 to 1916 period make this the most desirable and expensive collectible niche in trade stimulators. These cast iron Mills Novelty Company machines are the ELK of 1905, left, and KING DODO (five-way) of 1903, right.

Don't be misled by the wooden cabinet and bronze working parts. This is the Mills Novelty 1926 redesign of the LITTLE PERFECTION to create the "Flat Top." Slab sides and top made production a lot easier and less expensive. Neatness reveals production engineering not evident in the 1890s.

Details: Name headers are usually the first victims of time, breaking off and getting lost. FAIREST WHEEL machines with the top marquee are worth considerably more than wheels without them. This is the Albert Pick private label FAIREST WHEEL of 1906 in the Barbara and Bill Huber collection, Greenville, Ohio.

Details: Nameplates can reveal a great deal. The Waddell Wooden Ware Works THE BICYCLE of 1901 is a valuable machine. But this only known example with its "Poole Bros., Manufacturers, Chicago." brass nameplate is a bump up in rarity and value. Alex Warschaw collection.

Geometric designs and Art Deco elements indicate a circa 1928-1936 counter game. Although aluminum castings often provide interesting detail in game cabinets, they have never attracted the collector interest or value of the even more detailed cast iron cabinets that preceded them.

Details: Many pre-1930 machines are found with scratched nameplates and company decals, presumably by distributors and operators who didn't want their locations to know where the machines came from. This is the A.B.T. 36 LUCKY SPOT of 1927. Tom O'Connor collection, Delmar, New York.

Knowledge: Scale has a lot to do with identification and desirability. Machine at left is the Pace DANDY VENDER of 1931 with its marvelous painted gum ball front. It is difficult to conceive how much smaller the H. C. Evans EVANS BABY BELL of 1928 at left is unless you put the two machines side by side. Tom Gustwiller collection.

Acquisition: Watch what you buy No. 1. There is little question that this is an early model of the Decatur Fairest Wheel Company FAIREST WHEEL. But what model? It appears to be the Style 1 of 1895 missing the wheel cabinet and coin chute. Other than interesting, its main functional purpose is that of a spare parts machine.

Acquisition: Watch what you buy No. 2. Another large Style 1 FAIREST WHEEL lash up. The wheel is right, but the cash box is wrong. It looks like something a merchant used to keep going once the original base was broken or missing. Regard it as collectible but not correct.

Acquisition: A very exciting find at the Chicagoland Show in fall 1994. Collector Tim LaGanke of Novelty, Ohio, at left looks on while Carl Lepiane of Los Gatos, California, right, shows off his discovery. Machine is the rare and unusual Silver King PENNY BASE BALL of 1932. Baseball theme makes it particularly desirable. Machine requires complete restoration.

Acquisition: Watch what you buy No. 3. Operator or merchant revamps weren't restricted to the early days of trade stimulators. This Groetchen Tool IMP of 1940 has been permanently mounted on a large wooden cash box to cut down on collect calls. It is not the way the machine was made, but it is certainly the way it was r

Consideration: German "Allwin" style wall layout slot machine from the 1920s. Actively collected in Europe and the U. K., European machines have not yet caught fire in the United States. They will as domestic machines get harder to find and their intrinsic values are recognized. Their features were recognized long ago by the American coin machine industry. Compare this play action to the Groetchen Tool ZOOM of 1941 illustrated on page 180. G. S. Brierley collection.

Be alert: An exciting and interesting machine that has design features similar to the Brunhoff Manufacturing Company spinning top trade stimulators of 1899. It might have been identified as such, except for that large coin slot marked "Pennies only." It only accommodates the large size British penny, therefore A GOOD TURN is a British machine.

Be alert: It's old and it's iron, but its reels give it away. This is an American R. J. White six-way THE TRADER NO. 1 of 1902 found in Australia. Numerous card machines of this type were brought into Australia early in the 20th century, surviving for years to be re-reeled with local strips. Machines of this type led to the A. O. Buchanan THE AUSTRALIA replication made in Sydney in 1937.

Consideration: The play features and graphics of British, French, and German wall machines of the 1900-1960 period are often colorful and attractive and offer new collectible opportunities. This is a British WIN AND PLACE horse race machine of the 1930s found in New Zealand. Rod Cornelius collection.

It's a toy, and not a trade stimulator, counter game, or slot machine. But it plays on a penny size token. Many collectors appreciate peripheral collectibles of this nature, so it is listed as a baby bell. Maker is Buffalo Toy and Tool Works of Buffalo, New York. JACK POT GAME No.265 is particularly attractive as its box is intact. Similar JACK POT GAME No. 266 was advertised in *The Billboard* coin machine trade publication in August 1940. Robert Vicic collection.

While it doesn't operate with a coin, it is still regarded as a cigar trade stimulator. The John M. Waddell Manufacturing Company ROOLO game of 1896 is played with steel balls that drop down over the most pin studded playfield ever put on a counter machine. Players pay the counterman a nickel for a steel ball, and drop it in the top to score below to win anywhere from 1 to 2, 5, or 10 cigars as shown by flip-up indicators. PLAY BALL version of 1897 has baseball graphics.

What in the world? APPAREIL ELECTRIQUE MUSICAL (ELECTRICAL MUSIC MACHINE) is early 1900s French, and combines a shocker, music maker, and a simulated dice shaker. The dice are in the dial at upper left. Is it a music box, arcade machine or trade stimulator? Take your pick and add it to that collection. Marty Roenigk collection.

What is it, who made it, when was it made and how does it work? This target machine was unearthed in 1995, and no one seems to have any idea what it is. Components from Hance Manufacturing Company bulk vending machines made in Westerville, Ohio, just outside of Columbus, seem to provide a clue to its origins. But there is no record of Hance ever making counter games, and none have shown up in the past. Gary Kothera collection.

RESOURCES/RESTORATION

If you are lucky, you will find trade stimulators and counter games that are in perfect (possibly restored) condition, and work with ease and great entertainment as soon as you put in a coin. If you're lucky. But as in all antiques, you get 'em like you find 'em, which in all likelihood means the machines you find on the open market will require maintenance and repair, and possibly restoration.

Suddenly you will need special resources. Emphasis on the word "special," as it usually takes specialists in the field to bring these games back to life. Maybe so. But if you pay attention to what you have, and learn a little about what it takes to get them back in shape and operating, you can save yourself a bundle and get even more joy out of your machines by doing the bulk of this work yourself. All it takes is a degree of understanding, the courage to tackle the job, and time. Collectors that maintain their own games take a special pride in their work, and unless they really screw things up (which is hard to do, because in the long run these machines are simple) they also enhance the value of their collection. It all starts with getting a game working and, in failing that, determining what is needed to get a machine back in shape and operating.

That is the primary dictum of coin machine collecting; the machines *must* work. Coin machines are not static collectibles (any more than electric trains, home appliances, classic cars, or guns are; they must be workable or they have little collectible value or interest) and have to be playable before they can be regarded as worthy of keeping.

A reasonable restoration project. This John M. Waddell THE BICYCLE WHEEL of 1896 is missing its software. The paper number strip that weaves between the spokes needs replacement, as does the award card and glass sides. The owner has the option of making this a THE BICYCLE DISCOUNT WHEEL model with the proper paper and replacement etched glass.

A fairly simple restoration project. This Automatic Manufacturing Company LUCKY DICE of 1893 is in pretty good shape. The only inconsistency is the modern short round globe. What is needed to get the machine back in condition is a tall vintage round top dome. In the absence of the real thing a vertical wall glass dome for dried flowers found in most craft shops will do the job nicely.

Rust, and just plain wear, are the biggest problems facing old trade stimulators and counter games. Paradoxically, the older the machine is the better off you are. Early mechanisms were made of cast bronze, machined brass, and formed tinwork parts. All are virtually rustproof and are subject to repair. But the moment you get into the 1899-1920 period you will find more cast iron, stamped steel, and galvanized parts, all of which tend to excessively rust. It is often far better to attempt to recreate a part out of new material—using the old part as a guide which you should keep forever in case you need to replace the part again—than attempt to force the damaged material back into shape and run the risk of breakage and loss of configuration.

Often cast iron parts that have been nickel plated for outside trim have lost their thin plating and look like rusty blobs. Try to envision their original appearance, and then take steps to recreate the past. If the parts are still strong, or can be welded if they are cracked, sandblasting or shotblasting and a new nickel plating job will make them look brand new and quickly bring your machine back to its original glory. Nickel was the favored plating material between 1885-1930, and with age it provides a marvelous patina. Top of the line coin-operated machines also had copper, bronze, or ormolu (false gold) plating over cast iron parts, but only the more expensive and complicated machines were treated in this manner. Few if any of the lowly trade stimulators were afforded such luxury.

If your machine was made prior to 1930 or thereabouts stay away from chrome plating; they didn't have it. It's something that came along later in the 1930s.

If rust and water damage are the problems, take heart. Very little is beyond restoration. This is the 1905 Kelley THE IMPROVED KELLEY on page 150 of the specialty reel section before restoration by owner Jack Freund of Springfield, Wisconsin.

The biggest internal problems are those of broken or missing parts. But these problems also have a solution. Broken parts are the easiest to handle. At least they are there, and you can usually fathom what goes where, and why. Then you can either repair, or recreate it. Cast iron parts can be welded, or recast. Stampings, or sheet or plate metal parts, can be reformed, bent back into shape or recreated. The pot metal parts favored in the 1930s are the toughest as they cannot be soldered. They have to be arc welded, glued (if the parts will hold), or bridged and pinned. It is often better to cut or recast a new part. Missing screws, even rivets, can easily be replaced. But use brass wherever possible to avoid any rust problems in the future.

Above all else, take heart. Unless the machines you find are basket cases, coming all disassembled or as rusted parts in a box, there is a good chance that you can get the device working.

Here are a few guidelines to workability:

Somewhere along the line someone took the promotional paper for the Bally BOSCO cigarette dice machine of 1933 and glued it on the cabinet of this Exhibit Supply JUNIOR of 1927 to convert it to cigarette awards. Removal of the paste-ons and proper award paper will quickly restore this machine, although it is interesting as it is.

Basic Maintenance and Repair Steps for Bartop Card Machines and Reel Counter Games.

1. Always keep the machine clean.

That means inside and out. Lint and dirt can jam any mechanism, and lead to forceful play that can often break brittle parts. Many of the working parts are cast iron and bronze that get brittle with age, and can't stand up to the strain that the steel parts of more modern machines handle with ease. Use a vacuum cleaner with a nozzle or small brush attachment both outside and in and you will get most of the loose dirt. Dipping the working parts in gasoline will also help clean some of the caked grease, but be careful not to get the gasoline on the wooden cabinet or paper as you can end up losing or degrading what you have.

2. Use Vaseline to lubricate the working parts.

There is a lot of metal-to-metal contact in a LITTLE PERFECTION or similar card machine. Where there is wear, that's where the parts tend to break. So keep them lubricated with white Vaseline. That is what the original operators used years ago. However, keep the Vaseline away from the reel strips or the front and inside paper as it will destroy them.

3. Get to understand the mechanism.

There is not much to it! If the machine doesn't work, take it out and look it over. But do this carefully. The reels often come out as a unit, usually held in place by a long wire guard at the right side looking in from the back. Lift and rotate this wire guard carefully as it may be brittle with age or rusted and be ready to break or disintegrate. If it shows such signs of wear you might consider having a replacement made. A little oil or "Liquid Wrench" at the right spot might help on a rusty machine. Then check all the parts against any pictures or knowledge you have about these machines, such as the photographs, manuals and patent papers. You might see what is missing, if anything. Most likely some springs are worn out, or have disappeared as they usually degrade or wear out first. If it's a missing or broken part, have it repaired or re-made.

Another rust bucket that turned out looking like a million bucks. This is the Caille REGISTER of 1906 as found by collector Bill Whelan of Daly City, California. It is deeply rusted, the reels are in shatters and the cash box base is missing. The final restored result can be seen on page 150.

One of the biggest trade simulator restoration jobs in recent years. This only known example of the Mills Novelty RELIABLE of 1903 was discovered in deplorable condition. Contemporary advertising and catalog pages, and careful study of the paper and decal remnants on the machine, led to the final restored results as seen on page 72.

4. Restore your card machine only if it needs it.

There is a lot to be said for having a nice looking, working machine. But there are problems to watch out for. Much restoration work is overkill, and when newly restored vintage coin machines end up looking so much better then their originals they begin to look unreal. LITTLE PERFECTION and other similar wooden cabinet card machines do not usually need that much work. It is better to err on the side of conservatism than go overboard. Start by cleaning the cabinet. Above all else, retain whatever original paper you may have on your machine. You will never be able to replace it. But you can replace the reels and reward cards if necessary. Trade Stimulator collector Bill Whelan has specialized in restoration paper items for these old machines and offers a wide selection of reels and reward cards that fit most of the machines of this class. You can write for a catalog and costs. Send to: Slot Dynasty/Restorations, 234 Palmdale Avenue, Daly City, CA 94015.

Basic Guidelines for Pedestal Card and Cast Iron Machines

The "Iron Card" pedestal card machines and cast iron counter machines were the cheap throw-away machines of their day. Basic rules of good vintage coin-op housekeeping will suffice, and keep your machines in running trim. Just follow a few short rules.

1. Keep the machine and its mechanism clean.

Use the small brush or nozzle on the vacuum cleaner. You'll be amazed at the junk you'll pull out of a pot-bellied "Iron Card." But check first, or even use a magnet. There

One of the great thrills of trade stimulator collecting is finding a machine as exciting as this one. The Almy Manufacturing Company of Chicago "Type J" AUTOMATIC CASHIER AND DISCOUNT MACHINE of 1915 is the final model of the Yale Wonder Clock Company version of 1905. Yale sold out to Almy. Other than the fact that a few advertising cards are missing, as are some of the lighting panels, this ancient treasure is ready for restoration and cabinet refinishing. Frank Zygmunt collection.

Can this machine be saved? No. 1: The answer is yes, but the restoration cost of this wooden cabinet Daval specialty reel RACES of 1941 may well far exceed the value of the machine. Although the game is rare its value isn't high, so restoration is a judgement call. Bill Nesnay collection.

may be set screws or springs on the bottom that you'll need, or even old coins.

2. Keep the mechanism well lubricated.

Use Vaseline. They did, and it works best. But stay away from award cards or other paper graphics or you'll ruin them with the grease.

3. Know your mechanism.

Study it. Keep it running smoothly. Try to fathom out any missing parts. Patent drawings can be helpful in that analysis.

4. Make your machine look nice.

Cleaning up well should do the job. Restore only if you must, or if you have a lot of rust or a cracked cabinet that needs welding. But try and keep your machine as original as possible. And that includes the paint.

General Instructions
For Maintaining Counter Games

Counter games of the 1920-1960 period range all the way from wooden cabinets with iron and tinwork mechanisms to completely aluminum cast cabinets with steel stamped mechanisms, often including pot metal components. They were not built to endure, yet they were made strong enough to withstand harsh customer play and handling. That was enough to make most of them virtually immortal, provided they are maintained. Here are a number of basic steps to assure that.

1. Lubricate.

Greasing and oiling of metal to metal contact points at regular short intervals prolongs the life of the game.

2. Replace missing parts.

If you feel a block in the working mechanism, don't force the machine. Study the mechanism until you are certain you know the cause of the stoppages. Then adjust, bend, file to remove a burr or obstruction, or replace the broken or worn parts with identical parts. But remember, unnecessary adjustment or overbending may put the machine out of timing which leads to bigger problems.

3. Check the springs.

Springs should never be shortened, tension changed or substituted unless they are a perfect fit for the machine. Have a selection of springs on hand in order to have the right one at the right time.

You get 'em like you find 'em No. 1: Although the back is missing, everything else is in place in this delicate and rare Columbian Automatic Card Machine Company AUTOMATIC CARD MACHINE made in New York City in 1898. The clockwork mechanism and card rollers are amazingly clear of dust and grime whereas the wooden cabinet and award paper on the front have degraded with the years. All very fixable. Steve Gronowski collection.

You get 'em like you find 'em No. 2: SAFE HIT of 1938 is one of the rarest of the Daval "Humpback" games from the period. The traditional "Baseball Green" cabinet with yellow sides shows through and the award paper is in good condition. All this machine needs for restoration is the proper repaint job. Bill Whelan collection.

4. Check reel spin at intervals.
Spinning too fast can cause excessive bouncing and vibrations, leading to wear. To change reel speed, tighten or loosen the brake wires by the wing nuts on the back of the mechanism if the mechanism is provided with this level of control.

5. Be careful.
Never ever use a sharp instrument to dislodge coins, or attempt to make the mechanism work. Use toothpicks or small pieces of wire. If you find jammed or mutilated coins, get rid of them (spend them!) right away and keep them away from the machine.

6. Protect your machine.
Do not allow players to shake the machine to make it work or dislodge coins. The mechanism is precision built and can be dislocated. When replacing the mechanism in its cabinet, be sure you have used the available locks to fasten it in place.

7. Assure Access.
If you are lucky enough to have a key, make a duplicate and put it in a safe place. If not, before you drill out the lock, try other keys. A good trick is to have a box of small keys on hand (pick up all the keys you can find where you got the machine, or get them from other collectors, or at antique shops and malls) as they sometimes work. If one does, make that duplicate.

8. Remove possible obstructions.
It's amazing how much dirt an old machine can collect. Get rid of it. Also remove all coins stuck in corners or in the mechanism. Check them carefully. You may find a coin that has been stuck there since the machine was originally put into operation. Many a rare old coin has been found inside a coin machine.

Keeping a machine working often isn't enough. Your old trade stimulator and counter games also have to look like they did when they were on the nation's bars and counters, which means that their cosmetics need to be retained, restored, or replaced. The biggest problem area is in the graphics and machine "paper." Painted instructions and paper panels are often missing, having worn off or been torn off years ago, in which case you need new originals or reproductions. The home computer, and Kinkos, can also come to your aid, as many of the missing graphic pieces so sorely needed by collectors years ago are now being generated on a screen and printed. The results look

Can this machine be saved No. 2: EXCELSIOR race games made by the Excelsior Race Track Company in Chicago in 1890 often show up in fragmentary condition. Even so, if you have this much you have the game. The damaged steel cabinet needs patching and repair, four new horses are required and the glass dome is missing. The key to cosmetic success is the elaborate original nameplate. This machine is a doable deal. Jack Kelly collection.

so good you can hardly tell the difference from an original. Old advertising, and duplicate machines, are used as the source for the new replacements. The machines from the late 1930s into the 1950s are frequently better off in this regard as the instructions for play were often cast into the cabinets with the paper retained under glass. They also have the advantage of being later designs, with springs holding the cabinets together and operating the mechanism. Unless critical parts are missing, or spring lugs are broken off, restoration can often be as simple as replacing the springs that are worn, and determining which springs are missing altogether that need to be replaced. All of these restoration tricks can be accomplished by you with the help of other resources at your disposal.

These restoration resources can range from a part found at a show, reproduced award cards and reels found in hobby magazine advertising, and complete restoration jobs undertaken by the experts in the field. Chances are if you become a serious collector of these machines you will, at one point or another, want to subscribe to one or more of the hobby "fanzines" for their articles and stories about finding old trade stimulators and counter games. But that's not the best part. The dealers, parts suppliers, and restorers are advertising in their pages. This advertising alone is a tremendous asset. Somewhere along the line you will want to take advantage of any or all of these resources. To speed up that process and save you the time of searching for all of it on your own, those listed on pages 207 and 208 are probably all you need at this stage.

Can this machine be saved No. 4: No name, a remnant award card and a patent date in the base casting were all that were available to lead the restorer to the final machine. The result is the John J. Watson COMBINATION CARD AND DICE MACHINE of 1894 seen in restoration on page 64 of the card machines section. Tom Gustwiller collection.

Can this machine be saved No. 3: Just because this only known example of the Automatic Machine Company EMPIRE drop card machine of 1888 has a cracked cast iron cabinet with portions missing doesn't mean it has to spend the rest of its days in ignominy. This machine is a prime candidate for restoration, with new castings to be created based on indications provided by the old parts. Stan Muraski collection.

Can this machine be saved No. 5: Drop card machines seem to suffer the most from wear, tear, and time. While the innards are a mess and the cabinet is deeply rusted, this Reliance Novelty Company RELIANCE of 1896 contributes an original award card marquee frame, often missing from card machines. Eminently restorable.

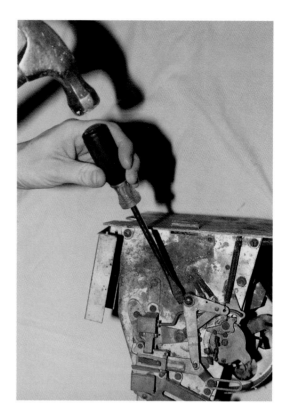

A big solid No! No! Never tackle a restoration job unless you know what you are doing. Never hit parts with a hammer. And never use blunt tools like the head of a screwdriver against metal. Counter games are delicate machines, and with proper repair and maintenance will deliver top notch performance for years—unless you bend, bash, or break something doing something dumb like this.

Can this machine be saved No. 6: Pieces, panels, and parts are missing from this Bat-A-Peny Corporation BAT-A-PENY of 1928. No big deal. It's a baseball game and worth restoration, and there are many ads and other examples to follow. The result will be a crisp, mint looking game.

Restoration conference between collectors Ed Smith, of Pecatonica, Illinois, left, and Stan Harris, right, of Philadelphia, Pennsylvania, ruminating over the course of a restoration for a Caille BON-TON missing the front casting and some of the internal parts. The final result can be seen on page 151 of the specialty reels section.

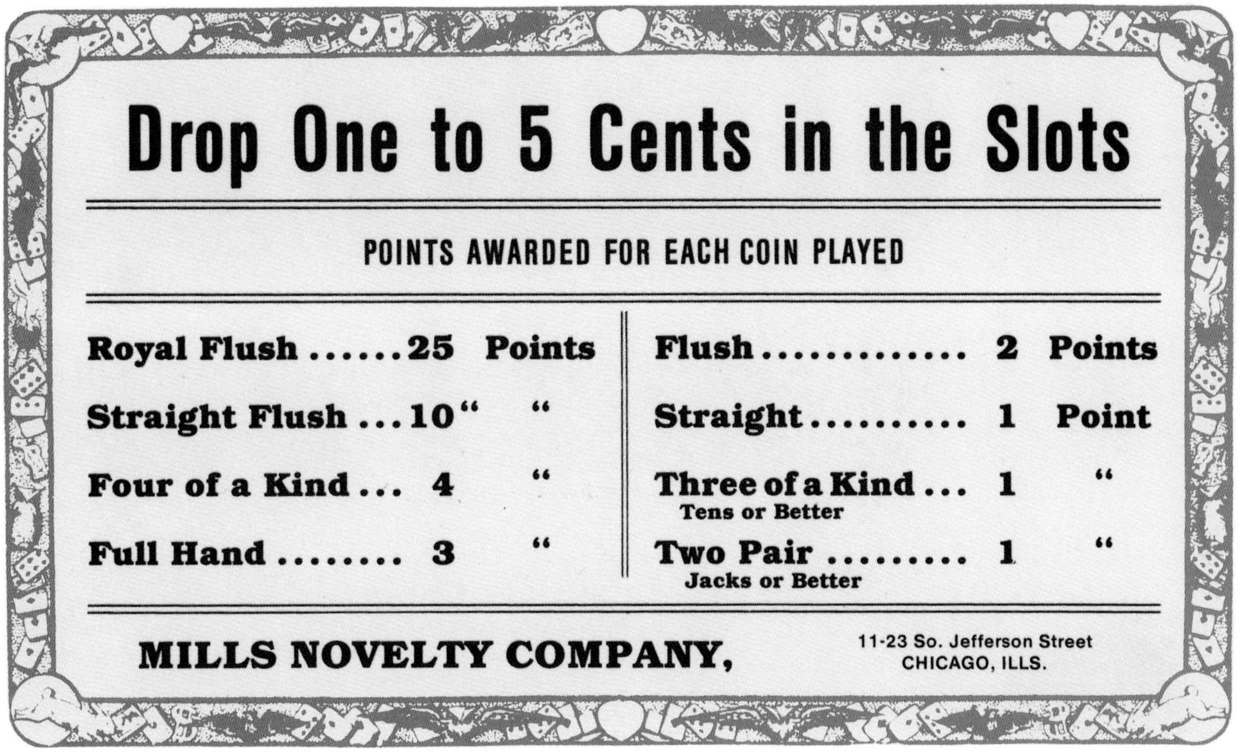

Drop One to 5 Cents in the Slots

POINTS AWARDED FOR EACH COIN PLAYED

Royal Flush ...... 25 Points		Flush .... 2 Points	
Straight Flush ... 10 "	"	Straight .... 1 Point	
Four of a Kind ... 4	"	Three of a Kind ... 1	"
		Tens or Better	
Full Hand 3	"	Two Pair 1	"
		Jacks or Better	

MILLS NOVELTY COMPANY, 11-23 So. Jefferson Street
CHICAGO, ILLS.

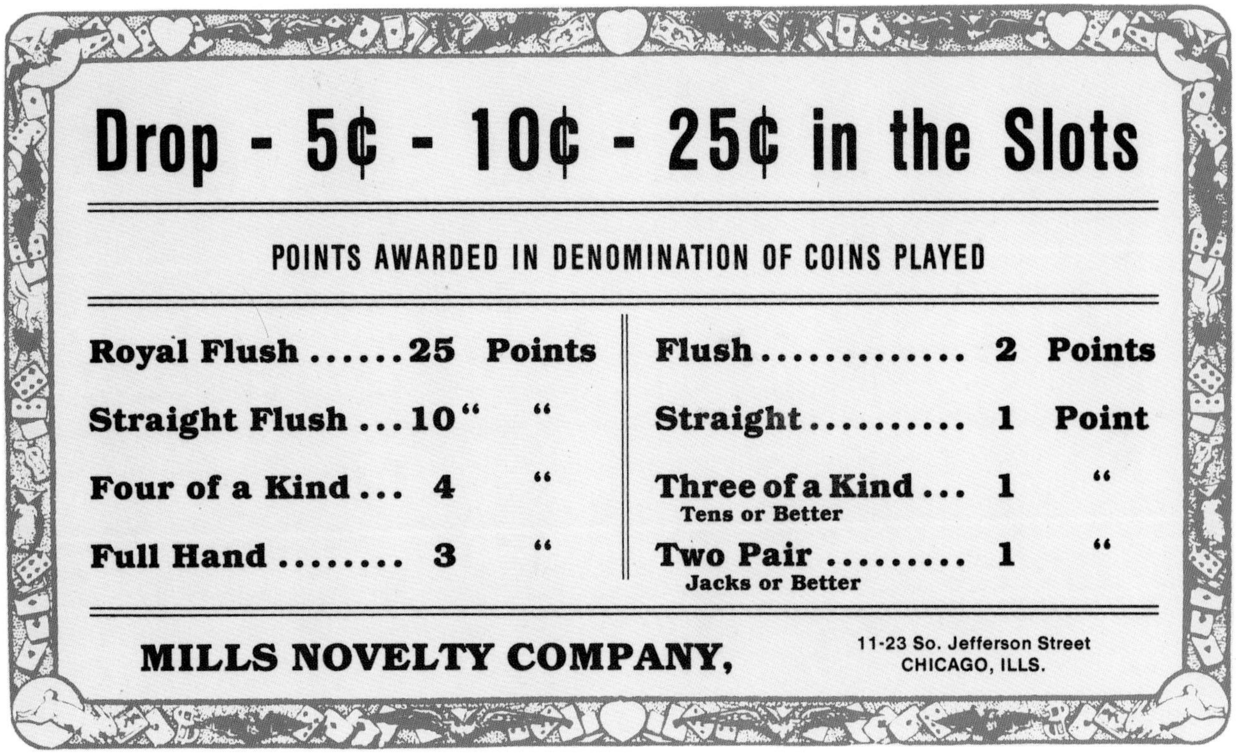

Drop - 5¢ - 10¢ - 25¢ in the Slots

POINTS AWARDED IN DENOMINATION OF COINS PLAYED

Royal Flush ...... 25 Points		Flush .... 2 Points	
Straight Flush ... 10 "	"	Straight .... 1 Point	
Four of a Kind ... 4	"	Three of a Kind ... 1	"
		Tens or Better	
Full Hand 3	"	Two Pair 1	"
		Jacks or Better	

MILLS NOVELTY COMPANY, 11-23 So. Jefferson Street
CHICAGO, ILLS.

Much of the missing paper and decals for award cards, marquees, and manufacturer identification has been recreated and reproduced for restoration work, particularly for machines that still survive in some numbers. These are reproduction award cards for the 5-way and 3-way versions of the Mills Novelty KING DODO of 1903. Source of supply is Slot Dynasty Restorations, 23 Palmdale Avenue, Daly City, CA 94015.

It may look grungy, but you are ninety percent of the way home for a new award paper panel for this American Automatic Machine Company AUTOMATIC DICE of 1893 because key copy is intact and readable. The remnant paper also reveals the type fonts and sizes. Some computer work and a dip in black coffee will recreate an almost perfect replacement to bring the machine up to snuff.

Auctions can be a good source of supply for hard to find machines. Here collector Ken Durham views a selection of 1930s counter games at an auction lineup prior to the beginning of bidding. This is the time to look closely and well at machines you'll be bidding on later.

Museum quality restorer Gary Taplin of Stamford, Connecticut, shows off some recent projects. On the table, foreground, is a Caille Brothers JOCKEY of 1905 with its unique front castings. Left is an early 1900s Caille PUCK floor slot machine with a Regina disk music box and behind is a restored Exhibit Supply arcade gun game marquee from the 1920s.

Author Dick Bueschel takes a good look at a Mooney and Goodwin TRADE STIMULATOR of 1946 that showed up at a Kane County Flea Market in St. Charles, Illinois, in 1989. Maybe this plug-in neon wall spinner isn't a truly classic piece, yet it is nevertheless significant, particularly since it is the only one to have shown up so far.

Showtime can be a good time for all. Here is a particularly tasty dealer display from a late 1980s Chicagoland Show. Machines of this caliber are now rarely seen in groups at shows and are snapped up in minutes. They are starting to show up at auctions as some major collections are disbanded.

Knowledge is power, and that comes out of study, books, and magazines. Book and counter game dealers Ken and Jackie Durham can be found at most major vintage coin machine shows, or can be reached by mail and the internet. Their addresses are included in this resources section.

Collector/dealer William E. "Willie" Howard of Akron, Ohio, shows off his wares at the Chicagoland show, offering a range of machines from ancient dicers to modern venders, and a lot in between. Show buying is the best opportunity to build a collection at realistic prices.

Major shows and events

Atlanta Gameroom Show
March
Gwinnett Civic & Cultural Center
6400 Sugarloaf Parkway, Duluth, GA
Contact: Bob Nelson
1-316/721-0836
FAX 1-316-263-1849

Chicagoland Antique Advertising, Slot-Machine & Jukebox Show
April and November
Pheasant Run Mega Center and Pavilion Building
1-800/999-3319
Route 64, North Avenue (2-1/2 miles west of Route 59), St. Charles, IL
Contact: Steve Gronowski
1-847/381-1234
FAX 1-847-382-5678

Coin-Op Super Show
March and September
Pasadena Exhibit Center
Green Street between Marengo and Euclid, Pasadena, CA
Contact: Rosanna Harris, Royal Bell, Ltd.
1-303/431-9266
FAX 1-303-431-6978
E-Mail rbltd@cris.com

Colorado Antique Slot Machine Jukebox Show and Sale
August
Denver Collector's Fair, National Western Stock Show Complex
46th and Humboldt, Denver CO
Contact: John or Kenna Joseffy
1-303/756-5369

Dallas Gameroom & Collectibles Show
September
Dallas Market Hall
2200 Stemmons Freeway, Dallas, TX
Contact: Be-Bop Jukeboxes
1-214/243-5725
E-Mail bebop@cyberramp.net
Internet http//pmadt.com/bebop

Fort Worth Gameroom Show
June
Amon G. Carter Exhibit Hall
Will Rogers Memorial Complex, Fort Worth, TX
Contact: Bob Nelson
1-316/721-0836
FAX 1-316-263-1849

Great Mid-Atlantic Antique Advertising & Coin-Op Show
February
Montgomery County Agricultural Center
Building 6, Fairgrounds, Gaithersburg, MD
1-410/329-2188
24 hour voice mail 1-800/803-1096

Hackensack Collectors Expo
October
Rotham Center, Fairleigh Dickinson University
100 University Plaza Drive, Hackensack, NJ
Contact: Bob Nelson
1-316/721-0836
FAX 1-316-263-1849

Philadelphia Gameroom Show
May
Fort Washington Expo Center
1100 Virginia Drive, Fort Washington, PA
Contact: Bob Nelson
1-316/721-0836
FAX 1-316-263-1849

Books with Trade Stimulators and Counter Games

Automatic Pleasures (1988)
By Nic Costa
(Out of print. Antique value $65.)
Kevin Francis Publishing Ltd., London, England

"Coin-Ops On Location" (1993)
By Richard M. Bueschel and Eric Hatchell
$35 postpaid
Wordmarque Design Associates
12644 Chapel Road, Suite 204ST, Clifton, VA 22024
1-703/968-9665
FAX 1-703-968-9667

Collectors Guide to Vintage COIN MACHINES (1995)
By Richard M. Bueschel
$44.95 postpaid
Schiffer Publishing, Ltd.
77 Lower Valley Road, Atglen, PA 19310
1-610/593-1777
FAX 1-610-593-2002

Collector's Treasury of Antique Slot Machines (and Trade Stimulators) From Contemporary Advertising (1980)
By Peter Bach
$29.95 Limited Availability
Post-Era Books
Arcadia, CA 91006

Drop Coin Here (1979)
By Ken and Fran Rubin
(Out of print. Antique value $175.)
Crown Publishers, Inc., New York

For Amusement Only (1995)
By Tom Gustwiller
$44.95 postpaid
William E. Howard Distribution
1114 Akron Drive, Akron, OH 44313
1-216/864-0844 or 1-216/376-3607

An Illustrated Price Guide to the 100 Most Collectible Trade Stimulators, Volume 1 (1978 and 1981)
By Richard M. Bueschel
(Out of print. Antique value $75.)
Coin Slot Books (Bill Harris), Denver, CO 80212

An Illustrated Price Guide to the 100 Most Collectible Trade Stimulators, Volume 2 (1981)
By Richard M. Bueschel
(Out of print. Antique value $60.)
Coin Slot Books (Bill Harris), Denver, CO 80212

Trade 2 Illustrated Guide to Collectible Trade Stimulators (1993)
By Richard M. Bueschel
$33 postpaid
Hoflin Publishing Ltd.
4401 Zephyr Street, Wheat Ridge, CO 80033-3299
1-303/934-5656

Silent Salesmen Too (1995)
By Bill Enes
$44.95 postpaid
Enes Publishing
8520 Lewis Drive, Lenexa, KS 66227
1-913/441-1492
FAX 1-913-441-1502

Book and Coin-Op Literature Dealers

Ken Durham
909 26 Street NW
Washington, D.C. 20037
1-202/338-1342
Internet http://Game Room Antiques.com

Rosanna Harris
Royal Bell Bookshelf
5815 W. 52nd Avenue
Denver, CO 80212
1-303/431-9266
FAX 1-303-431-6978

Hobby Publications

Antique Amusements Magazine
Antique Amusement Company, Mill Lane,
Swaffham, Bulbeck
Cambridge CB5 ONF. ENGLAND
Monthly. £23 per year US and Canada (£19 Europe, £25 Australia, Pacific Rim)
011-44-1223-813041. FAX, same.

Antiques Amusements, Slot Machine & Jukebox Gazette
909 26 Street NW, Washington, D.C. 20037
Semi-annual newspaper/2 issues per year.
$10 for four issues
Sample copy $5
1-202/338-1342
Internet http://Game Room Antiques.com

Chicagoland Program
414 N. Prospect Manor Avenue,
Mt. Prospect, IL 60056-2046
Semi-annual show program. Spring and Fall. $2 each.
1-847/253-0791
FAX 1-847-253-7919

Coin Drop International
5815 W. 52nd Avenue, Denver, CO 80212
Bi-monthly. $15 per year US
($US21 Canada, $US40 offshore)
Sample copy $3
1-303/431-9266
FAX 1-303-431-6978

Coin-Op Classics
17844 Toiyabe Street, Fountain Valley, CA 92708
Bi-monthly/ 5 issues per year. $39 per year US ($US54 Canada, $US70 Europe, $US75 Pacific Rim)
Sample copy $9
1-714/756-8746
FAX 1-714-963-1716
Internet http://www.coin-op-classics-com

Coin-Op Newsletter
909 26 Street NW, Washington, D.C. 20037
Bi-monthly. $15 per year US
Sample copy $5
1-202/338-1342
Internet http://Game Room Antiques.com

GameRoom
P. O. Box 41, Keyport, NJ 07735-0041
Monthly. $28 per year US ($US35 Canada, $US53
Europe Surface/Europe Air $US87, $US57 Pacific Rim
Surface/Pacific Rim Air $US93)
Sample copy $4
1-908/739-1955
FAX 1-908-739-2834
e-mail: GameRmMag@aol.com

Vintage Coin Machine Identification and Evaluation

Richard M. Bueschel
414 N. Prospect Manor Avenue
Mt. Prospect, IL 60056-2046
1-847/253-0791
FAX 1-847-253-7919
email: buschlhist@aol.com

Ken Durham
909 26 Street NW
Washington, D.C. 20037
1-202/338-1342
Internet http://Game Room Antiques.com

Slot Machine Parts and Trade Stimulator and Counter Game Restorations

Bernie Berten
Springs and Parts
9420 S. Trumbull Avenue
Evergreen Park, IL 60642
1-708/499-0688
FAX 1-708-499-5979

Fred De Baugh
222 E. Thomas Avenue
Baltimore, MD 21225
1-410/789-4811
Evening 1-410/592-2936
FAX 1-410-789-4638

John M. Fignar
Coin Machine Restoration and Repair
12 Leaward Way
Saratoga Springs, NY 12866
1-508/252-9711
Weekends 1-518/584-0268

Tom Kolbrener
St. Louis Slot Machine Company
2111-R, South Brentwood
St. Louis, MO 63144
1-314/961-4612
FAX 1-314-961-3846

Tom Krahl
Antique Slot Machine Part Company
140 N. Western Avenue
Carpentersville, IL 60110
1-847/428-8476
FAX 1-847-428-4471

Gary Taplin
Penny Arcade Restorations
28 Southfield Avenue
Stamford, CT 06902
1-203/357-1913

Reel Strips, Decals and Award Cards:

Evans & Frink
2977 Eager
Howell, MI 48843

Barbara Larks
The Gumball King's Decals & More
8444 Lawndale Avenue
Skokie, IL 60076
1-847/679-4765

Bill Whelan
Slot Dynasty Coin Machine Restorations
23 Palmdale Avenue
Daly City, CA 94015
1-415/756-1189

Collectors Who May Want to Buy Machines

Collector's Arcade
Allan B. Pall & Susan Pall
1118 N. Harlem Avenue
River Forest, IL 60305
1-708/771-7446

Ken Durham
909 26 Street NW
Washington, D.C. 20037
1-202/338-1342
Internet http://Game Room Antiques.com

Jack Freund
Slots Of Fun
P. O. Box 4
Springfield, WI 53176
1-414/642-3655
FAX 1-414-642-2632

Tom Gustwiller
116 W. Main Street
Ottawa, OH 45875
Day 1-419/523-6395
Evenings 1-419/523-6556

William E. Howard
1114 Akron Drive
Akron, OH 44313
Day 1-216/376-3607
Evenings 1-216/864-0844
FAX 1-216-376-3603

Alex Warschaw
424 Honey Oak Lane
San Antonio, TX 78253
1-210/679-6151
FAX 1-210-679-8550

Bill Whelan
23 Palmdale Avenue
Daly City, CA 94015
1-415/756-1189

Dealers Who May Have Machines to Sell

Collector's Arcade
Allan B. Pall & Susan Pall
1118 N. Harlem Avenue
River Forest, IL 60305
1-708/771-7446

Ken Durham
909 26 Street NW
Washington, D.C. 20037
1-202/338-1342
Internet http://Game Room Antiques.com

Jack Freund
Slots Of Fun
P. O. Box 4
Springfield, WI 53176
1-414/642-3655
FAX 1-414-642-2632

Pete Levin's Scrap Book
1916 Piedmont Circle NE
Atlanta, GA 30324
1-404/875-7212

St. Louis Slot Machine Company
2111-R, South Brentwood
St. Louis, MO 63144
1-314/961-4612
FAX 1-314-961-3846

Alan Sax
3239 Victorian RFD
Long Grove, IL 60047
1-847/438-5900

Frank Zygmunt
P. O. Box 542
Westmont, IL 60559
1-630/985-2742
FAX 1-630-985-5151

Collectible Trade Stimulators (1870~1919) and Counter Games (1920~1996)

(dated by manufacturers)

In the over 120 years since their inception thousands of different trade stimulators and counter games have been created and introduced, with most produced in short runs while others were produced in great volume. It is the nature of mechanical, and particularly coin-op collectibles, that a number of these machines will never be found. Yet at least two or three and often more of the "undiscovereds" have continually surfaced every year since vintage coin machine collecting became a popular interest in the early 1960s. These newly found machines are generally known in the hobby as "Fantastic Finds," providing a reward for those who continue to seek the unknown.

Knowledge is power, and knowing the makers and some of the characteristics of the undiscovered games aids in their discovery. To aid you in that search, as well as identify and date the known machines, trade stimulator and counter game manufacturers are listed in alphabetical order and geographic location with their machines placed in chronological order as they appeared. This listing was prepared from known existing examples, contemporary trade media advertising and other documentation such as legal papers, interviews with producers, articles and other mentions. Approximately two-thirds of these machines have been found, with the remaining one-third constituting the exciting undiscovered inventory of trade stimulators and counter games. Many of these machines had a long manufacturing and service life with their production and usage dates often extending considerably beyond their introduction dates.

The description of a vintage coin-operated device as a "Trade stimulator" or a "Counter Game" does little to define the actual machine or its working characteristics. Unlike slot machines, arcade machines, pinball machines, vending, scales or music machines, the vintage trade stimulators and counter games were produced for a narrow window of application, with a variety of small points differentiating one class of machine from another. Often the choice of reels, be they playing cards, fruit, cigarette, color spot, numbers, fortune, dice or other graphic, had a significant effect on the operation of the machine, with different machine names often applied.

The differences between trade stimulators and counter games is more chronological than physical, with their classification primarily based on when they were made and how they were used as the determinant factor, even if they looked very much alike. Some counter games of the 20s and 30s are virtually identical to the trade stimulators that proceeded them in the tens and teens, yet they differ in fabrication and materials, game awards and location placement. Trade stimulators of the 1870-1919 period were primarily saloon, cigar counter and country store pieces that offered "Free Drinks" and "Free Cigars" in their saloon placements, and an opportunity to get equal value or an increase in merchandise over the money spent on the machine in stores, more often in general and country stores in rural areas rather than urban.

With the coming of Prohibition and the automobile as an aftermath of World War I, eliminating the saloons and enlarging the offerings of the stores, counter games replaced the trade stimulators as soft-core gambling devices that paid off in merchandise, cigarettes or over-the-counter cash, with the games surviving in numbers into the 1950s. It was the automobile, supermarkets and suburban shopping malls that finally applied the kiss of death to the Mom and Pop and cigar stores of the past, increasingly moving family shopping into much larger retail outlets dedicated to volume sales. With their growth and the operation of these new outlets by hired management, replacing the small shop and the owner's need to make a penny-nickel-dime-quarter gain in income by means of often illegal gambling devices, as well as the disappearance of checkout counter space, the counter game was eliminated. By the 1960s they were nowhere to be found. It has only been in recent years, surviving in privately owned bars, restaurants and truck stops—holdouts against the chains—that the games, often in video formats, have returned. As often as not they have returned to their original role as trade stimulators. Perhaps there is an upcoming Golden Age of trade stimulators and counter games on the horizon. But until that is a reality the games of the past are the ones that must provide the joys and game playing rewards of these remnants of a simpler age.

For purposes of clarification in the chapters of this collectors guide, the following distinct classifications and formats have been established for identification purposes:

Race Game, Roulette, Wheel, Pointer, Card Machine, Dial, Dice, Coin Drop, Target, Single Reel, Baby Bell, Cigarette, Specialty Reel, Novelty/Skill.

The machines are described in the same manner in the tables.

United States

Manufacturer and Location

Format	Name	Date

A.B.C. Coin Machine Company, Inc., Chicago, Illinois
Coin Drop	BELL SKILL	1933
Target	CENTURY BELL	Sept. 1933
Target	LINE EM UP	1933
Specialty Reel	JOCKEY CLUB (Horses)	Nov. 1933
Specialty Reel	JOCKEY CLUB GUM VENDER (Horses)	Jan. 1934
Cigarette	JOCKEY CLUB	1934

ABCO Novelty Company, Chicago, Illinois
Dice	HAPPY DAY	1944
Dice	IMP	1944
Card Machine	21 GAME	1944
Baby Bell	CUB	1944
Cigarette	CUB	1944
Card Machine	ACE	1944
Cigarette	YANKEE	1944
Baby Bell	YANKEE	1944
Cigarette	IMP	1944
Baby Bell	IMP	1944
Cigarette	WINGS	1944
Card Machine	POK-O-REEL	1944
Specialty Reel	KLIX (Numbers)	1944
Cigarette	GINGER	1944
Cigarette	PENNY PACK	1944
Cigarette	MERCURY	1944
Specialty Reel	SPARKS CHAMPION (Sports)	1944
Cigarette	LIBERTY	1944
Baby Bell	LIBERTY BELL	1944
Specialty Reel	FREE PLAY (Numbers)	1944
Cigarette	BUDDY	1944
Baby Bell	AMERICAN EAGLE	1944
Baby Bell	NON-COIN AMERICAN EAGLE	1944
Cigarette	MARVEL	1944
Cigarette	NON-COIN MARVEL	1944
Cigarette	SPARKS	1944
Cigarette	NON-COIN SPARKS	1944
Cigarette	ZEPHYR	1944
Novelty/Skill	TILT TEST	May 1950
Novelty/Skill	HIT-A-HOMER	July 1950
Novelty/Skill	TOUCHDOWN	1953

Bernard Abel and Company, New York, New York
Dice	SQUARE DEAL	Apr. 1893

A.B.T. Manufacturing Company, Chicago, Illinois
Novelty/Skill	PLAY BALL	Dec. 1924
Roulette	36 ROULETTE	1927
Dice	36 LUCKY SPOT	1927
Novelty/Skill	MIRROR OF FORTUNE	1928
Novelty/Skill	FINGER TEST	Feb. 1932
Novelty/Skill	THE SIX HORSEMEN	June 1932
Novelty/Skill	IMP GOLF	1933
Card Machine	ROYAL REELS	1934
Card Machine	ROYAL REELS GUM VENDOR	1934
Wheel	WAGON WHEELS	Nov. 1935
Wheel	SPORTLAND	Nov. 1935
Dial	THREE CADETS	Jan. 1936
Wheel	SMOKE-UP	1936
Wheel	PROSIT	1936
Wheel	A.B.C.	1936
Wheel	CIGA-ROLA	1936
Wheel	GRAND PRIZE	1936
Dial	SKILL DRAW	1936
Race Game	HALF MILE	1939
Wheel	3-IN-A-ROW	1939
Dice	TOP-ITS	1939
(Unknown)	SHOOT-A-PAK	1939

Acme Game Company, Benton Harbor, Michigan
Dial	MATCH COLOR	Feb. 1940
Dial	MATCH COLOR GUM VENDER	Feb. 1940
Dial	DIZZY DISKS	Mar. 1940
Dial	DIZZY DISKS GUM VENDER	Mar. 1940

Acme Novelty Company, Detroit, Michigan
(Unknown)	(UNKNOWN)	Nov. 1928

Acme Novelty Works, Detroit, Michigan
Roulette	ROULETTE	1891

Acme Sales Company, Toledo, Ohio
Dice	LITTLE WINNER	1922
Dice	THE ACME	1923

Acme Sales Company, Brooklyn, New York
Novelty/Skill	SINK OR SWIM	Oct. 1936

Adams Gum Company
Novelty/Skill	FIVE STAR	1915

The Ad-Lee Company, Inc., Chicago, Illinois
Card Machine	QUINTETTE	1925
Dice	TRY IT	1927
Dice	TRY IT BALL GUM VENDER	1927
Target	TARGET PRACTICE	1928
Dice	CRYSTAL GAZER (Cigarette)	1932
Baby Bell	1933 AD-LEE VENDER	Sept. 1932
Dice	4-WAY FROLIC	Sept. 1932
Dice	CRYSTAL DESTINY	1933
Dice	CUBA	1933
Race Game	KING'S HORSES	1933
Wheel	COIN OPERATED WHEEL	Apr. 1933
Coin Drop	RADIO	1934
Dice	SPIN-EM	1935
Wheel	ZIP	1935
(Unknown)	TNT	1935
Race Game	1935 KING'S HORSES	1935
Dice	LEAP FROG	1935
Dice	ROYAL SPORT DICE GAMER	1935
Roulette	ROYAL SPORT MARBLE GAME	1935
Coin Drop	ZIG-ZAG	1939

Advance Electric Company, Los Angeles, California
(Unknown)	(UNKNOWN)	May 1935
(Unknown)	(UNKNOWN)	June 1935
Race Game	(UNKNOWN)	Oct. 1938

Advance Machine Company, Chicago, Illinois
Coin Drop	PIN FORTUNE	1927
Coin Drop	COINSKILL	June 1928
Novelty/Skill	ADVANCE BAT-A-PENY	1931
Target	SKILSHOT	Apr. 1931
Novelty/Skill	MINIATURE BASEBALL WORLD CHAMPION	1931
Coin Drop	MARVEL	Dec. 1933

A. and F. Engineering Company, Chicago, Illinois
Dice	TRI-COLOR DICE	Oct. 1945

The Alb Company, Belleville, Illinois
Race Game	DERBY	Jan. 1927

Albany Novelty Works, Albany, New York
Wheel	(UNKNOWN)	Dec. 1893
Pointer	(UNKNOWN)	Oct. 1899

The All-In-One Company, St. Louis, Missouri
Novelty/Skill	ALL-IN-ONE	1914

Almy Manufacturing Company, Chicago, Illinois
Novelty/Skill	AUTOMATIC CASHIER AND DISCOUNT MACHINE	Mar. 1915

American Automatic Machine Company, New York, New York
Roulette	AUTOMATIC ROULETTE	1892
Dice	AUTOMATIC DICE	Jan. 1893
Dice	AUTOMATIC DICE SHAKING MACHINE	1893
Roulette	IMPROVED AUTOMATIC ROULETTE	1894

American Automatics, Elmont, Long Island, New York
Dice	JITTER CUBE	Nov. 1944

American Manufacturing and Sales Company, Portland, Oregon
Cigarette	HAV-A-SMOKE	1934
(Unknown)	FLTEL	1935

American Mechanical Toy Company, New York, New York
Race Game	(UNKNOWN)	1888

American Novelty Company, Cincinnati, Ohio
Card Machine	SUCCESS	1894

American Sales Company, Kansas City, Missouri
Coin Drop	PENNY SKILLO	1938

American Specialty Manufacturing Company, Buffalo, New York
Card Machine	SUCCESS	1896

American Supply Company, Philadelphia, Pennsylvania
(Unknown)	(UNKNOWN)	1907

Amusement Coin Machine Manufacturing Company, Pittsburgh, Pennsylvania
Novelty/Skill	COCO-NUTS	Jan. 1933

Amusement Enterprises, Houston, Texas
Dice	DICE SHAKER	June 1946
Novelty/Skill	PITCHEM	June 1946
Card Machine	POKER JR.	1947

Amusement Machine Company, Jersey City, New Jersey
Race Game	(UNKNOWN)	Oct. 1889

Card Machine	CARD MACHINE	Apr. 1890	

Amusement Machine Company, New York, New York

Card Machine	CARD MACHINE "Iron Card"	Dec. 1890
Specialty Reel	POLICY (Numbers)	1891
Card Machine	BABY CARD MACHINE	1891
Card Machine	COUNTER IRON CARD	1891
Card Machine	STANDARD "Iron Card"	1891
Card Machine	COUNTER CARD "Two Hand"	1891
Card Machine	CARD MACHINE "Two Hand"	1891
Card Machine	SUCCESS "Iron Card"	1891
Pointer	ARROW	1892
Coin Drop	COMBINATION JACK POT	1892

Amusement Machine Company, Oakland, California

(Unknown)	BASEBALL	1915

Amusement Novelty Company, Toledo, Ohio

Novelty/Skill	5¢ A PUSH 5¢	July 1931

Amusement Products Corporation, Jersey City, New Jersey

Dice	SWEEPSTAKES	Aug. 1932

A. H. Andrews and Company, Chicago, Illinois

(Unknown)	(UNKNOWN)	Jan. 1890

The Anthony (Cigar) Company, Cincinnati, Ohio

Pointer	ECLIPSE	1892

Anthony and Smith, Cincinnati, Ohio

Pointer	ECLIPSE	1893

Arcade Supply Company, Chicago, Illinois

Novelty/Skill	THE CLOWN	Sept. 1929
Novelty/Skill	SKEE-SHOT (POCKET BALL)	1929

Arkansas Novelty Company, Magnolia, Arkansas

Dice	3-A-LIKE	1947

Arlington Heights Machine Works, Arlington Heights, Illinois

Baby Bell	BABY BELL	1936

ASCO, Newark, New Jersey

Coin Drop	POISON THIS RAT (Hitler)	May 1942

Aspin and Furry, Wilmington, Delaware

Target	O-TO-GO	1898

Associated Amusements, Inc., Boston, Massachusetts

Novelty/Skill	BASKETBALL STARS	Mar. 1947

Associated Distributors, Detroit, Michigan

Card Machine	DEALER'S CHOICE	Dec. 1933

Atlas Games, Cleveland, Ohio

Coin Drop	CIRCUS	Dec. 1941
Coin Drop	CIVILIAN DEFENSE	Apr. 1942
Novelty/Skill	TILT TEST	1947

Atlas Indicator Works, Chicago, Illinois

Coin Drop	PICK A WINNER	1931
Target	BASE BALL	Feb. 1931
Target	BASE BALL	Mar. 1931

Atlas Manufacturing Company, Cincinnati, Ohio

Roulette	ROULETTE	1895

Atlas Manufacturing Company, Kaukauna, Wisconsin

Roulette	MIDGET ROULETTE	1926
Dice	MIDGET	1926
Roulette	3-IN-1	Nov. 1927
Dice	3-IN-1	Nov. 1927

Atlas Manufacturing Company, New York, New York

Coin Drop	FIVE ON	Aug. 1936

Atlas Novelty Company, Oakland, California

Dice	26 GAME	1908

Atlas Novelty Company, San Francisco, California

Card Machine	(UNKNOWN)	1907
Dice	(UNKNOWN)	1908
Dice	26 GAME	1926

August Grocery Company, Richmond, Virginia

Wheel	THE HOO DOO	Apr. 1901

Auto-Bell Novelty Company, Chicago, Illinois

Cigarette	(UNKNOWN)	1951
Novelty/Skill	VICTORY	1951
Novelty/Skill	HIT-A-HOMER	1951
Novelty/Skill	TOUCHDOWN	1953
Novelty/Skill	TILT TEST	1957
Novelty/Skill	PLAY BALL	Mar. 1958
Novelty/Skill	LUCKY HOROSCOPE	Jan. 1959

Auto-Dice Company, Columbus, Ohio

Dice	AUTO-DICE AMUSEMENT TABLE	1930

Auto-Vender Company, Chicago, Illinois

Wheel	DANDY VENDER	Apr. 1911
Wheel	IMPROVED DANDY	Nov. 1911

Automat Games Company, Chicago, Illinois

Coin Drop	TEXAS LEAGUER	July 1935

Automatic Amusement, Chicago, Illinois

Roulette	POCKET POOL	Mar. 1938

Automatic Cash Discount Register Company, Chicago, Illinois

Pointer	THE PROFIT SHARER (non-coin)	1902

Automatic Coin Device Company, Cincinnati, Ohio

(Unknown)	(UNKNOWN)	1914

Automatic Coin Machine Company, Chicago, Illinois

Dice	CRYSTAL GAZER	Oct. 1931
Race Game	TURF	May 1932

Automatic Corporation, Chicago, Illinois

(Unknown)	DICELESS DICE	1932

Automatic Games Company, Inglewood, California

Card Machine	NEW LARK POKER	1931
Baby Bell	NEW LARK BABY BELL	1931

Automatic Industries, Inc., Youngstown, Ohio

Roulette	JINGLE	Mar. 1932

Automatic Machine Company, New York, New York

Card Machine	EMPIRE	1888
Card Machine	AUTOMATIC POKER PLAYER	1889
Dice	AUTOMATIC DICE	1892

Automatic Machine and Tool Company, Chicago, Illinois

Card Machine	SUCCESS	1901
Card Machine	JUMBO SUCCESS	1901
Card Machine	JOCKEY	1901
Card Machine	JOCKEY CABINET	1901
Card Machine	MUSICAL JOCKEY	1901

Automatic Manufacturing Company, New York, New York

Dice	POPE DICE MACHINE	Feb. 1892
Dice	AUTOMATIC DICE SHAKING MACHINE	1892
Dice	LUCKY DICE	Jan. 1893

Automatic Novelty Company, Kansas City, Missouri

Dice	(UNKNOWN)	1905

Automatic Novelty Company, New York, New York

(Unknown)	(UNKNOWN)	1907

Automatic Novelty Company, Philadelphia, Pennsylvania

Dice	(UNKNOWN)	June 1893

Automatic Novelty Company, San Antonio, Texas

Race Game	STEEPLE-CHASE	Sept. 1933

Automatic Novelty Machine Company, New York, New York

Dice	AUTOMATIC DICE	1888

Automatic Trading Company, New York, New York

Coin Drop	AUTOMATIC TRADER	June 1908

Automatic Vending Machine Company, New York, New York

(Unknown)	(UNKNOWN)	1900

Alfred R. Babcock, Chicago, Illinois

Specialty Reel	(UNKNOWN)	Nov. 1937

Louis Badaracco, San Francisco, California

Card Machine	(UNKNOWN)	1905

Badger Bay Company, Green Bay, Wisconsin

Dice	(UNKNOWN)	1935

Baker Machine Company, Fort Worth, Texas

Dice	(UNKNOWN)	1932
Wheel	(UNKNOWN)	1932
Cigarette	MISSOURI MULE	1933

Baker Novelty Company, Chicago, Illinois

Dice	PICK-A-PACK	Jan. 1939
Wheel	LUCKY STRIKE	Jan. 1941
Novelty/Skill	KICKER AND CATCHER (5 Balls)	June 1941
Novelty/Skill	SKILL-A-RETTE	Feb. 1942
Coin Drop	BOMB HIT	Apr. 1942
Novelty/Skill	KICKER AND CATCHER (3 Balls)	1946
Novelty/Skill	KICKER AND CATCHER	1947

F. H. Waldie, Seattle, Washington

(Unknown)	(UNKNOWN)	1903

Bally Manufacturing Company, Chicago, Illinois

Dial	BALLY	1932
Dice	BOSCO	1933
Dice	BEER BOSCO	Aug. 1933
Dice	BALRICKY	1933
Dice	DICETTE	1933
Wheel	CUB	1933
(Unknown)	(UNKNOWN)	1934
Dice	NATURAL	1934
(Unknown)	IMP	1934
(Unknown)	TUNNEL	1934
(Unknown)	SLUGGER	1934
Novelty/Skill	FLICKER	1934
(Unknown)	LONE STAR	1935
(Unknown)	HUNCH	1935

Pointer	SPINNER		1935
(Unknown)	WIN OR LOSE		1935
(Unknown)	TEASER		1936
Cigarette	BALLY BABY		1936
Specialty Reel	BALLY BABY (Novelty)		1936
Specialty Reel	BALLY BABY (Numbers)		1936
Baby Bell	BALLY BABY		1936
(Unknown)	REDWOOD		1936
(Unknown)	AUTOMAT		1937
(Unknown)	SPARK-O-LITE		1937
Dice	CONGO		1937
Dice	NATURAL		1937
(Unknown)	MATCH IT		1937
(Unknown)	PLUS OR MINUS		1937
Pointer	1937 SPINNER		1937
(Unknown)	MATCH-EM		1937
Specialty Reel	THE NUGGET (Numbers)		May 1937
Specialty Reel	GOLD RUSH (Numbers)		1937
(Unknown)	EAGLE EYE		1937
(Unknown)	VENDOR		1937
Dice	BALLY BOOSTER		1937
Specialty Reel	SUM FUN (Novelty)		June 1937
Pointer	SPINNERINO		1937
(Unknown)	DOG HOUSE		1937
Novelty/Skill	LITE-A-PAX		Dec. 1937
Novelty/Skill	MILLWHEEL		1938
Novelty/Skill	CIGARETTE MILLWHEEL		1938
(Unknown)	TOPSY		1938
Specialty Reel	BABY RESERVE (Numbers)		1938
Baby Bell	BABY RESERVE		1938
Cigarette	BABY RESERVE		1938
(Unknown)	POPPY		1938
Novelty/Skill	PONIES		1938
(Unknown)	SWEETHEART		1938
Novelty/Skill	PUNCH BOARD		1938
Wheel	SPIRAL		1938
Cigarette	WAMPUM		1938
(Unknown)	RAINBOW		1939
Wheel	PENCIL VENDER		1939
Novelty/Skill	LINE-A-BASKET		1947
Baltimore Vending Machine Company, Baltimore, Maryland			
Dice	AUTOMATIC SHOW CASE		June 1895
Banner Specialty Company, Philadelphia, Pennsylvania			
Coin Drop	LEADER		1922
Card Machine	BANNER PERFECTION		1924
Specialty Reel	BANNER PURITAN (Numbers)		1924
Target	BANNER TARGET PRACTICE		1924
Target	NEW TARGET PRACTICE		1924
Barco Products Company, Chicago, Illinois			
Coin Drop	PENNY PITCH		1936
Monroe Barnes Manufacturer, Bloomington, Illinois			
Wheel	FAIREST WHEEL (Later renamed due to lawsuit)		July 1895
Wheel	BONUS WHEEL		Apr. 1897
Wheel	CRESCENT (CIGAR WHEEL)		Apr. 1897
Barr and Company, Chicago, Illinois			
Roulette	(UNKNOWN)		Apr. 1906
Barr Novelty Company, Shamokin, Pennsylvania			
Target	DOUBLE TARGET		1924
W. R. Bartley, Butte, Montana			
Card Machine	SUCCESS		1904
Bartley and McFarland, Seattle, Washington			
Card Machine	SUCCESS		1903
Bat-A-Peny Corporation, Rochester, New York			
Novelty/Skill	BAT-A-PENY		Jan. 1926
Novelty/Skill	1928 BAT-A-PENY		Jan. 1928
Baxter and Ellis, Cincinnati, Ohio			
(Unknown)	(UNKNOWN)		1904
Bay City Novelty Company, Bay City, Michigan			
(Unknown)	(UNKNOWN)		1897
Baynard Novelty and Machine Works, Denver, Colorado			
(Unknown)	(UNKNOWN)		1900
B.B. Novelty Company, Milwaukee, Wisconsin			
Coin Drop	THE NEXT		1926
Henry A. Behn, Union Hill (Union City), New Jersey			
Race Game	(UNKNOWN)		May 1889
Belk, Schafer and Company			
Wheel	HUMMER (CIGAR SELLER)		Oct. 1896
Jonas D. Bell and Company, Chicago, Illinois			
Single Reel	NICKELSCOPE		Jan. 1897
Single Reel	PENNYSCOPE		Jan. 1897
Single Reel	HOW IS YOUR LUCK		1898
Single Reel	(WRIGLEY) TRY YOUR LUCK		1899
Single Reel	WRIGLEY'S SLOT MACHINE		1899
Coin Drop	(WRIGLEY'S) DEWEY (PIN MACHINE)		1899
Single Reel	VICTORY TRADE MACHINE		1903
Bell Fruit Vending Company, Streater, Illinois			
(Unknown)	COIN-GETTER		Dec. 1913
Bennett and Company, Kalamazoo, Michigan			
Wheel	STAR GREEN (H. Vantongeren)		1900
Wheel	SHENANDOAH (Stuckey Cigar)		1900
Wheel	COURT HOUSE (George W. Stattler)		1900
Paul Bennett and Company, Chicago, Illinois			
Cigarette	LUCKY PACK		1938
Card Machine	DEUCES WILD		1938
Card Machine	DEUCES WILD VENDER		1939
Cigarette	TOKETTE		1940
Specialty Reel	DOUGH BOY (Numbers)		Mar. 1940
Cigarette	DOUGH BOY		Mar. 1940
Specialty Reel	SKILL TEST (Numbers)		1940
Paul E. Berger Manufacturing Company, Chicago, Illinois			
Card Machine	SUCCESS		1899
Card Machine	JUMBO SUCCESS		1899
Roulette	MONTE CARLO		1899
Card Machine	PERFECTION CARD		1901
Berger Manufacturing Company, Chicago, Illinois			
Novelty/Skill	CATCH-N-MATCH		1936
Best Novelty Company, Hartford, Connecticut			
Coin Drop	WIZARD		1904
Beyer and Company, Seattle, Washington			
(Unknown)	(UNKNOWN)		1901
B. and F. Sales and Manufacturing Company, Fort Wayne, Indiana			
Dice	DIXIE		1935
A. C. Bindner Company, Inc., Chicago, Illinois			
Novelty/Skill	THE AUTOMATIC CHAMP		Dec. 1927
Bingo Sales Company, Toledo, Ohio			
Coin Drop	BINGO		June 1929
Binks Industries, Inc., Chicago, Illinois			
Coin Drop	ZIPPER		Apr. 1954
Coin Drop	WHIZ-BOWLER		1954
Biscayne Manufacturing Company, Miami, Florida			
Roulette	ROLL A POINT SENIOR		1933
Dice	ROLL A POINT JUNIOR		1933
Charles C. Bishop and Company, St. Louis, Missouri			
Dice	THE TRIOGRAPH		Nov. 1889
Frederick W. Bishop, Los Angeles, California			
Coin Drop	TEN POCKET		July 1893
Coin Drop	NINE POCKET		1893
Blake Manufacturing Company, Holly, Michigan			
Novelty/Skill	JIGGER		Sept. 1941
Novelty/Skill	"V"		Oct. 1941
Novelty/Skill	JUMPER		Oct. 1941
Novelty/Skill	HU-LA		Feb. 1942
Blanchard			
Roulette	ROLLING POKER		1935
Blue Bird Sales Corporation, Kansas City, Missouri			
Target	RED, WHITE AND BLUE GUM TARGET		1926
Target	AUTOMATIC GUM TARGET		1926
Target	TARGET		1927
Target	BASE BALL TARGET		1927
Target	AUTOMATIC GUM TARGET		1927
Dice	TRY IT		1927
Dice	TRY IT BALL GUM VENDER		1927
Specialty Reel	BABY BELL (Novelty)		Dec. 1927
Specialty Reel	BABY BELL VENDER (Novelty)		June 1928
Baby Bell	BABY BELL VENDER		June 1928
Baby Bell	1929 BABY VENDER		Dec. 1928
Specialty Reel	SIX IN ONE (Novelty)		1929
Blue Bird Products Company, Kansas City, Missouri			
Target	AUTOMATIC GUM TARGET		1931
B. and M. Products Company, Chicago, Illinois			
Baby Bell	BABY VENDER		1927
Novelty/Skill	PLAY BALL		1927
Boardman Rubber Stamp Works, Toledo, Ohio			
Dice	SLOT DICE		Nov. 1892
Boduwil Company, Inc., Chicago, Illinois			
Novelty/Skill	PROMOTION SALES REGISTER/PSA		July 1931
Issac T. Bomar, Campbellsville, Kentucky			

Novelty/Skill	DOLL PITCH	Mar. 1895	

Bonus Advertising System, Los Angeles, California

Novelty/Skill	JACK-POT BONUS ?	1946	
Novelty/Skill	? JACK-POT BONUS ?	Oct. 1947	

Bonus Sales Company, Lawrence, Massachusetts

Coin Drop	BASEBALL	1938	
Pointer	SKILO	1938	

Booth Games Company, Minneapolis, Minnesota

Coin Drop	PLAY HORSESHOES	1935	

Bowman Specialty Company, Cleveland, Ohio

Dice	DIXIE DICE	1932	
Roulette	DIXIE	1932	

Boyce Coin Machine Amusement Corporation, Tuckahoe, New York

Novelty/Skill	OVER THE TOP	Feb. 1925	
Novelty/Skill	JUGGLER	June 1925	
Coin Drop	PENNY BACK GUM MACHINE	June 1925	
Race Game	TWO PENNY RACING MACHINE	Oct. 1925	
Coin Drop	WEE-GEE	Oct. 1925	
Novelty/Skill	THE PATIENCE DEVELOPER	Nov. 1925	
Coin Drop	RUNABOUT	Mar. 1927	

James A. Bracewell, Tulsa, Oklahoma

Specialty Reel	(UNKNOWN) (Sports)	1936	

W. A. Bradford Company, San Francisco, California

Dice	(UNKNOWN)	1901	

Bradford Novelty Machine Company, San Francisco, California

Dice	THE LARK	1907	

Bradford Novelty Company, Providence, Rhode Island

Coin Drop	LITTLE GEM FORTUNE TELLER	1913	
Coin Drop	LITTLE GEM	1913	

Bradley Industries, Chicago, Illinois

Dice	7-GRAND	1948	

Charles Brewer and Sons, Chicago, Illinois

Novelty/Skill	(UNKNOWN)	1930	

Charles A. Breyfogle, Allentown, Pennsylvania

Coin Drop	(UNKNOWN)	Jan. 1912	

M. Brodie Company, Long Beach, California

Target	LUCKY ROLL	1939	
(Unknown)	STEVE BRODIE	1940	

Brunhoff Manufacturing Company, Cincinnati, Ohio

Wheel	FIVE CIGARS	Apr. 1898	
Wheel	AUTOMATIC VOTE RECORDER AND CIGAR SELLER	Apr. 1898	
Dice	CRAZY	Nov. 1898	
Dice	CIGAR CUTTER	1899	
Novelty/Skill	DIAMOND EYE	1899	
Novelty/Skill	SPINNING TOP	May 1899	
Wheel	SLOTLESS (CIGAR CUTTER) (non-coin)	1903	
Coin Drop	DAISY ("Hump Back")	1907	

Brunswick Manufacturing Company, Chicago, Illinois

Coin Drop	PEP	Oct. 1932	
Novelty/Skill	STEEPLECHASE	Nov. 1932	
Dice	PAY DAY	1936	

Bryant Pattern and Novelty Company, Detroit, Michigan

Wheel	(UNKNOWN)	Feb. 1902	

L.H. Buchanan and Company, Pasadena, California

Coin Drop	THE PYRAMID	1892	

Buckley Manufacturing Company, Chicago, Illinois

Target	PIN BOARD TARGET PRACTICE	1931	
Target	JACK POT TARGET PRACTICE	1931	
Target	POKER	1931	
Baby Bell	PURITAN BABY BELL MODEL 1	Dec. 1931	
Baby Bell	PURITAN BABY BELL MODEL 2	Dec. 1931	
Baby Bell	PURITAN BABY BELL MODEL 3	Dec. 1931	
Baby Bell	PURITAN BABY BELL MODEL 4	Dec. 1931	
Baby Bell	PURITAN BABY BELL MODEL 5	Dec. 1931	
Baby Bell	JACK POT PURITAN BABY BELL MODEL 6	Dec. 1931	
Specialty Reel	PURITAN BABY VENDER (Numbers)	Jan. 1932	
Specialty Reel	PURITAN BABY VENDER (Fortune)	Jan. 1932	
Baby Bell	PURITAN BABY VENDER MODEL 1	1932	
Baby Bell	PURITAN BABY VENDER MODEL 2	1932	
Baby Bell	PURITAN BABY VENDER MODEL 3	1932	
Baby Bell	PURITAN BABY VENDER MODEL 4	1932	
Baby Bell	PURITAN BABY VENDER MODEL 5	1932	
Baby Bell	RESERVE JACK-POT PURITAN VENDER	1933	
Baby Bell	JACK-POT ATTACHMENT	Sept. 1933	
Dice	BABY SHOES	Sept. 1933	
Card Machine	PILGRIM VENDER	Sept. 1933	
Specialty Reel	PURITAN BELL (Fortune)	1933	
Specialty Reel	PURITAN BELL (Beer)	1933	
Baby Bell	PURITAN BELL MODEL 1	1933	
Baby Bell	PURITAN BELL MODEL 2	1933	
Baby Bell	PURITAN BELL MODEL 3	1933	
Baby Bell	PURITAN BELL MODEL 4	1933	
Baby Bell	PURITAN BELL MODEL 5	1933	
Baby Bell	JACK POT PURITAN BELL MODEL 1	1933	
Baby Bell	JACK POT PURITAN BELL MODEL 2	1933	
Baby Bell	JACK POT PURITAN BELL MODEL 3	1933	
Baby Bell	JACK POT PURITAN BELL MODEL 4	1933	
Baby Bell	JACK POT PURITAN BELL MODEL 5	1933	
Specialty Reel	IMPROVED PURITAN VENDOR (Fortune)	1933	
Baby Bell	IMPROVED PURITAN VENDOR	1933	
Specialty Reel	IMPROVED PURITAN VENDOR (Beer)	1934	
Specialty Reel	JACK POT IMPROVED PURITAN VENDOR (Fortune)	1934	
Baby Bell	JACK POT IMPROVED PURITAN VENDOR	1934	
Specialty Reel	JACK POT IMPROVED PURITAN VENDOR (Beer)	1934	
Card Machine	DRAW POKER	Apr. 1934	
Card Machine	DRAW POKER GUM VENDER	Apr. 1934	
(Unknown)	SEVEN AND ONE HALF	Apr. 1934	
Cigarette	CENT-A-PACK MODEL A	Sept. 1935	
Cigarette	CENT-A-PACK MODEL B	Sept. 1935	
Cigarette	SPIN-A-PACK	1936	
Cigarette	ROLL-A-PACK	1936	
Cigarette	WIN-A-PACK	1936	
Specialty Reel	HORSES (Horses)	Apr. 1936	
Specialty Reel	GOLDEN HORSES (Horses)	July 1936	
Dice	BOTTOMS UP	1936	
Dice	CHAMPION	1936	
Cigarette	DELUXE CENT-A-PACK	Feb. 1937	
Baby Bell	ALWIN	Mar. 1937	
Specialty Reel	BINGO (Numbers)	1937	
Specialty Reel	MUTUEL HORSES (Horses)	1937	
Specialty Reel	NUMBERS (Numbers)	1938	
Cigarette	1940 CENT-A-PACK	1940	
Cigarette	PURITAN BELL	1942	
Cigarette	PURITAN VENDOR	1942	
Cigarette	CENT-A-PACK	1942	
Cigarette	LUCKY PACK	1942	
Cigarette	ZIP	1942	

Bucyrus Manufacturing Company, Bucyrus, Ohio

Dice	DICE BOX	Oct. 1891	
Dice	ELECTRIC DICE	Oct. 1891	
Dice	ECLIPSE	Oct. 1891	
Dice	RIVAL	1892	
Dice	AUTOMATIC DICE SHAKER	1902	

Buddick Engineering Company, Inc., New York, New York

Novelty/Skill	SKE-BAL-ETE	Aug. 1931	
Novelty/Skill	PAR-KET	1932	

Buddy Sales Corporation, Brooklyn, New York

Novelty/Skill	BUDDY BALL GUM VENDER	Sept. 1931	

Budin's Specialties, Inc., Brooklyn, New York

Coin Drop	BUDDY GUM VENDER	1929	

Buffalo Toy and Tool Works, Buffalo, New York

Baby Bell	JACK-POT-GAME No. 265	1940	
Baby Bell	JACK-POT-GAME No. 266	1940	

Burnham Gum Machine Works, Chicago, Illinois

Baby Bell	BABY VENDER	1927	
Baby Bell	QUARTER BABY VENDER	1927	

Burnham and Mills, Chicago, Illinois

Baby Bell	BABY VENDER	1927	
Novelty/Skill	PLAY BALL	1927	
Baby Bell	IMPROVED BABY VENDER	1928	

Burton Machine Company, Chicago, Illinois

Dice	MONTE CARLO	1932	

Herbert H. Buxbaum, Philadelphia, Pennsylvania

Dice	(UNKNOWN)	June 1893	

The Caille Brothers Company, Detroit, Michigan

Wheel	BUSY BEE No.1	1901	
Wheel	BUSY BEE No.2	1901	
Wheel	BUSY BEE No.3	1901	
Wheel	BUSY BEE No.4	1901	
Wheel	BUSY BEE No.5	1901	
Card Machine	SUCCESS	July 1901	
Card Machine	COUNTER SUCCESS	July 1901	
Card Machine	JUMBO SUCCESS	July 1901	
Card Machine	ROYAL JUMBO	July 1901	
Card Machine	PERFECTION	1901	
Card Machine	QUINTETTE	1901	

Wheel	SEARCHLIGHT	Apr. 1902
Wheel	(TRADE OR CASH) SEARCHLIGHT	Apr. 1902
Coin Drop	SUNBURST	1903
Wheel	CALIFORNIA BEAR	May 1904
Card Machine	SENSATIONAL	1904
Card Machine	HY-LO (COUNTER)	1904
Card Machine	HY-LO	1904
Card Machine	HY-LO GUM VENDER	1904
Wheel	WASP	1904
Wheel	WASP (COUNTER)	1904
Card Machine	GOOD LUCK	1904
Specialty Reel	GOOD LUCK (Numbers)	1904
Card Machine	GOOD LUCK (SWIVEL BASE)	1904
Specialty Reel	GOOD LUCK (SWIVEL BASE) (Numbers)	1904
Card Machine	GOOD LUCK SPECIAL	1904
Specialty Reel	GOOD LUCK SPECIAL (Numbers)	1904
Single Reel	ELK	1905
Single Reel	IMPROVED AUTOMATIC CHECK PAYING CARD MACHINE	June 1905
Card Machine	IMPROVED SUCCESS	June 1905
Card Machine	IMPROVED COUNTER SUCCESS	June 1905
Card Machine	IMPROVED JUMBO SUCCESS	June 1905
Card Machine	COUNTER JUMBO	June 1905
Specialty Reel	PURITAN (Numbers)	1905
Specialty Reel	PURITAN VENDOR	1905
Specialty Reel	CLIPPER (Numbers)	1905
Card Machine	JOCKEY	1905
Card Machine	JOCKEY (CABINET)	1905
Card Machine	BANKER	1906
Card Machine	BANKER (SWIVEL BASE)	1906
Card Machine	(PLAIN) BANKER	1906
Card Machine	DRAW POKER	1906
Card Machine	(COUNTER) DRAW POKER	1906
Specialty Reel	REGISTER (Numbers)	1906
Card Machine	GLOBE	1906
Card Machine	(COUNTER) GLOBE	1906
Card Machine	RELIANCE	1906
Single Reel	SPECIAL (SPECIAL AUTOMATIC CHECK PAYING CARD MACHINE)	1906
Single Reel	TIGER	1907
Single Reel	TIGER	1907
Specialty Reel	HIAWATHA (Numbers)	1907
Card Machine	HIAWATHA JR.	1907
Dice	WINNER DICE	1907
Specialty Reel	BON TON (Numbers)	1907
Specialty Reel	BON TON SIDE VENDER (Numbers)	1907
Baby Bell	MERCHANT	1908
Card Machine	PILGRIM STYLE A	1908
Baby Bell	PILGRIM STYLE B	1908
Specialty Reel	PILGRIM STYLE C (Numbers)	1908
Specialty Reel	PILGRIM STYLE D (Numbers)	1908
Single Reel	SPECIAL TIGER	1908
Single Reel	SPECIAL TIGER (Novelty)	1908
Single Reel	SPECIAL TIGER GUM VENDER	1909
Specialty Reel	NEW PURITAN (Numbers)	1909
Specialty Reel	CHECK-PAY PURITAN (Numbers)	1909
Specialty Reel	JUMBO PURITAN (Numbers)	1910
Specialty Reel	MATADOR (Dice)	1910
Card Machine	HIAWATHA	1910
Specialty Reel	LA WA-WO-NA (Numbers)	1910
Card Machine	MAYFLOWER STYLE A	1910
Baby Bell	MAYFLOWER STYLE B	1910
Specialty Reel	MAYFLOWER STYLE C (Numbers)	1910
Specialty Reel	MAYFLOWER STYLE D (Numbers)	1910
Card Machine	MAYFLOWER	1910
Coin Drop	LITTLE DREAM	1910
Single Reel	IMPROVED SPECIAL TIGER	1910
Single Reel	LE TIGRE (3-Way)	1910
Single Reel	LE TIGRE (5-Way)	1910
Single Reel	BASEBALL ("The Tiger")	1910
Coin Drop	INDIAN PIN POOL	1911
Single Reel	NEW SPECIAL TIGER (Side Handle)	1911
Single Reel	JEWEL	May 1911
Single Reel	IMPROVED BASE-BALL	1911
Single Reel	THE COMET	1911
Single Reel	LE COMETE 3-WAY	1911
Single Reel	LE COMETE 5-WAY	1911
Wheel	MASCOT	1912
Wheel	DANDY GUM VENDER	1912
Wheel	LINCOLN (CIGARS)	1912
Wheel	LINCOLN (MONEY)	1912
Wheel	LINCOLN (TRADE)	1912
Single Reel	BIG STAR SIX	1912
Novelty/Skill	CLOWN	1924
Novelty/Skill	JESTER	1925
Baby Bell	PURITAN BELL	1926
Baby Bell	PURITAN BELL VENDER	1926
Baby Bell	PENNY BALL GUM VENDER	1926
Baby Bell	PURITAN GUM VENDER	1926
Baby Bell	FORTUNE BALL GUM VENDER	1927
Baby Bell	JUNIOR BELL	1928
Card Machine	GOOD LUCK	Jan. 1933

Caille-Richards Company, Union City, Michigan

Coin Drop	LITTLE WONDER	1902
Novelty/Skill	LION, JR.	1902

Caille-Schiemer Company, Detroit, Michigan

Wheel	(LITTLE) BUSY BEE	1901

California Machine Company, San Francisco, California

Card Machine	TUXEDO	1906

California Sales Company, Chicago, Illinois

Target	TARGET PRACTICE	1924

J. W. Calvert and Company, Baltimore, Maryland

Target	BASEBALL	Jan. 1931

Calvert Manufacturing Company, Baltimore, Maryland

Novelty/Skill	INDIAN SHOOTER	1929
Target	ALL ABOARD	1929
Target	SPOOK HOUSE	1930
Target	GEM	1930
Target	GEM CONFECTION VENDER	1930
Target	(UNKNOWN)	1931

Calvert Novelty Company, Baltimore, Maryland

Card Machine	TOT	1941

Camco Products Company, Grand Rapids, Michigan

Dice	HI-LO	1939
Dice	SNOOKY	1939
Dice	LUCKY STAR	1939

Leo Canda Company, Cincinnati, Ohio

Card Machine	MODEL CARD MACHINE	Aug. 1893
Card Machine	NEW CARD MACHINE	1893
Specialty Reel	POLICY MACHINE (Numbers)	1893
Specialty Reel	DICE MACHINE (Dice)	1893
Coin Drop	TRADE VENDING MACHINE	1893
Card Machine	GIANT (CARD)	1894
Card Machine	GIANT COUNTER CARD	1894
Specialty Reel	GIANT POLICY (Numbers)	1894
Specialty Reel	COUNTER GIANT POLICY (Numbers)	1894
Specialty Reel	GIANT DICE (Dice)	1894
Specialty Reel	COUNTER GIANT DICE (Dice)	1894
Pointer	GIANT ARROW	1894
Pointer	COUNTER GIANT ARROW	1894
Coin Drop	THE EAGLE	1894
Card Machine	NEW CARD MACHINE	1894
Specialty Reel	NEW POLICY MACHINE (Numbers)	1894
Specialty Reel	NEW DICE MACHINE (Dice)	1894
Card Machine	BONANZA	1895
Card Machine	SUCCESS CARD MACHINE	1895
Card Machine	SUCCESS	1896
Card Machine	COUNTER SUCCESS	1896
Specialty Reel	FIGARO (Numbers)	Jan. 1897
Card Machine	ACME	1897
Card Machine	EXCELSIOR	1897
Card Machine	CHECK EXCELSIOR	1897
Card Machine	COUNTER EXCELSIOR	1897
Card Machine	COUNTER CHECK EXCELSIOR	1897
Roulette	IMPROVED ROULETTE	1897
Card Machine	THE SHUFFLER	1897
Card Machine	SUCCESS	1898
Card Machine	COUNTER SUCCESS	1898
Card Machine	PERFECTION CARD ("Round Top")	1898
Specialty Reel	PERFECTION FIGARO (Numbers)	1898
Card Machine	JUMBO SUCCESS	1898
Card Machine	COUNTER JUMBO SUCCESS	1898
Card Machine	UPRIGHT CARD MACHINE	1898
Card Machine	ROYAL CARD MACHINE	1898
Card Machine	COUNTER PERFECTION	1898
Specialty Reel	UPRIGHT FIGARO (Numbers)	1898
Specialty Reel	FIGARO CHECK (Numbers)	1898
Specialty Reel	UPRIGHT POLICY (Numbers)	1898

Card Machine	JUMBO		1898
Card Machine	CHECK JUMBO		1898
Card Machine	COUNTER JUMBO		1898
Card Machine	COUNTER CHECK JUMBO		1898
Card Machine	JUMBO GIANT		1898
Card Machine	CLOVERLEAF		1899
Card Machine	SKYSCRAPER		Jan. 1900
Card Machine	CANDA CARD MACHINE		Jan. 1900

Leo Canda Manufacturing Company, Cincinnati, Ohio

Card Machine	SUCCESS	May 1902
Card Machine	JUMBO SUCCESS	May 1902
Card Machine	LITTLE PERFECTION	May 1902
Card Machine	UPRIGHT CARD	1902
Specialty Reel	FIGARO (Numbers)	1902
Card Machine	JUMBO	1902
Card Machine	COUNTER JUMBO	1902
Card Machine	JUMBO GIANT	1902
Card Machine	QUINTETTE	1902
Card Machine	HAMILTON	1903
Specialty Reel	DICE MACHINE (Dice)	1903
Card Machine	CLOVER (Pinochle)	1903
Roulette	ROULETTE	1903
Card Machine	AUTOMATIC CARD MACHINE	1903

Cardinal Company, Dallas, Texas

Dice	TUMBLER (Fruit)	1934
Dice	ROLL 'EM	1934
Dice	NEW ROLL 'EM	1935

Carldick Manufacturing Company, Manitowoc, Wisconsin

Wheel	(UNKNOWN)	1934

Carney Manufacturing Company, Miami, Florida

Dice	MAE AND HER PALS	May 1935

Joseph Carpenter, Kenton, Ohio

Coin Drop	(UNKNOWN)	1926

Cato Novelty Works, Lakeview, Michigan

(Unknown)	(UNKNOWN)	1893

Caudle and McCrary Manufacturing Company, Phoenix, Arizona

Coin Drop	NAVAJO	1957

Cawood Novelty Company, Danville, Illinois

Wheel	PANAMA CANAL	1913
Wheel	PLAY BALL	1913

Central City Novelty Company, Syracuse, New York

Race Game	GEM	1892

Central Manufacturing Company, Chicago, Illinois

Novelty/Skill	HI FLY	Aug. 1946
Novelty/Skill	HIT A HOMER	1947

Century Games Company, Chicago, Illinois

Roulette	ROLLETTO	1934
Roulette	ROLETTO JR.	1934

Century Manufacturing Company, Chicago, Illinois

(Unknown)	PENNY SKILLO	1938

Champion Manufacturing Company, Beverly, Massachusetts

Novelty/Skill	BASKETBALL	Jan. 1948

C. and F. Manufacturing Company, Chicago, Illinois

Coin Drop	PLAY THE FIELD	1933

C. Charle and Company, Springfield, Missouri

Novelty/Skill	DING THE DINGER	1933

Chase Vending Machine Company, St. Louis, Missouri

Pointer	SPIN-O	Oct. 1940

The Checker Sales Company, Youngstown, Ohio

Roulette	JINGLES	June 1932

Checkerboard Corporation, Henderson, North Carolina

Novelty/Skill	CHECKER MATCH	1933

Chicago Coin Machine Company, Chicago, Illinois

Dice	SHOOTEM	Aug. 1933
Dice	SHOOTEM (Fruit)	Aug. 1933
Dial	PIPE-EYE	Oct. 1934
Dice	SNAKE-EYES	Oct. 1934
Dial	LUCKY EYES	Oct. 1934
Single Reel	(UNKNOWN)	Feb. 1936

Chicago Coin Machine Exchange, Chicago, Illinois

Dice	SHOOTEM	Mar. 1933

Chicago Mint Company, Chicago, Illinois

Baby Bell	PURITAN CONFECTION VENDER	1928
Baby Bell	ALL-WAYS CONFECTION VENDER	1929
Baby Bell	PURITAN CONFECTION VENDER	1931
Baby Bell	JACK POT PURITAN CONFECTION VENDER	1933

Chicago Nickel Works, Chicago, Illinois

Race Game	AUTOMATIC RACE TRACK	1889

Chicago Slot Machine Exchange, Chicago, Illinois

Target	FLIP TARGET	June 1927
Baby Bell	WONDER VENDER	June 1927

Chicle Chocle Products Company

Pointer	CHICLE-CHOCLE	1929

Cincinnati Automatic Machine Company, Cincinnati, Ohio

(Unknown)	(UNKNOWN)	1895

Cincinnati Novelty Manufacturing Company, Cincinnati, Ohio

(Unknown)	(UNKNOWN)	1892

Clark Novelty Manufacturing Company, Detroit, Michigan

(Unknown)	(UNKNOWN)	1889

Clawson Machine Company, Newark, New Jersey

Coin Drop	LIVELY CIGAR SELLER	Nov. 1897
Wheel	FAIREST WHEEL	1898
Card Machine	PERFECTION CARD	1898
Card Machine	SUCCESS	1899
Card Machine	JUMBO SUCCESS	1899
Card Machine	JUMBO	1899
Card Machine	JUMBO GIANT	1899
Card Machine	STANDARD	1899
Card Machine	CLOVERLEAF (PINOCHLE SUCCESS)	1899
Card Machine	THE MONARCH	1899

Clawson Slot Machine Company, Newark, New Jersey

Dice	DICE TOSSER No.1	1889
Dice	DICE TOSSER No.2	1889
Dice	HOO DOO CIGAR CUTTER	1889
Dice	AUTOMATIC DICE (SHAKER)	Aug. 1890
Dice	AUTOMATIC FORTUNE TELLER	1890
Dice	(COUNTER) AUTOMATIC FORTUNE TELLER	1890
Dice	AUTOMATIC DICE SHAKER	1890
Dice	(COUNTER) AUTOMATIC DICE SHAKER	1890
Dice	TRY YOUR LUCK	1891
Dice	TRY YOUR FORTUNE	1891
Dice	DICE MACHINE	1892
Coin Drop	FAIR-SELLING MACHINE	Sept. 1892
Coin Drop	PERFECT SELLING MACHINE	1892
Coin Drop	HAPPY THOUGHT	1893
Dice	AUTOMATIC SALESMAN	June 1893
Coin Drop	LIVELY CIGAR SELLER	June 1893
Coin Drop	HEADS AND TAILS	Oct. 1893
Coin Drop	LIVELY CIGAR SELLER No.2	Oct. 1893
Coin Drop	TEN TO ONE	Dec. 1893
Card Machine	SUCCESS No.1	1894
Roulette	THREE BALL	Apr. 1895
Roulette	ONE BALL	Aug. 1895
Novelty/Skill	THE NEWARK RAINBOW	1896

W. H. Clune Manufacturer, Los Angeles, California

Card Machine	VICTOR	1900
Card Machine	COMMERCIAL	1900

Coast Novelty Company, San Francisco, California

(Unknown)	(UNKNOWN)	1903

Coin Auto Company, Hammond, Indiana

(Unknown)	(UNKNOWN)	1909

Coin Machine Company of America, Indianapolis, Indiana

Coin Drop	BOMB HITLER	Jan. 1942

Coin Machine Service Company, Adrian, Michigan

Dial	HEADS OR TAILS	June 1936

Coin Sales Corporation, New York, New York

Target	MILLARD'S MINIATURE BASEBALL	1927

Cointronics, Mountain View, California

Novelty/Skill	BALL/WALK	1968
Novelty/Skill	ELECTRIC BALL/WALK	Apr. 1969

Colby Specialty Supply Company, Chicago, Illinois

Dice	COMBINATION LUNG TESTER	1892

Merriam Collins and Company, Decatur, Illinois

Roulette	PEERLESS ADVERTISER (THE CODE)	Apr. 1897

Coleman Novelty Company, Rockford, Illinois

Target	THE GOLD MINE	1932

Colonial Manufacturing and Sales Company, Kansas City, Missouri

Coin Drop	5 STAR FINAL	Apr. 1939

Columbia Manufacturing Company, Rochester, New York

Card Machine	SUCCESS	1900

Columbia Novelty Manufacturing Company, Chicago, Illinois]

Novelty/Skill	20TH CENTURY PROSPECTOR	Feb. 1900

Columbian Automatic Card Machine Company, New York, New York

Card Machine	AUTOMATIC CARD MACHINE	1898

Columbian Automatic Machine Company, New York, New York

Card Machine	AUTOMATIC POKER PLAYER	1901

Columbian Machine Company, New York, New York

Card Machine	POKER CARD MACHINE	1901

Columbine Novelty Company, Denver, Colorado
(Unknown)	(UNKNOWN)	1902

Comet Industries Inc., Chicago, Illinois
Cigarette	NON-COIN MARVEL	1949
Cigarette	MARVEL	1949
Baby Bell	NON-COIN AMERICAN EAGLE	1949
Baby Bell	AMERICAN EAGLE	1949
Baby Bell	CUB	1949
Cigarette	CUB	1949
Card Machine	ACE	1949
Cigarette	BUDDY	1949
Cigarette	NON-COIN COMET	1950
Cigarette	COMET	1950
Baby Bell	NON-COIN METEOR	1950
Baby Bell	METEOR	1950
Baby Bell	MITE	1950
Cigarette	MITE	1950
Card Machine	KING	1950
Cigarette	CIGGY	1950
Baby Bell	CIGGY	1950

Comstock Novelty Works, Fort Wayne, Indiana
Coin Drop	THE PERFECTION	Dec. 1897
Pointer	PERFECTION WHEEL	1898

Condon and Company, Vinalhaven, Maine
Coin Drop	GAME O'SKILL	Oct. 1903

Consolidated Automatic Machine Company, Los Angeles, California
Coin Drop	ADD YOUR SCORE	1931
Novelty/Skill	INDOOR STRIKER	Jan. 1932

The Consolidated Coin Control Company, Chicago, Illinois
Coin Drop	200	1894

Consolidated Service Bureau, Chicago, Illinois
Dice	KENTUCKY DERBY (non-coin)	Dec. 1934

Continental Novelty Manufacturing Company, Williamsville, New York
(Unknown)	(UNKNOWN)	1908

Convex Sign Company, Chicago, Illinois
Race Game	(UNKNOWN)	1925

S. A. Cook and Company, Medina, New York
Coin Drop	HOWARD'S FAVORITE	1896

Ralph B. Cooley, Brooklyn, New York
Pointer	AUTOMATIC REGISTERING BANK	Jan. 1891

Henry A. Cordray, Brenham, Texas
Race Game	(UNKNOWN)	Dec. 1886

J. Edward Cowles and Company, New York, New York
Pointer	PILOT	1899

Cowper Manufacturing Company, Chicago, Illinois
Wheel	THE MIDGET	Sept. 1897
Wheel	MASCOT	1897
Race Game	MINIATURE RACE TRACK	1897
Coin Drop	NEW DROP CASE	1897
Coin Drop	DONKEY	1897
Dice	DICE MACHINE	1897
(Unknown)	FIRE EAGLE	1898
Roulette	WINNER ROULETTE	1898
Card Machine	PERFECTION CARD	1902
Pointer	STAR POINTER (non-coin)	1902
Card Machine	DRAW POKER	1906
Card Machine	LITTLE DUKE	1906
Specialty Reel	PURITAN (Numbers)	1906
Coin Drop	THE IDEAL	1906
Card Machine	THE ELK	1907

Coyle and Rogers, Washington, District of Columbia
Dice	ELECTRICAL DICE	Sept. 1888
Dice	AUTOMATIC DICE	Oct. 1888
Dice	AUTOMATIC ADVERTISER	Oct. 1888
Dice	AUTOMATIC DICE VENDING MACHINE	Apr. 1889
Race Game	AUTOMATIC RACE COURSE	Jan. 1890

Craft Engineering Company, Grand Rapids, Michigan
Novelty/Skill	NERVE SCALE	July 1941

Crooks and Crooks, San Francisco, California
Card Machine	2-PLAYER POKER	Oct. 1896
Card Machine	3-PLAYER POKER	Apr. 1897

Crown Machine Company, Chicago, Illinois
Novelty/Skill	MASTER	1938

M. T. Daniels, Wichita, Kansas
Coin Drop	BALL-QUET	Mar. 1932
Target	LUCKY COIN TOSSER	1932
Coin Drop	ODD PENNY MAGNET	1933

Daval (Manufacturing) Company, Chicago, Illinois
Baby Bell	DAVAL GUM VENDOR (Style 1)	1932
Baby Bell	DAVAL GUM VENDOR (Style 2)	1932
Baby Bell	ULTRA MODERN DAVAL GUM VENDOR (Style 3)	1933
Baby Bell	ULTRA MODERN DAVAL GUM VENDOR JACKPOT (Style 4)	Mar. 1933
Baby Bell	ULTRA MODERN DAVAL GUM VENDOR JACKPOT (Style 5)	May 1933
Card Machine	CHICAGO CLUB HOUSE	June 1933
Card Machine	JACK POT CHICAGO CLUB HOUSE	June 1933
Card Machine	CHICAGO EXPRESS	1933
Baby Bell	DAVAL GUM VENDOR REGULAR	Oct. 1933
Baby Bell	DAVAL GUM VENDOR JACKPOT	Oct. 1933
Baby Bell	NEW DAVAL GUM VENDOR	1934
Baby Bell	NEW DAVAL GUM VENDOR JACKPOT	1934
Card Machine	CHICAGO CLUB HOUSE (Model No.1)	1933
Baby Bell	CHICAGO CLUB HOUSE (Model No.2)	1933
Card Machine	JACKPOT CHICAGO CLUB HOUSE (Model No.3)	1933
Baby Bell	JACKPOT CHICAGO CLUB HOUSE (Model No.4)	1933
Card Machine	JACKPOT CHICAGO CLUB HOUSE (Model No.5)	1933
Baby Bell	JACKPOT CHICAGO CLUB HOUSE (Model No.6)	1933
Card Machine	CHICAGO CLUB HOUSE (Model No.7)	1933
Baby Bell	CHICAGO CLUB HOUSE (Model No.8)	1933
Card Machine	CHICAGO CLUB HOUSE "Gold Medal" (Model No.1)	1934
Baby Bell	CHICAGO CLUB HOUSE "Gold Medal" (Model No.2)	1934
Card Machine	JACKPOT CHICAGO CLUB HOUSE "Gold Medal" (Model No.3)	1934
Baby Bell	JACKPOT CHICAGO CLUB HOUSE "Gold Medal" (Model No.4)	1934
Card Machine	JACKPOT CHICAGO CLUB HOUSE "Gold Medal" (Model No.5)	1934
Baby Bell	JACKPOT CHICAGO CLUB HOUSE "Gold Medal" (Model No.6)	1934
Card Machine	CHICAGO CLUB HOUSE "Gold Medal" (Model No.7)	1934
Baby Bell	CHICAGO CLUB HOUSE "Gold Medal" (Model No.8)	1934
Specialty Reel	DERBY (Model No.9) (Horses)	1934
Baby Bell	DAVAL GUM VENDOR (Model No. 10)	1934
Specialty Reel	JACK POT DERBY (Model No. 11) (Horses)	1934
Baby Bell	JACK POT DAVAL GUM VENDOR (Model No. 12)	1934
Specialty Reel	JACK POT DERBY (Model No. 13) (Horses)	1934
Baby Bell	JACK POT DAVAL GUM VENDOR (Model No. 14)	1934
Specialty Reel	DERBY (Model No. 15) (Horses)	1934
Baby Bell	DAVAL GUM VENDOR (Model No. 16)	1934
Wheel	MATCH-A-BALL	1935
Specialty Reel	SEVEN COME ELEVEN (Dice)	1935
Dice	TRUE-DICE	1935
Dice	TRUE-DICE (Poker)	1935
Dice	TRUE-DICE (Alphabet)	1935
Cigarette	PENNY PACK	June 1935
Cigarette	PENNY PACK REGISTER	June 1935
Cigarette	WIN-A-SMOKE	1935
Cigarette	WIN-A-PACK	1936
Cigarette	SPIN-A-PACK	1936
Cigarette	CENTA-SMOKE	1936
Cigarette	CENTA-SMOKE REGISTER	1936
Cigarette	CENTA-SMOKE COIN DIVIDER	1936
Specialty Reel	TIT-TAT-TOE	1936
Specialty Reel	TIT-TAT-TOE REGISTER	1936
Specialty Reel	CLEARING HOUSE (Numbers)	1936
Specialty Reel	CLEARING HOUSE REGISTER (Numbers)	1936
Specialty Reel	'RITHMATIC (Numbers)	1936
Specialty Reel	'RITHMATIC REGISTER (Numbers)	1936
Specialty Reel	REEL 21 (Numbers)	1936
Novelty/Skill	AUTO-PUNCH	June 1936
Specialty Reel	RACES (Horses)	Oct. 1936
Specialty Reel	REEL DICE (Dice)	Nov. 1936
Cigarette	TRI-O-PACK	Jan. 1937
Specialty Reel	RED 'N' BLUE (Colors)	1937
Card Machine	DOUBLE DECK	1937
Race Game	DAVAL DERBY	1937
Card Machine	REEL SPOT	1937
Baby Bell	BELL SLIDE	Jan. 1938
Baby Bell	JACKPOT BELL SLIDE	Jan. 1938
Cigarette	1938 PENNY PACK	1938
Specialty Reel	TRACK REELS (Horses)	1938
Card Machine	JOKER (JOKER WILD)	June 1938

Card Machine	JOKER GUM VENDOR	1938
Cigarette	SMOKE REELS	July 1938
Cigarette	SMOKE REELS GUM VENDOR	1938
Specialty Reel	SAFE HIT	1938
Specialty Reel	TALLY (Numbers)	Oct. 1938
Cigarette	TALLY	Oct. 1938
Baby Bell	TALLY	Oct. 1938
Cigarette	1939 PENNY PACK	Dec. 1938
Novelty/Skill	STEP-UP	1939
Cigarette	1940 PENNY PACK	1939
Cigarette	1940 PENNY PACK DIVIDER	1939
Cigarette	X-RAY	Jan. 1940
Specialty Reel	X-RAY VISIBILITY (Cigarette)	Jan. 1940
Specialty Reel	X-RAY (Beer)	Jan. 1940
Specialty Reel	X-RAY VISIBILITY (Beer)	Jan. 1940
Cigarette	COMET	1940
Cigarette	JIFFY	Jan. 1940
Specialty Reel	HEADS OR TAILS	1940
Cigarette	MARVEL	May 1940
Cigarette	MARVEL BALL GUM VENDOR	May 1940
Specialty Reel	BEER MARVEL (Beer)	1940
Cigarette	MARVEL VISIBILITY	1940
Cigarette	MARVEL VISIBILITY BALL GUM VENDOR	1940
Baby Bell	AMERICAN EAGLE	July 1940
Baby Bell	AMERICAN EAGLE GUM VENDOR	July 1940
Specialty Reel	STAR AMERICAN EAGLE	1940
Specialty Reel	STAR AMERICAN EAGLE GUM VENDOR	1940
Baby Bell	AMERICAN EAGLE VISIBILITY	1940
Baby Bell	AMERICAN EAGLE VISIBILITY GUM VENDOR	1940
Baby Bell	GOLD AWARD AMERICAN EAGLE	1940
Baby Bell	GOLD AWARD AMERICAN EAGLE GUM VENDOR	1940
Card Machine	ACE	Nov. 1940
Baby Bell	CUB	Nov. 1940
Specialty Reel	CUB (Numbers)	Nov. 1940
Cigarette	CUB	Nov. 1940
Specialty Reel	21 (Numbers)	Mar. 1941
Specialty Reel	DEFENSE AMERICAN EAGLE (Military)	May 1941
Specialty Reel	DEFENSE AMERICAN EAGLE GUM VENDOR (Military)	May 1941
Specialty Reel	AMERICAN FLAGS	June 1941
Cigarette	LUCKY SMOKES	June 1941
Specialty Reel	RACES (Horses)	Aug. 1941
Cigarette	REX	1941
Cigarette	NON-COIN MARVEL	1941
Cigarette	NON-COIN MARVEL BALL GUM VENDOR	1941
Cigarette	NON-COIN AMERICAN EAGLE	1941
Cigarette	NON-COIN AMERICAN EAGLE GUM VENDOR	1941

Daval Products Corporation, Chicago, Illinois

Cigarette	MARVEL	May 1946
Cigarette	MARVEL GUM VENDOR	May 1946
Cigarette	NON-COIN MARVEL	May 1946
Cigarette	NON-COIN MARVEL GUM VENDOR	May 1946
Cigarette	AMERICAN EAGLE	May 1946
Cigarette	AMERICAN EAGLE GUM VENDOR	May 1946
Cigarette	NON-COIN AMERICAN EAGLE	May 1946
Cigarette	NON-COIN AMERICAN EAGLE GUM VENDOR	May 1946
Card Machine	ACE	1946
Baby Bell	CUB	1946
Baby Bell	GUSHER	June 1946
Cigarette	BUDDY	July 1946
Specialty Reel	FREE-PLAY (Numbers)	Oct. 1946
Baby Bell	FREE-PLAY	Oct. 1946
Cigarette	FREE-PLAY	Oct. 1946
Baby Bell	ELECTRIC AMERICAN EAGLE	1947
Novelty/Skill	BEST HAND	Feb. 1947
Novelty/Skill	MEXICAN BASEBALL	Feb. 1947
Novelty/Skill	OOMPH	Feb. 1947
Cigarette	TAX FREE MARVEL	Sept. 1947
Baby Bell	TAX FREE MARVEL	Sept. 1947

John Henry Davis, Chicago, Illinois

Pointer	THE DEWEY SALESMAN	May 1897

Davis Novelty Company, Manistee, Michigan

(Unknown)	(UNKNOWN)	1897

Daycom Incorporated, Dayton, Ohio

Single Reel	REEL-O-BALL	Nov. 1932

Dealer's Choice, Inc., Collingswood, New Jersey

Card Machine	DEALER'S CHOICE	1966

Dean Novelty Company, Tulsa, Oklahoma

Card Machine	LITTLE PERFECTION	1924
Target	TARGET PRACTICE	1924
Wheel	EVER READY TRADE BOARD	1930
Coin Drop	VAUDETTE	Nov. 1932
Coin Drop	NEW VAUDETTE	Jan. 1933
Card Machine	PENNY DRAW	1934
Card Machine	PENNY ANTE	Mar. 1934

Decatur Fairest Wheel Company, Decatur, Illinois

Wheel	FAIREST WHEEL (Style 1)	Feb. 1895
Wheel	DISCOUNT WHEEL (Style 1)	Feb. 1895
Wheel	IMPROVED FAIREST WHEEL	Jan. 1896
Wheel	IMPROVED DISCOUNT WHEEL	Jan. 1896

Decatur Fairest Wheel Works, Decatur, Illinois

Wheel	FAIREST WHEEL "Large Wheel"	Apr. 1894
Wheel	FAIREST WHEEL "Small Wheel"	Dec. 1894
Wheel	IMPROVED FAIREST WHEEL (Style 1)	Mar. 1896
Wheel	IMPROVED DISCOUNT WHEEL (Style 1)	Mar. 1896
Wheel	FAIREST WHEEL No.2	1897
Wheel	FAIREST WHEEL No.3	1899

Decatur Novelty Works, Decatur, Illinois

Wheel	(UNKNOWN)	1896

George Deddens Distillery Company, Cincinnati, Ohio

(Unknown)	(UNKNOWN)	1908

Reinhold F. DeGrain, Washington, District of Columbia

Specialty Reel	(UNKNOWN) (Dice)	Apr. 1890
Specialty Reel	(UNKNOWN) (Dice)	Dec. 1892

Victor P. DeKnight, Washington, District of Columbia

Wheel	(UNKNOWN)	Apr. 1894

Del Norte Specialty Company, Elgin, Illinois

Baby Bell	PURITAN BABY BELL	1933

William Dennings, National Military Home (Dayton), Ohio

Wheel	GAME WHEEL	Feb. 1882
Wheel	GAME WHEEL No.2	Aug. 1885

Denver Novelty Works, Denver, Colorado

(Unknown)	(UNKNOWN)	1890

Dependable Enterprises, Barberton, Ohio

Dice	DICE-MAT (5-column) (non-coin)	Jan. 1938
Dice	DICE-MAT (10-column) (non-coin)	1948

Detroit Brass and Iron Novelty Company, Detroit, Michigan

(Unknown)	(UNKNOWN)	1904

Detroit Coin Machine Company, Detroit, Michigan

Specialty Reel	PURITAN (Numbers)	Mar. 1905

Detroit Manufacturing Novelty Company, Detroit, Michigan

(Unknown)	(UNKNOWN)	1895

Devices Manufacturing Sales Company, Chicago, Illinois

Specialty Reel	PURITAN BABY VENDOR (Fortune)	1939
Baby Bell	PURITAN BABY VENDOR	1939

Sidney Diamant, New York, New York

Coin Drop	(UNKNOWN)	Jan. 1925

Diamond Novelty Company, Syracuse, New York

Card Machine	PERFECTION CARD	1904

Dicta-Card Inc., Mishawaka, Indiana

Novelty/Skill	DICTA-CARD	1936
Novelty/Skill	DICTA-RACE	1936

William Diebel, Philadelphia, Pennsylvania

Wheel	(UNKNOWN)	May 1894

Dixie Manufacturing Company, Portland, Oregon

Baby Bell	DIXIE BELL	Sept. 1934

Dixie Music Company, Miami, Florida

Dice	DIXIE	1931

S. H. Dixon, Chicago, Illinois

Coin Drop	ROLL SKILL	1936

Charles L. Dobrick, New York, New York

Wheel	(UNKNOWN)	Nov. 1891

Doraldina Corporation, Rochester, New York

Novelty/Skill	ROCK-IT	Feb. 1932

Douglass Specialties, Inc., New York, New York

Coin Drop	THE FAVORITE	June 1924

A. S. Douglis and Company, Chicago, Illinois

Baby Bell	DANDY VENDER	1930
Baby Bell	COIN DIVIDER DANDY VENDER	1931
Novelty/Skill	PENNY BASEBALL	Feb. 1931
Novelty/Skill	SPIRAL GOLF	Feb. 1931
Baby Bell	PURITAN VENDOR	1932

Douglis Machine Company, Chicago, Illinois

Baby Bell	PURITAN VENDOR	Apr. 1932
Baby Bell	JACKPOT PURITAN VENDOR	Sept. 1932
Baby Bell	DAVAL GUM VENDOR	1932

	Baby Bell	ARISTOCRAT	1932

Albert S. Drais, San Francisco, California

	(Unknown)	(UNKNOWN)	1898

George Draper and Sons, Hopedale, Massachusetts

	Wheel	(UNKNOWN)	Jan. 1893

E. F. Driver, Chicago, Illinois

	Novelty/Skill	PLAY POOL	1929

Drobisch Brothers and Company, Decatur, Illinois

	Coin Drop	ADVERTISING REGISTER	June 1896
	Wheel	BONUS WHEEL	Dec. 1896
	Pointer	VICTOR	Jan. 1897
	Wheel	(UNKNOWN)	Mar. 1897
	Pointer	STAR ADVERTISER	Mar. 1897
	Dice	No.5 MONARCH DICE MACHINE	Apr. 1897
	Wheel	(UNKNOWN)	Apr. 1897
	Pointer	THE LEADER	1897

Peter Drummer, Corning, New York

	Dice	(UNKNOWN)	Nov. 1890

J. D. Drushell Company, Chicago, Illinois

	Novelty/Skill	ROLL SKILL	Feb. 1936

David W. Dunn, Ashland, Kentucky

	Wheel	BICYCLE RIDER	July 1915

Dunn Brothers, Anderson, Indiana

	Novelty/Skill	PERFECTION (Round Globe)	1905
	Novelty/Skill	PERFECTION (Capped Globe)	1905
	Dice	WRIGLEY DICE MACHINE	1905
	Novelty/Skill	WRIGLEY PERFECTION (Straight Globe)	1906

Dutch (Manufacturing Company)

	Coin Drop	SMOKEMASTER	Dec. 1930
	Coin Drop	DRINKMASTER	Dec. 1930

Eagle Amusement Machine/E.A.M.

	Dice	EAGLE	1899

Eagle Manufacturing Company, Detroit, Michigan

	Dice	EAGLE	1892

Henry J. Eastman, San Francisco, California

	(Unknown)	(UNKNOWN)	1905

Raphael E. Ebersole, Roanoke, Indiana

	Coin Drop	DAISY	July 1897

Eclipse Novelty Works, Denver, Colorado

	(Unknown)	(UNKNOWN)	1904

William Edge, Orange, New Jersey

	Dice	(UNKNOWN)	Oct. 1892

Electra Corporation, Chicago, Illinois

	Novelty/Skill	DIZZY DIALS	Feb. 1933
	Novelty/Skill	MILL RACE	Feb. 1933
	Novelty/Skill	WATCHEM	Feb. 1933

Electrical Supply Company, Sacramento, California

	(Unknown)	(UNKNOWN)	1904

Electro Ball Company, Inc., Dallas, Texas

	Dice	PAK-O-CIGS (Cigarette)	Jan. 1936

Elm City Novelty Sales Company, New Haven, Connecticut

	Coin Drop	AMERICAN SPORT	1928

Elston Sales Company, Chicago, Illinois

	Race Game	RAILBIRD	1933

Martin Elzas, Los Angeles, California

	Dice	(UNKNOWN)	Apr. 1907

Henry T. Emeis, Salt Lake City, Utah

	Coin Drop	(UNKNOWN)	Sept. 1897

Engel Manufacturing Company, New Holland, Michigan

	Baby Bell	IMP	Nov. 1927

William A. Etter, El Paso, Texas

	Pointer	(UNKNOWN)	Mar. 1938

Ennis and Carr, Syracuse, New York

	Dice	PERFECTION	1904

Erickson, Portland, Oregon

	Dice	LOG CABIN	1898

Erie Machine Company, Cleveland, Ohio

	Novelty/Skill	BOMBER	1939

Erie Manufacturing Company, Hartford, Connecticut

	Novelty/Skill	THE EMCO (NERVE EXERCISE)	Dec. 1928

Eureka Novelty Sales Company, Eureka, California

	(Unknown)	EUREKA	1902

H. C. Evans and Company, Chicago, Illinois

	Wheel	STAR POINTER (non-coin)	June 1903
	Race Game	LITTLE BILLIKIN	1907
	Roulette	CUBE ROULETTE	1907
	Race Game	RACE TRACK	1919
	Race Game	CANDY RACE TRACK	1921
	Race Game	MINIATURE MUTUEL RACE TRACK	1923
	Race Game	MINIATURE RACE COURSE	1923
	Baby Bell	EVAN'S BABY BELL	1928
	Race Game	CHICAGO DERBY	1932
	Race Game	SARATOGA SWEEPSTAKES ("6 Horse")	1932
	Race Game	SARATOGA SWEEPSTAKES ("8 Horse")	1932
	Race Game	PARI-MUTUEL SARATOGA SWEEPSTAKES	1933
	Race Game	SARATOGA SWEEPSTAKES SPECIAL	1933
	Novelty/Skill	BY-A-BLADE	Feb. 1938
	Pointer	POCKET EDITION GALLOPING DOMINOS	1939
	Race Game	SARATOGA SWEEPSTAKES	1941
	Pointer	WIN-O	Oct. 1942
	Dice	AFRICAN GOLF	1944
	Dice	HI-LO CHUCK LUCK	1944
	Dice	MIAMI COLOR GAME	1944
	Dice	MONTE CARLO	1944
	Dice	CROWN & ANCHOR	1944
	Pointer	DOMINO JR.	1944

Evans and Son Specialty Company, Geneva, Ohio

	Roulette	(UNKNOWN)	1930

Excelsior Race Track Company, Chicago, Illinois

	Race Game	EXCELSIOR	1890

Exhibit Supply Company, Chicago, Illinois

	Dice	DICE FORTUNE TELLER	June 1925
	Target	BULLS EYE BALL GUM VENDER	Aug. 1925
	Dice	OPERATORS DICE MACHINE	Feb. 1926
	Target	PLAY BALL (Wood)	Mar. 1926
	Target	PLAY BALL (Aluminum)	Aug. 1926
	Dice	JUNIOR	Dec. 1927
	Dice	KROMO KOLORED KUBE	Dec. 1927
	Dice	ARCADE DICE FORTUNE TELLER	1928
	Target	PLAY BALL GUM VENDER	1928
	Target	PLAY BALL PEANUT VENDER	Oct. 1928
	Novelty/Skill	BATTER-UP	Oct. 1928
	Novelty/Skill	GOLDEN BELL	Dec. 1928
	Target	SMILING JOE	1929
	Target	HIT THE COON	1929
	Novelty/Skill	STEEPLECHASE	Aug. 1932
	Dice	THE BOOSTER	1933
	Dice	THE BOOSTER (Fruit)	1933
	Wheel	SWEET SALLY	Jan. 1934
	Wheel	SWEET SALLY (Horses)	Jan. 1934
	Dice	THE BOOSTER (Number)	Mar. 1934
	Dice	THE BOOSTER (Fruit)	Mar. 1934
	Dice	SELECT-EM	1934
	Dice	SELECT-EM (Cigarette)	1934
	Dice	I OWE YOU (Alphabet)	Jan. 1935
	Dice	SELECT-EM GUM VENDOR	1935
	Dice	SELECT-EM GUM VENDOR (Number)	1935
	Dice	SELECT-EM GUM VENDOR	1935
	Dice	THE BOOSTER (Novelty)	1935
	Dice	THE BOOSTER (Number)	1935
	Dice	THE BOOSTER (Fruit)	1935
	Dice	THE BOOSTER	1935
	Dice	HORSE SHOES	1935
	Dice	GET-A-PACK (Number)	1935
	Dice	PAK-O-CIGS (Novelty)	1936
	Roulette	BEAT IT	1936
	Coin Drop	DOUBLE DICE	1936
	Coin Drop	MAKE TWENTY ONE	1936
	Coin Drop	THE BOUNCER	1936
	Coin Drop	STAR CIGARETTE MERCHANDISER	1936
	Dice	BASE BALL (Novelty)	1936
	Dice	TANGO (Novelty)	1936
	Novelty/Skill	TIC-TAC-TOE	1936
	Dice	KEY HOLE (Number)	1936
	Dice	WIN (Novelty)	1936
	Dice	WIN (Alphabet)	1936
	Dice	OLD AGE PENSION	1936
	Dice	FREE PLAY (Number)	1937
	Dice	GOAL LINE (Novelty)	1937
	Dial	SKILL DRAW	1937
	Dial	RED DOG	1937
	Dial	TURF TIME	1938
	Dice	HORSE PLAY (Novelty)	1938
	Dice	36 GAME (Number)	1938
	(Unknown)	HONEY	1938
	(Unknown)	SUSIE Q	1938
	Dice	INDIAN DICE	1938

The Factories Sales Company, Dallas, Texas

	(Unknown)	WHIM	1934

C. D. Fairchild Inc., Syracuse, New York
	Novelty/Skill	BAT-A-BALL	Mar. 1931

Fairgames Manufacturing Company, Niles, Michigan
	(Unknown)	(UNKNOWN)	1939

A. Feinberg Company, Rochester, New York
	Card Machine	SUCCESS	1904

Charles Fey and Company, San Francisco, California
	Wheel	SKILL MACHINE	1895
	Pointer	THREE SPINDLE	1896
	Pointer	KLONDIKE	1897
	Specialty Reel	POLICY (Numbers)	1897
	Specialty Reel	4-11-44 (Numbers)	1897
	Drop Cards	PAYING TELLER	1897
	Card Machine	THE DUKE	1899
	Card Machine	DRAW POKER	1905
	Single Reel	ELK	1905
	Dice	ON THE SQUARE	1907
	Dice	ON THE LEVEL	1907
	Roulette	SKILL ROLL	1907
	Dice	AUTOMATIC DICE BOX	1909
	Roulette	TRIPLE ROULETTE	1909
	Roulette	36 ROULETTE	1920
	Dice	36 LUCKY SPOT MIDGET	1920
	Roulette	TRIPLE ROLL	1924
	Roulette	TRIPLE ROULETTE	1924
	Dice	MIDGET WITH SALESBOARD (Novelty)	1926
	Dice	MIDGET 36	1926
	Roulette	MIDGET	1926
	(Unknown)	LITTLE VENDER	1927
	Dice	3-IN-1	1927
	Dice	NEW 36 GAME	1927
	Roulette	PEE-WEE	1927
	Pointer	FAIR-N-SQUARE ("The Cigarette Salesman")	1934
	Dice	FAIR-N-SQUARE ("Blue Eagle Model")	1934
	Dice	ROLL 'EM	1935
	Dial	SKILL-DRAW (Style 1)	1935
	Dial	SKILL-DRAW (Style 2)	1936
	Dial	THREE CADETS	1936
	Dial	PLAY AND DRAW	1936

Edmund Fey, San Francisco, California
	Dice	THE ACE	Feb. 1928
	Novelty/Skill	INDOOR STRIKER	June 1930

Fey Automatic Machine Company, San Francisco, California
	Target	SHOOT-A-PAK	1939

F. F. Sales Company, Chicago, Illinois
	Baby Bell	MINIATURE BELL	Dec. 1950

Field Manufacturing Corporation, Peoria, Illinois
	Baby Bell	RELIABLE PURITAN BABY VENDER MODEL 1	1930
	Baby Bell	RELIABLE PURITAN BABY VENDER MODEL 2	1930
	Baby Bell	RELIABLE PURITAN BABY VENDER MODEL 3	1930
	Baby Bell	RELIABLE PURITAN BABY VENDER MODEL 4	1930
	Baby Bell	JACK POT RELIABLE PURITAN BABY VENDER MODEL 5	1930
	Baby Bell	PURITAN BABY BELL VENDER	1931
	Baby Bell	2-IN-1 BABY VENDER	June 1931
	Specialty Reel	CLEARING HOUSE	June 1931
	Baby Bell	BABY BELL	1931
	Baby Bell	JACK POT BABY BELL	1931
	Specialty Reel	DIAMOND (Baseball)	1931
	Specialty Reel	BASEBALL BABY VENDER (Baseball)	1931
	Novelty/Skill	WHOOPEE BALL	Sept. 1931
	Novelty/Skill	POKER PLAY	Sept. 1931
	Novelty/Skill	VEST POCKET BASEBALL	Sept. 1931
	Novelty/Skill	INDOOR SPORTS	1932
	Novelty/Skill	DING THE DINGER	1932
	Novelty/Skill	CHURCHILL DOWNS	1932
	Novelty/Skill	POKERENO	Apr. 1932
	Novelty/Skill	WORLD'S FAIR MODEL A	1933
	Novelty/Skill	WORLD'S FAIR MODEL B	1933
	Novelty/Skill	WORLD'S FAIR MODEL C	1933
	Dial	WHIRLWIND (Electric)	1933
	Dial	WHIRLWIND (Mechanical)	1933
	(Unknown)	HULLABALOO	1934
	(Unknown)	K'RAZY	1934
	(Unknown)	BLACK BEAUTY	1934
	Dice	THE SIZZLER	1935

Field Paper Products Company, Peoria, Illinois
	Baby Bell	GYPSY FORTUNE TELLER	Oct. 1928
	Baby Bell	PURITAN BABY	Jan. 1929
	Baby Bell	PURITAN BABY BALL GUM VENDER	Jan. 1929
	Baby Bell	1929 GYPSY FORTUNE TELLER	Feb. 1929
	Baby Bell	BABY VENDER	1929
	Baby Bell	BABY JACK-POT VENDER	1929
	Baby Bell	BABY SKILL VENDER	1929
	Baby Bell	BABY JACK-POT SKILL VENDER	1929
	Baby Bell	2-IN-1 BABY VENDER	1929
	Baby Bell	2-IN-1 BABY JACK-POT VENDER	1929
	Baby Bell	2-IN-1 BABY SKILL VENDER	1929
	Baby Bell	2-IN-1 BABY JACK-POT SKILL VENDER	1929
	Specialty Reel	CHIEF (Fortune)	1931

Filmascope Manufacturing Company, Chicago, Illinois
	Novelty/Skill	KENTUCKY DERBY	Feb. 1932

A. J. Fisher and Company, Pittsburgh, Pennsylvania
	Target	THE LEGAL	1908
	Coin Drop	(ORIGINAL) PREMIUM	1910

Five Boro Machine Manufacturing Company, Brooklyn, New York
	Coin Drop	TURTLE SOUP	1936
	Coin Drop	DUCK SOUP	June 1937

Flatbush Gum Company, Inc., Brooklyn, New York
	Target	KING BASE BALL MACHINE	1930
	Novelty/Skill	PREMO MERCHANDISER	Feb. 1931

Flour City, Minneapolis, Minnesota
	Race Game	AUTOMATIC RACE COURSE	1889

O. H. Flower, Cincinnati, Ohio
	(Unknown)	(UNKNOWN)	1912

Flying Fun Company, Sheboygan, Wisconsin
	(Unknown)	FLYING FUN	July 1938

Flying Skill Machine Company, New York, New York
	Target	FLYING SKILL	1930

J. L. Foley Manufacturing Company, Chicago, Illinois
	Card Machine	DRAW POKER	1906

Fort Wayne Novelty Manufacturing Company, Fort Wayne, Indiana
	Dice	LITTLE JOE	1934

Fortune Machine Works, Chicago, Illinois
	Coin Drop	FORTUNE TELLER	1903

Fortune Sales Company, Los Angeles, California
	Novelty/Skill	LUCKY ROLL	Jan. 1939

Four Jacks Corporation, Chicago, Illinois
	Specialty Reel	CLUB JACK (Numbers)	1937
	Specialty Reel	BASEBALL MUTUEL	1937
	Specialty Reel	FOOTBALL MUTUEL	1937
	Specialty Reel	PUT AND TAKE (Numbers)	1937
	Specialty Reel	CRAPS (Dice)	1937

Benjamin F. Fowler, Minneapolis, Minnesota
	(Unknown)	(UNKNOWN)	Feb. 1901

Richard K. Fox, New York, New York
	Race Game	FRENCH RACE GAME (non-coin)	Oct. 1889
	Race Game	EXCELSIOR (AUTOMATIC)	Oct. 1889
	Dice	IMPROVED DICE GAME	Apr. 1890

Sidney Frankel, Pittsburgh, Pennsylvania
	Race Game	(UNKNOWN)	Mar. 1930

J. F. Frantz Manufacturing Company, Chicago, Illinois
	Novelty/Skill	BAKER KICKER AND CATCHER	1952
	Novelty/Skill	LITTLE LEAGUE	1952
	Novelty/Skill	KICKER AND CATCHER	1954
	Novelty/Skill	BASE BALL	1954
	Coin Drop	BULLS EYE	1954
	Novelty/Skill	POT OF GOLD	1955
	Novelty/Skill	BASKETBALL	1955
	Novelty/Skill	LITTLE LEAGUE	1955
	Novelty/Skill	BIG TOP	1960
	Novelty/Skill	DOUBLE HEADER	1962
	Novelty/Skill	LONG SHOT	1963
	Novelty/Skill	SAVE OUR BUSINESS	1966
	Novelty/Skill	BI-CENTENNIAL KICKER AND CATCHER	1976

Charles A. French, Boston, Massachusetts
	Wheel	(UNKNOWN)	July 1894

James P. Frey, Modesto, California
	Novelty/Skill	(UNKNOWN)	Oct. 1932

Bill Frey, Inc., Miami, Florida
	Dice	JITTER-ROLL 12"	1939
	Dice	JITTER-ROLL 18"	1939
	Dice	COLOR-ROLL 12"	1939
	Dice	COLOR-ROLL 18"	1939
	Dice	CHUCK-A-ROLL 12"	1939
	Dice	CHUCK-A-ROLL 18"	1939
	Dice	NUMBER-ROLL	1939

Dice	TWIN-ROLL	1939
Dice	JITTERBUG	1940
Dice	AFRICAN GOLF	1947
Dice	MONTE CARLO	1947
Dice	BING	1950
Dice	BEAT THE DEALER	1950

Friedman and Company, Chicago, Illinois

Coin Drop	SONG DICK	1898

Charles J. Froeleich Novelty Company, Utica, New York

(Unknown)	(UNKNOWN)	1891

Froom Laboratories, Inc., Youngstown, Ohio

Dice	PENNY PLAY	1937

Warren Fuhrmann, Keansburg, New Jersey

Race Game	(UNKNOWN)	1937

Fun Industries, East Moline, Illinois

Novelty/Skill	LITTLE LEAGUE	1993
Novelty/Skill	LONG SHOT	1993
Novelty/Skill	KICK 'N CATCH	1993
Novelty/Skill	COPY CAT	1993

Game Of Games Company, Cleveland, Ohio

Roulette	ROLL'ETTO	1932

Gamemasters, Norcross, Georgia

Dice	BONES	1989

Games, Inc., Chicago, Illinois

(Unknown)	(UNKNOWN)	1954

Ganss, Clarke and Dengler, East Saginaw, Michigan

Novelty/Skill	GAME TABLE (Non-coin)	Oct. 1889

Garden City Novelty Company, Chicago, Illinois

Coin Drop	LEAPING LENA	1932
Novelty/Skill	PICK A PACK	1933
Roulette	WHIRL-SKILL	1933
Roulette	JACKPOT WHIRL-SKILL	1933
Roulette	DELUXE VENDER	1933
Novelty/Skill	CHERRY JITTERS	1934
Wheel	PAIR-IT	1934
Dial	MATCH-A-PAK	Nov. 1934
Pointer	SPIN-O	1935
Pointer	PLA-PAX	1935
Cigarette	GEM	1936
Specialty Reel	BAR BOY (Beer)	1936
Cigarette	THREE OF A KIND	1936
Specialty Reel	TURF (Horses)	1936
Specialty Reel	ARMY 21 GAME (Numbers)	1936
Specialty Reel	BABY JACK (Numbers)	1937
Cigarette	PRINCE	1937

J. H. Gasser, Webster, Massachusetts

(Unknown)	(UNKNOWN)	1899

Gatter Novelty Company, Philadelphia, Pennsylvania

Coin Drop	NOVELTY GEM	1921
Dice	THE LITTLE WONDER	Sept. 1927

Gatter Manufacturing Company, Philadelphia, Pennsylvania

Roulette	PANSY	1932

Gayton Novelty Company, Providence, Rhode Island

Coin Drop	(UNKNOWN)	1906

Genco, Inc., Chicago, Illinois

Novelty/Skill	SPIRAL GOLF	Feb. 1931
Novelty/Skill	HEARTS	Mar. 1931
Novelty/Skill	SOCCER	Mar. 1931
Novelty/Skill	DEUCES WILD	Sept. 1931
Novelty/Skill	MERRY-GO-ROUND	1931
Novelty/Skill	RACE HORSE	Dec. 1931
Novelty/Skill	DICE	Dec. 1931
Novelty/Skill	POKER	Dec. 1931
Novelty/Skill	SWEET SIXTEEN	Dec. 1931
Novelty/Skill	SEVEN ELEVEN	Dec. 1931
Novelty/Skill	BASKET BALL	Dec. 1931
Novelty/Skill	RITZ	1934
Novelty/Skill	HOOPS	1936
Novelty/Skill	PUNCH A BALL (non-coin)	1939
Novelty/Skill	PUNCH A BALL (Baseball) (non-coin)	1939

General Novelty Manufacturing Company, Chicago, Illinois

Coin Drop	WIN-A-PACK	Feb. 1933
Novelty/Skill	KENTUCKY DERBY	Feb. 1933

William Gent Vending Machine Company, Cleveland, Ohio

Coin Drop	RELIABLE NERVE AND EYE TESTER	May 1928

Mark Gerard Company, Fort Lee, New Jersey

Novelty/Skill	MARK I TIC TAC TOE	May 1961

G. R. Gibson, Denver, Colorado

(Unknown)	(UNKNOWN)	1907

Giles and Simpkins, Inc., Inglewood, California

Baby Bell	HOLLYWOOD	Jan. 1930

Gillespie Games Company, Long Beach, California

(Unknown)	KONTEST BOMBER	1947
(Unknown)	KONTEST POKER	1947

Gillet, Hunter and Company, Springfield, Illinois

(Unknown)	(UNKNOWN)	1887

Gillet Novelty Company, Detroit, Michigan

(Unknown)	(UNKNOWN)	1903

Gisha Company, Anderson, Indiana

(Unknown)	(UNKNOWN)	1912

Gist Cabinet Company, Kansas City, Missouri

Roulette	THE RUMBA WHEEL	1933
Wheel	T.N.T.	1933

Theodore C. Glaser, East Orange, New Jersey

Card Machine	(UNKNOWN)	Jan. 1935

J. F. Gleason and Company, Chicago, Illinois

Baby Bell	PURITAN BELL	1930

Globe Novelty Company, Chicago, Illinois

Dice	GALLOPING CUBES	1936
Dice	BOTTOMS UP	1936

G. and M. Machinery Company, Jacksonville, Florida

Baby Bell	CHERRY ROLL	1933

G. and M. Manufacturing Company, Minneapolis, Minnesota

Dice	FORTUNE MATCHES	1940

G. M. Laboratories, Inc., Chicago, Illinois

Specialty Reel	(UNKNOWN)	Mar. 1936

John Goodbody, Rochester, New York

Novelty/Skill	WHOOPEE BALL	May 1931
Novelty/Skill	THE WILDCAT	Dec. 1931

D. Gottlieb and Company, Chicago, Illinois

Novelty/Skill	BASEBALL ROLL-ET	1930
Novelty/Skill	ROLL-ET	1930
Novelty/Skill	MINIATURE BASEBALL	Jan. 1931
Novelty/Skill	CHUCK-O-LUCK HORSE RACING	Nov. 1931
Novelty/Skill	CHUCK-O-LUCK FOOTBALL	Nov. 1931
Novelty/Skill	CHUCK-O-LUCK DICE	Nov. 1931
Novelty/Skill	CHUCK-O-LUCK POKER	Nov. 1931
Novelty/Skill	CHUCK-O-LUCK 21 BLACK JACK	Nov. 1931
Novelty/Skill	STOP AND SOCK	Dec. 1931
Target	SKILL SHOT	1936
Dice	DAILY RACES JR.	1938
Dice	TOBACCO PAK	1938
Dice	INDIAN DICE	1938
Novelty/Skill	LUCKY STAR	1938
Novelty/Skill	MIDGET RACES	Oct. 1938
Novelty/Skill	STEEPLE RACES	Oct. 1938
(Unknown)	MATCH-IT	1939

L. C. Graham Company, Albany, New York

Dice	MIDGET	1926
Roulette	MIDGET ROULETTE	1926
Dice	3-IN-1	1927
Roulette	3-IN-1	1927

Grand National Sales Company, Chicago, Illinois

Cigarette	SPIN-A-PACK	1939
Coin Drop	ROLL-A-CENT	1941

Grand Rapids Novelty Manufacturing Company, Grand Rapids, Michigan

(Unknown)	(UNKNOWN)	1893

Grand Rapids Slot Machine Company, Grand Rapids, Michigan

Pointer	ECLIPSE	1894

Great States Manufacturing Company, Kansas City, Missouri

Coin Drop	FOOT BALL PRACTICE	1932
Baby Bell	PURITAN BABY VENDOR	1932
Dice	BUCK-A-DAY	1933
Target	BEER TARGET	1933
Target	JACK POT BEER TARGET	1933
Target	HAPPY DAYS	1933
Target	JACK POT HAPPY DAYS	1933
Target	CIGARETTE TARGET	1933
Target	JACK POT CIGARETTE TARGET	1933
Target	CANDY TARGET	1933
Target	JACK POT CANDY TARGET	1933
Target	CIGAR TARGET	1933
Target	JACK POT CIGAR TARGET	1933
Race Game	SANDY'S HORSES	1936
Coin Drop	FLIP FLOP FLUZZEE	1939
Coin Drop	DROP IT	1939
Coin Drop	FLIP-A-KOPPER	1939

Great Western Products Company, Kansas City, Missouri
(Unknown)	(UNKNOWN)	1917
Dice	THREE-WAY DIVIDEND PRODUCER	1927

M. O. Griswold and Company, Rock Island, Illinois
Pointer	LAYOUT (non-coin)	1889
Dice	DICE MACHINE	1892
Card Machine	CARD MACHINE	1892
Wheel	WHEEL OF FORTUNE	Oct. 1893
Wheel	THE BLACK CAT	1895
Roulette	ROULETTE	1899
Wheel	NEW IDEA	1900
Wheel	THE BIG THREE	1901
Wheel	STAR	1902

Griswold Manufacturing Company, Rock Island, Illinois
Wheel	STAR	Mar. 1905
Wheel	SELF PAY	1916
Wheel	NEW STAR	1919
Wheel	NEW STAR (Aluminum)	1922

Groetchen Tool (and Manufacturing) Company, Chicago, Illinois
Baby Bell	DANDY VENDER	1930
Baby Bell	NEW DANDY VENDER	1932
Baby Bell	NEW DANDY VENDER JAK POT	1932
Baby Bell	NEW DANDY VENDER CONCEALED JAK POT	1932
Baby Bell	DANDY VENDER	1932
Baby Bell	COIN DIVIDER DANDY VENDER	1932
Baby Bell	VEEDER COUNTER DANDY VENDER	1932
Baby Bell	DANDY VENDER MODEL 1933	Sept. 1932
Card Machine	POK-O-REEL	Feb. 1932
Card Machine	POK-O-REEL GUM VENDER	Apr. 1933
Baby Bell	SILENT DANDY VENDER	1933
Cigarette	SILENT DANDY VENDER	1933
Dial	SOLITAIRE	Sept. 1933
Dial	SOLITAIRE GUM VENDER	Oct. 1933
Dial	GOLD RUSH	1934
Dial	GOLD RUSH CALENDER	1934
Card Machine	POK-O-REEL TRIPLEX	1934
Card Machine	POK-O-REEL TRIPLEX JAK-POT	1934
Baby Bell	POK-O-REEL TRIPLEX	1934
Baby Bell	POK-O-REEL TRIPLEX JAK-POT	1934
Specialty Reel	POK-O-REEL TRIPLEX	1934
Specialty Reel	POK-O-REEL TRIPLEX JAK-POT	1934
Card Machine	21 VENDER	Mar. 1934
Cigarette	ZIG-ZAG	July 1934
Specialty Reel	CIGARETTES DANDY VENDER	1934
Specialty Reel	JACKPOT CIGARETTES DANDY VENDER	1934
Specialty Reel	BEER DANDY VENDER (Beer)	1934
Specialty Reel	JACKPOT BEER DANDY VENDER (Beer)	1934
Specialty Reel	FREE PLAY DANDY VENDER (Numbers)	1934
Specialty Reel	JACKPOT FREE PLAY DANDY VENDER (Numbers)	1934
Dice	DICE-O-MATIC VENDER	Oct. 1934
Dice	DICE-O-MATIC VENDER (Poker)	1934
Dice	DICE-O-MATIC VENDER (Fruit)	1934
Dice	DICE-O-MATIC VENDER (Number)	1934
Dice	ROTO-MATIC	1934
Specialty Reel	TURF FLASH (Horses)	1935
Cigarette	PENNY CIGARETTE MACHINE	1935
Cigarette	PENNY CIGARETTE MACHINE GUM VENDER	1935
Specialty Reel	TAVERN VENDER STYLE A (Beer)	1935
Cigarette	TAVERN VENDER STYLE B	1935
Specialty Reel	THE FORTUNE TELLER	1935
Cigarette	PENNY SMOKE	1935
Cigarette	PENNY SMOKE VEEDER COUNTER	1935
Novelty/Skill	PUNCHETTE	1936
Specialty Reel	HIGH STAKES (Horses)	Aug. 1936
Specialty Reel	TWENTY ONE (Numbers)	Aug. 1936
Card Machine	POK-O-REEL	Aug. 1936
Specialty Reel	HIGH TENSION (Horses)	Aug. 1936
Cigarette	SMOKE HOUSE	Aug. 1936
Cigarette	ZEPHYR	Jan. 1937
Baby Bell	ZEPHYR	May 1937
Cigarette	GINGER	June 1937
Baby Bell	GINGER	1937
Specialty Reel	GINGER FORTUNE	1937
Card Machine	ROYAL FLUSH	1937
Specialty Reel	GINGER BASEBALL (Sports)	1937
Specialty Reel	DIXIE DOMINOES (Dice)	1937
Specialty Reel	DIXIE SPELLING BEE (Alphabet)	1937
Cigarette	SPARKS DELUX	1938
Specialty Reel	SPARKS DELUX (Beer)	1938
Specialty Reel	SPARKS DELUX (Horses)	1938
Cigarette	SPARKS JACKPOT	1938
Specialty Reel	SPARKS JACKPOT (Beer)	1938
Specialty Reel	SPARKS JACKPOT (Horses)	1938
Novelty/Skill	BLUE BONNET	1938
Specialty Reel	ZEPHYR FORTUNE (Fortune)	1938
Cigarette	MERCURY	1939
Cigarette	MERCURY DISCREET	1939
Novelty/Skill	GINGER PUNCHBOARD	Feb. 1939
Novelty/Skill	SPARKS PUNCHBOARD	Feb. 1939
Novelty/Skill	MERCURY PUNCHBOARD	Feb. 1939
Cigarette	MOTORIZED SPARKS	Dec. 1939
Cigarette	MOTORIZED MERCURY	Dec. 1939
Cigarette	IMP	Jan. 1940
Specialty Reel	IMP (Numbers)	1940
Baby Bell	IMP	1940
Baby Bell	1940 LIBERTY	June 1940
Cigarette	1940 DELUXE MERCURY	July 1940
Cigarette	1940 DELUXE MERCURY BALL GUM VENDER	1940
Cigarette	1940 DELUXE MERCURY DISCREET	1940
Cigarette	1940 DELUXE MERCURY DISCREET BALL GUM VENDER	1940
Baby Bell	1940 LIBERTY BALL GUM VENDER	1940
Specialty Reel	LIBERTY SPORTS PARADE (Sports)	1940
Specialty Reel	LIBERTY SPORTS PARADE DISCREET (Sports)	1940
Specialty Reel	LIBERTY SPORTS PARADE DISCREET BALL GUM VENDER (Sports)	1940
Cigarette	LIBERTY	1940
Cigarette	LIBERTY BALL GUM VENDER	1940
Cigarette	LIBERTY DISCREET	1940
Cigarette	LIBERTY DISCREET BELL BALL GUM VENDER	1940
Baby Bell	LIBERTY	1940
Baby Bell	LIBERTY BALL GUM VENDER	1940
Baby Bell	LIBERTY DISCREET	1940
Baby Bell	LIBERTY DISCREET BALL GUM VENDER	1940
Cigarette	SPARKS CHAMPION	Sept. 1940
Cigarette	SPARKS CHAMPION DISCREET	1940
Specialty Reel	SPARKS CHAMPION (Numbers)	1940
Specialty Reel	SPARKS CHAMPION DISCREET (Numbers)	1940
Specialty Reel	SPARKS CHAMPION (Beer)	1940
Specialty Reel	SPARKS CHAMPION DISCREET (Beer)	1940
Specialty Reel	SPARKS CHAMPION (Horses)	1940
Specialty Reel	SPARKS CHAMPION DISCREET (Horses)	1940
Specialty Reel	SPARKS CHAMPION SPORTS (Sports)	1940
Sports Reels	SPARKS CHAMPION SPORTS DISCREET	1940
Novelty/Skill	PIKES PEAK	Jan. 1941
Novelty/Skill	ZOOM	Jan. 1941
Novelty/Skill	SKILL SHOT	Jan. 1941
Baby Bell	SPARKS CHAMPION BELL	Jan. 1941
Baby Bell	SPARKS CHAMPION BELL DISCREET	Jan. 1941
Dial	DIAL-IT	Jan. 1941
Card Machine	POK-O-REEL	1941
Specialty Reel	KLIX (Numbers)	1941
Cigarette	WINGS	1941
Novelty/Skill	NEW PIKES PEAK	Apr. 1941
Cigarette	YANKEE	1941
Baby Bell	YANKEE	1941
Novelty/Skill	PIKE'S PEAK	Feb. 1941
Coin Drop	POISON THIS RAT (Hitler)	1943
Coin Drop	KILL THE JAP	1943
Coin Drop	(Mussolini)	1943
Baby Bell	1946 LIBERTY BELL	1946
Cigarette	1946 LIBERTY BELL	1946
Baby Bell	1947 IMP	June 1946
Cigarette	1947 IMP	May 1947
Baby Bell	ATOM	June 1949
Cigarette	ATOM	June 1949
Cigarette	WINGS	June 1949
Card Machine	POK-O-REEL	June 1949
Cigarette	YANKEE	June 1949
Baby Bell	YANKEE	June 1949
Specialty Reel	KLIX ("21" BLACK JACK) (Numbers)	June 1949

Grove Brothers, Philadelphia, Pennsylvania
Specialty Reel	POKER-DICE (Dice)	Apr. 1893

L. G. Grund, Philadelphia, Pennsylvania
Card Machine	CARD MACHINE	1903

Gumatic Corporation, St. Louis, Missouri
Coin Drop	MELODY GUM	1938

Hamilton Manufacturing Company, Hamilton, Ohio

Card Machine	THE HAMILTON (SUCCESS)		1902
Card Machine	THE HAMILTON		1903
Card Machine	COUNTER HAMILTON		1903
Coin Drop	DAISY NO BLANK ("Bread Loaf Top")		1907
Coin Drop	DAISY ("Diamond Top")		1910
Race Game	(UNKNOWN)	Oct.	1933

Hammond and Jones, Baltimore, Maryland
Race Game	HORSE RACE		1888

Happy Jack Company, Glendale, California
Dial	HAPPY JACK		1932

Harkness-Miller Company, San Antonio, Texas
Dice	PENN-E-WIZE	Feb.	1933

Harlich Manufacturing Company, Chicago, Illinois
Coin Drop	BIG SIX	Feb.	1935

Lorenzo W. Harris, Asbury Park, New Jersey
Pinfield	(UNKNOWN)	Sept.	1932

Hawkeye Novelty Company, Des Moines, Iowa
Coin Drop	POT-O-GOLD	Aug.	1936
Dial	WAGON WHEELS		1937
Coin Drop	WHIRL POOL	Feb.	1947

Hawes, Butman and Company, Boston, Massachusetts
Race Game	(UNKNOWN)	Mar.	1876

Andrew T. Hayashi, Ventnor, New Jersey
(Unknown)	(UNKNOWN)	May	1930

Charles W. Heeg, St. Louis, Missouri
(Unknown)	(UNKNOWN)		1891

G. Henry and Company, Chicago, Illinois
Dice	(UNKNOWN)		1895

Hercules Manufacturing Company, Fort Worth, Texas
Cigarette	MUTUAL		1934

Hercules Novelty Company, Chicago, Illinois
Novelty/Skill	MIDGET BASEBALL	Jan.	1931
Dial	TRIANGLES		1933
Coin Drop	NEW MOON		1933
Coin Drop	BUY AMERICAN		1933

Homer A. Herr Mechanical Engineer, Philadelphia, Pennsylvania
Dice	(UNKNOWN)	June	1893

Hiawatha Amusement Company, Los Angeles, California
Target	TARGET PRACTICE		1927

Hiawatha Manufacturing Company, Detroit, Michigan
Specialty Reel	HIAWATHA		1907
Card Machine	HIAWATHA CARD		1907

Hillsboro Wooden Ware Company, Hillsboro, Ohio
Coin Drop	THE HILLSBORO		1897

G. F. Hochriem and Company, Chicago, Illinois
Race Game	BOOSTER	June	1932
Dice	26 GAME		1940

The Hollingsworth Corporation, Little Rock, Arkansas
Novelty/Skill	AFRICAN GOLF	Jan.	1932

Holly Manufacturing Company, Detroit, Michigan
Coin Drop	CIRCUS	Jan.	1941

A. L. Holt Company, Philadelphia, Pennsylvania
Novelty/Skill	POCKET-BALL	May	1931

T. F. Holtz and Company, San Francisco, California
Card Machine	CARD MACHINE		1896
Card Machine	BROWNIE		1897

T. F. Holtz Novelty Machine Works, San Francisco, California
Card Machine	CARD MACHINE		1895

J. and E. Homan Machinists, New York, New York
Dice	(UNKNOWN)	Oct.	1893

Home Novelty Company, Cincinnati, Ohio
(Unknown)	(UNKNOWN)		1899

Home Novelty Company Limited, Detroit, Michigan
(Unknown)	(UNKNOWN)		1907

Hot Shot Sales Company, Toledo, Ohio
Target	HOT-SHOT	July	1932

Howard Novelty Company, Detroit, Michigan
(Unknown)	(UNKNOWN)		1909

Howard Novelty Company, St. Louis, Missouri
(Unknown)	(UNKNOWN)	May	1912

Howard Sales Company
Baby Bell	BRODI STANDARD (non-coin)		1939
Baby Bell	BRODI DELUXE (non-coin)		1939
Baby Bell	BRODI LIBERTY (non-coin)		1940

Hub Manufacturing Company, Milwaukee, Wisconsin
Wheel	FLYING HEELS		1935
Wheel	CORK TIP		1935

Hudson Moore Company, New York, New York
Race Game	RACE		1889

Dice	POKER DICE	July	1891
Dice	AUTOMATIC DICE SHAKING MACHINE		1892
Card Machine	AUTOMATIC POKER PLAYER		1893
Roulette	AUTOMATIC ROULETTE		1893
Roulette	IMPROVED AUTOMATIC ROULETTE		1894
Card Machine	PERFECTION		1894
Card Machine	UPRIGHT PERFECTION		1894

Charles D. C. Huestis Manufacturer, Seattle, Washington
Coin Drop	COMSTOCK		1898

Huffman Novelty Company, New Haven, Connecticut
Coin Drop	(UNKNOWN)		1914

Hunt and Company, Chicago, Illinois
Wheel	ZULU		1913

Hunts Club, Inc., Chicago, Illinois
Card Machine	ROYAL FLUSH		1939

Hutchison Engineering Company, Nashville, Tennessee
Novelty/Skill	ZIG-ZAG		1929
Novelty/Skill	BLUE STREAK (STEEPLECHASE)		1931

I Will Novelty Company, Chicago, Illinois
Dice	I WILL		1894

Idea Novelty Manufacturing Company, Grand Rapids, Michigan
Coin Drop	THE TOWER		1894

Ideal Amusement Machine, Chicago, Illinois
Novelty/Skill	CHURCHILL DOWNS		1933

Ideal Manufacturing Company, Chicago, Illinois
(Unknown)	(UNKNOWN)		1889

Ideal Manufacturing Company, Los Angeles, California
Card Machine	(UNKNOWN)		1947

Ideal Toy Company, Chicago, Illinois
Card Machine	SAMPLE EXHIBITOR	Apr.	1890
Card Machine	CARD EXHIBITING MACHINE	Apr.	1890
Dice	(UNKNOWN)	May	1891
Card Machine	IMPROVED SAMPLE EXHIBITOR	Oct.	1891
(Unknown)	(UNKNOWN)	Apr.	1892

Illinois Axle Skein and Nut Lock Company, Pana, Illinois
(Unknown)	(UNKNOWN)		1906

Martin G. Imbach, Brooklyn, New York
Dice	(UNKNOWN)	June	1891

Imperial Manufacturing Company, Chester, Pennsylvania
Dice	IMPERIAL		1912
Dice	IS IT ANY OF YOUR BUSINESS?		1924

In and Out-Door Games Company, Chicago, Illinois
Novelty/Skill	EPSOM-DOWNS		1932

Indian Head Novelty Company, Los Angeles, California
Novelty/Skill	INDOOR STRIKER	Apr.	1931
Novelty/Skill	INDOOR JUNIOR	May	1931
Novelty/Skill	THE CHIEF	May	1931

Industry Novelty Company, Chicago, Illinois
Card Machine	PILGRIM		1916
Specialty Reel	PURITAN		1916
Single Reel	BASE BALL		1916
Coin Drop	PREMIUM TRADER		1918
Coin Drop	SILENT SALESMAN		1918
Target	TARGET PRACTICE		1918
Baby Bell	THE TRADER		1918
Single Reel	1918 BASEBALL		1918

LeGrand Ingersoll, Denver, Colorado
(Unknown)	(UNKNOWN)		1891

International Automatic Machine Company, Cincinnati, Ohio
Coin Drop	(UNKNOWN)		1892

International Amusement Company, Chicago, Illinois
Race Game	SPEEDWAY	Sept.	1932

International Mutoscope Reel Company, Inc., New York, New York
Novelty/Skill	JUGGLING CLOWN	Oct.	1930
Novelty/Skill	TIP THE BELL HOP	Sept.	1931
Novelty/Skill	TINY TIM GOLF	Sept.	1931

Iowa Novelty Company, Cedar Rapids, Iowa
Target	TARGET PRACTISE		1923
Card Machine	LITTLE PERFECTION		1923

Iowa Paper Company, Waterloo, Iowa
(Unknown)	(UNKNOWN)		1910

The Jackson Company, New Castle, Indiana
Coin Drop	NEVER LOSE		1928
Coin Drop	CAN'T LOSE		1930
Coin Drop	HAVE SOME FUN		1932

Conrad Jackson Desk Company, Cincinnati, Ohio
Wheel	GAME O'SKILL		1902

Jacobs Novelty Company, Stevens Point, Wisconsin
(Unknown)	(UNKNOWN)		1929

Jaeger Automatic Machine Company, Philadelphia, Pennsylvania
  (Unknown)   (UNKNOWN)   1899

Harry S. Jarboe, Trenton, New Jersey
  Pointer   THE ADVERTISER   Sept. 1901

Jayess Novelty Company, Syracuse, New York
  Novelty/Skill   SNAP-A-BALL   Apr. 1931

O. D. Jennings and Company, Chicago, Illinois
Type	Name	Date
Target	THE TARGET	1926
Target	FAVORITE GUMBAL	1926
Target	FAVORITE PEANUTS	1926
Target	FAVORITE CANDY	1926
Target	FAVORITE BALL GUM	1928
Baby Bell	PURITAN GIRL	1928
Specialty Reel	PURITAN GIRL (Numbers)	1928
Cigarette	WIN-A-PACK	1934
Cigarette	LITTLE MERCHANT	1934
Specialty Reel	LITTLE MERCHANT (Numbers)	1934
Baby Bell	LITTLE MERCHANT	1934
Baby Bell	REBATER	1934
Card Machine	21 VENDER	1934
Card Machine	CARD MACHINE	1934
Roulette	LITTLE MYSTERY	1934
Pointer	ARROW	1934
Dice	DICETTE	1934
Card Machine	POKER GAME	June 1935
Card Machine	21 BLACK JACK	June 1935
Cigarette	CLUB VENDER	1936
Cigarette	STAR PENNY PLAY	1936
Cigarette	STAR (PENNY PLAY)VENDER	1936
Baby Bell	STAR VENDER	1936
Specialty Reel	TREASURY BANK (Numbers)	1936
Specialty Reel	TREASURY BANK VENDER (Numbers)	1936
Baby Bell	CANDY	1936
Baby Bell	CANDY VENDER	1936
Cigarette	GRANDSTAND	1937
Specialty Reel	GRANDSTAND (Baseball)	1937
Cigarette	PENNY CLUB	1938
Pointer	SWEET MUSIC	1940

Jess Manufacturing Company
  Dice   SHAKE RATTLE 'N' ROLL (Poker)   1954

Johnston Products Company, East Moline, Illinois
Type	Name	Date
Novelty/Skill	KICK 'N CATCH	1979
Novelty/Skill	LONG SHOT	1979
Novelty/Skill	LITTLE LEAGUE	1979

William C. Jones Machine Shop, Niantic, Illinois
Type	Name	Date
Novelty/Skill	AUTOMATIC TRADE CLOCK	Feb. 1903
Novelty/Skill	AUTOMATIC WIZARD CLOCK	Sept. 1904

Jones Novelty Company, Rochester, New York
  Card Machine   JONES CARD MACHINE   1892

Jones Novelty Company, Danville, Illinois
  (Unknown)   (UNKNOWN)   1919

Jorgensen Manufacturing Company, Detroit, Michigan
  (Unknown)   (UNKNOWN)   1905

E. E. Junior Manufacturing Company, Los Angeles, California
Type	Name	Date
Novelty/Skill	PLAY BASKETBALL ("3 Shots")	Jan. 1931
Novelty/Skill	PLAY FOOTBALL	1931
Novelty/Skill	PLAY BASKETBALL ("2 Shots")	July 1931
Novelty/Skill	(NEW) PLAY BASKETBALL ("3 Shots")	Aug. 1931

Kalamazoo Automatic Music Company, Kalamazoo, Michigan
Type	Name	Date
Dice	KALAMAZOO	1934
Dice	'LEVEN COME SEVEN	1935
Dice	KAZOO	1935
Dice	KAZOO-ZOO	1935
Dice	SEVEN COME ELEVEN	1935
Poker Dice	LITTLE SILENT POKER	1935

Kalamazoo Shutter Company, Kalamazoo, Michigan
  (Unknown)   (UNKNOWN)   1902

K. and E. Novelty Company, San Jose, California
  Dice   (UNKNOWN)   1932

Keane Novelty Company, Chicago, Illinois
Type	Name	Date
Dice	SQUARE DEAL (non-coin)	May 1891
Coin Drop	CIGAR CUTER	Apr. 1892

Keeney and Sons, Inc., Chicago, Illinois
Type	Name	Date
Specialty Reel	BABY VENDER	1927
Novelty/Skill	ELECTRIC BASEBALL	Apr. 1931
Novelty/Skill	CHUCK-O-LUCK HORSE RACING	Dec. 1931
Novelty/Skill	CHUCK-O-LUCK FOOTBALL	Dec. 1931
Novelty/Skill	CHUCK-O-LUCK DICE	Dec. 1931
Novelty/Skill	CHUCK-O-LUCK POKER	Dec. 1931
Novelty/Skill	CHUCK-O-LUCK 21 BLACK JACK	Dec. 1931
Novelty/Skill	STEEPLECHASE	Oct. 1932
Novelty/Skill	NEW STEEPLECHASE	Nov. 1932
Race Game	DERBY VENDER	Apr. 1933
Pointer	MAGIC CLOCK (Numbers)	1933
Pointer	MAGIC CLOCK (Fruit symbols)	1933

J. H. Keeney and Company, Chicago, Illinois
Type	Name	Date
Pointer	SPINNER WINNER (Dice)	Jan. 1939
Pointer	SPINNER WINNER (Numbers)	Jan. 1939
Pointer	SPINNER WINNER (Cigarettes)	Jan. 1939
Dice	JITTER-BONES (non-coin)	1939
Novelty/Skill	SCRAMBALL	1941

Henry M. Keith, Spokane, Washington
  Dial   (UNKNOWN)   May 1938

Philip Keller, Springfield, Massachusetts
  Dice   (UNKNOWN)   1900

Kelley Manufacturing Company, Chicago, Illinois
Type	Name	Date
Novelty/Skill	FLIP FLAP	1901
Wheel	BICYCLE	1902
Card Machine	COUNTER PERFECTION	1903
Specialty Reel	THE KELLEY (Numbers)	1903
Specialty Reel	THE NEW IMPROVED KELLEY (Numbers)	1905

W. H. Kelly Company, Tulsa, Oklahoma
  Novelty/Skill   PENNY JITTER BALL   1938

Kelly Novelty Company, Chicago, Illinois
  Novelty/Skill   TOPSY TURVY   Feb. 1932

Kellogg and Company, New York, New York
Type	Name	Date
Pointer	WHEEL OF FORTUNE (Horses)	1888
Pointer	WHEEL OF FORTUNE (Numbers)	1888

Kemo Novelty Company, West Allis, Wisconsin
  Race Game   NECK AND NECK   1932

Clarence M. Kemp, Baltimore, Maryland
  Wheel   VOTING MACHINE   Nov. 1889

Kennedy and Diss, Brooklyn, New York
Type	Name	Date
Race Game	AUTOMATIC RACE TRACK	Oct. 1889
Card Machine	CARD MACHINE	Apr. 1890

Kentucky Gum Company, Louisville, Kentucky
  Dice   LUCKY BOY   Dec. 1934

Kenyon, Inc.
Type	Name	Date
Target	PITCH-A-PENNY	1939
(Unknown)	PLEASURE ISLAND	1939

George Kern, Peru, Illinois
Type	Name	Date
Pointer	(UNKNOWN)	Mar. 1916
Pointer	(UNKNOWN)	June 1916

D. Kernan Manufacturing Company, Chicago, Illinois
Type	Name	Date
Card Machine	THE SHUFFLER	1897
Roulette	IMPROVED ROULETTE	1897
Dice	DICE SLOT MACHINE	1897
Wheel	MIDGET	Dec. 1898
Card Machine	SUCCESS	1901
Card Machine	JUMBO SUCCESS	1901

Keystone Automatic Company, Philadelphia, Pennsylvania
  (Unknown)   (UNKNOWN)   1902

Keystone Engineering Company, St. Louis, Missouri
  Novelty/Skill   THE KICKO ARISTOCRAT   Apr. 1933

Keystone Novelty Manufacturing Company, Chicago, Illinois
Type	Name	Date
Roulette	ROULETTE POKER	1932
Roulette	COUNTER ROULETTE POKER	1932
Roulette	SPOT POKER	1933

Keystone Novelty and Manufacturing Company, Philadelphia, Pennsylvania
Type	Name	Date
Coin Drop	SILENT SALESMAN	1920
Dice	WINNER	1924
Dice	CHUCK-O-LUCK	1927

Keystone Novelty and Sales Company, Indianapolis, Indiana
Type	Name	Date
Baby Bell	KEYSTONE PURITAN BELL	Jan. 1928
Baby Bell	PURITAN BELL GUM VENDER	Jan. 1928
Specialty Reel	PURITAN BELL FORTUNE	Jan. 1928
Dice	IMPROVED DICE MACHINE	Jan. 1928

Keystone Sales Company, Chicago, Illinois
  Dice   KEYSTONE   1931

Kimsey Manufacturing Company, Spokane, Washington
  Novelty/Skill   DIALIT   1947

King and Company, Chicago, Illinois
  Novelty/Skill   HOLD-A-BALL   1960

A. R. Kiser Company, Charlotte, North Carolina
  Novelty/Skill   PUT AND TAKE   Aug. 1931

The Klondike Slot Machine Company, Cincinnati, Ohio
  Coin Drop   THE KLONDIKE   1899

Klondyke Prospector Manufacturing Company, Chicago, Illinois
    Coin Drop   KLONDYKE PROSPECTOR   Mar. 1900

Oscar G. Klugel, Indianapolis, Indiana
    Dice   (UNKNOWN)   Mar. 1892

Howard Knight, Trenton, New Jersey
    Wheel   (UNKNOWN)   Oct. 1892

Knight Novelty Company, Marblehead, Massachusetts
    (Unknown)   (UNKNOWN)   1905

Knight Novelty Company, Cliftondale, Massachusetts
    Coin Drop   LEADER   1922

Koplo Sales and Supply Company, Chicago, Illinois
    Coin Drop   ROLL-A-CENT   1940
    Dice   SEVEN GRAND   1941
    Coin Drop   ROLL-A-CENT NO.2   1941

Kramer Manufacturing Company
    Roulette   POCKET POOL   1938

A. H. F. Kruse, Portland, Oregon
    Card Machine   (UNKNOWN)   1904
    Wheel   (UNKNOWN)   1909
    Pointer   (UNKNOWN)   1914

K. and S. Sales Company, Chicago, Illinois
    Roulette   BELL-SKILL   1932
    Dice   BOUNCING BONES   1933

LaBuff Manufacturing Company, Rochester, New York
    Card Machine   IMPROVED CARD MACHINE   1892

Joseph J. Lane, Providence, Rhode Island
    (Unknown)   PENNY SLOT MACHINE   Oct. 1903

Lark Distributing Company, Inglewood, California
    Specialty Reel   THE LARK (Numbers)   1927
    Baby Bell   THE LARK   1928

Latimer and Company, San Francisco, California
    Target   GAME O'SKILL   July 1893

Lehigh Novelty Company, Philadelphia, Pennsylvania
    Dice   PRIZE PEPSIN GUM VENDING MACHINE   1897

Care Leonardo Novelty Company, New York, New York
    Coin Drop   AMERICAN SPORT   1928

Chas. Leonhardt, Jr. and Company, San Francisco, California
    Card Machine   MONARCH CARD MACHINE   1894
    Card Machine   IMPROVED MONARCH CARD MACHINE   1895
    Coin Drop   TWO FOR ONE SKILL   1895
    Card Machine   DRAW POKER   1895

C. C. Letts and Company, Chicago, Illinois
    Specialty Reel   (UNKNOWN)   1901

Levitt Manufacturing Corporation, Rochester, New York
    Novelty/Skill   THE KNOCKOUT   Oct. 1931

William T. Lewis, Buffalo, New York
    Dice   (UNKNOWN)   Oct. 1892

Lewis Manufacturing Company, Minneapolis, Minnesota
    Race Game   (UNKNOWN)   1889

Lewis and Strobel, Philadelphia, Pennsylvania
    Card Machine   POKER SOLITAIRE   Feb. 1891
    Specialty Reel   4 CARD (Numbers)   Feb. 1891

Liberty Manufacturing Company, Council Bluffs, Iowa
    Dice   SHIMMER DICE   1939
    Dice   TWINS WIN   1940

Liberty Manufacturing Company, Ltd., Kalamazoo, Michigan
    (Unknown)   (UNKNOWN)   1901

Norman Lichty Manufacturing Company, Des Moines, Iowa
    Wheel   AUTOMATIC SALESMAN AND PHRENOLOGIST 1893

C. R. Light and Company, San Francisco, California
    (Unknown)   (UNKNOWN)   1901

James A. Lighthipe, San Francisco, California
    (Unknown)   (UNKNOWN)   1897

John Lighton Machine Company, Syracuse, New York
    Dice   DICE SHAKER   1892

Lilliput Manufacturing Company, New York, New York
    Novelty/Skill   LILLIPUT GOLF   Feb. 1931

Linbrite Specialty Company, Cleveland, Ohio
    Roulette   WHIRL-O-BALL   1927

Lincoln Novelty Company, Chicago, Illinois
    (Unknown)   (UNKNOWN)   1936

Lion Manufacturing Company, Chicago, Illinois
    Baby Bell   LION PURITAN BABY VENDOR 1931
    Baby Bell   PURITAN BABY BELL   1931
    Cigarette   PURITAN BABY BELL   1931

Lincoln Novelty Company, Toledo, Ohio
    Dice   (UNKNOWN)   1926

Robert H. Little, Chicago, Illinois
    Coin Drop   (UNKNOWN)   May 1891

Little Casino Amusement Company, Rochester, New York
    Card Machine   SUCCESS   1894

Little Giant Manufacturing Company, New Haven, Connecticut
    Coin Drop   LITTLE GIANT   1887

Little Nut Vendor Company, Lansing, Michigan
    Dice   GOLD NUT   1935

Loheide Manufacturing Company, St. Louis, Missouri
    Novelty/Skill   WIZARD CLOCK ("2 column")   1907
    Novelty/Skill   WIZARD CLOCK ("4 column")   1908

Loss (Novelty Company), Grand Rapids, Michigan
    Novelty/Skill   SKILLIARD   Nov. 1903
    Novelty/Skill   GAME-O-SCIENCE   Apr. 1905

Loudon Novelty Company, Galesburg, Illinois
    Card Machine   (UNKNOWN)   1924

Lowe Vending Machine Company, Cleveland, Ohio
    Novelty/Skill   PLAY GOLF   July 1931

Ludington Novelty Works, Ludington, Michigan
    (Unknown)   (UNKNOWN)   1897

Lu-Kat Novelty Company, San Francisco, California
    Novelty/Skill   LUCAT ("The Lucky Cat")   1934

Edwin J. Lumley, Washington, District of Columbia
    Race Game   HORSE RACE   Aug. 1889

Lyon Novelty Manufacturing Company, Long Beach, California
    Novelty/Skill   DRAW FIVE   Jan. 1932

Mabey Electric and Manufacturing Company, Indianapolis, Indiana
    Dice   HIGH DICE   1927

Mac Clatchie Manufacturing Company, Compton, California
    Novelty/Skill   DRAW POKER   1934

B. Madorsky, Brooklyn, New York
    Novelty/Skill   FOOT BALL   Feb. 1930
    Novelty/Skill   TOWER WONDER WORKER   Feb. 1930
    Novelty/Skill   BASKEE BALL   Mar. 1931

Maitland Manufacturing Company, Chicago, Illinois
    Race Game   MISTIC DERBY   1947

Majestic Manufacturing and Sales Company, San Antonio, Texas
    Pointer   PICK-A-PLUM   Nov. 1933
    Dial   SHOOT-THE-MOON   Aug. 1934
    Dial   AMERICA'S SMOKE   1934
    Dial   PLAY A HAND   1935

Malcolm and Tratsch, Chicago, Illinois
    (Unknown)   (UNKNOWN)   1916

Charles T. Maley Novelty Company, Cincinnati, Ohio
    Dice   AUTOMATIC DICE SHAKING SLOT MACHINE 1893
    Card Machine   MODEL CARD MACHINE   1893
    Specialty Reel   THE DART (Numbers)   1893
    Card Machine   AUTOMATIC CARD MACHINE   1893
    Dice   AUTOMATIC DICE MACHINE   1893
    Dice   COMMON SENSE DICE MACHINE 1893
    Coin Drop   NICKEL TICKLER No.1   1893
    Coin Drop   NICKEL TICKLER No.2   1893
    Coin Drop   NICKEL TICKLER No.3   1894
    Pointer   ECLIPSE   1894
    Coin Drop   PENNY TICKLER   1894
    Pointer   DIAL   1894
    Coin Drop   CASHIER   1895
    Card Machine   UPRIGHT PERFECTION   1897
    Card Machine   PERFECTION CARD   1897
    Card Machine   EXCELSIOR   1897
    Pointer   TWO ARROW   1898

Horace A. Manley and Company, Boston, Massachusetts
    Wheel   (UNKNOWN)   Aug. 1893

Mansfield Brass Foundry, Mansfield, Ohio
    Roulette   ROULETTE   1893

George E. Maple, Great Falls, Montana
    Card Machine   CARD MACHINE   1893

Marathon Specialty Company, Wausau, Wisconsin
    Coin Drop   PENNY ANDY   1947

M. M. Marcus and Company, Cleveland, Ohio
    Novelty/Skill   SCOTCH GOLF   Oct. 1931

Josiah T. Marean, Brooklyn, New York
    Race Game   RACE COURSE   Nov. 1885

Marion Manufacturing Company, Chicago, Illinois
    Dice   (UNKNOWN)   Oct. 1891

The Markepp Company, Cleveland, Ohio
    Target   FLIPPER   Aug. 1941

Marvel Manufacturing Company, Chicago, Illinois
    Novelty/Skill   POP-UP   1946
    Novelty/Skill   DIAMOND   1947
    Novelty/Skill   SLUGGER   1948

Novelty/Skill	LUCKY HOROSCOPE	Jan. 1959	

George Mason and Company, Chicago, Illinois
Race Game	MINIATURE RACE COURSE MACHINE	1888

Mason Manufacturing Company, Chicago, Illinois
Coin Drop	(UNKNOWN)	June 1899

W. A. Mason, Philadelphia, Pennsylvania
(Unknown)	(UNKNOWN)	1894

Mascot Machine Company, Chicago, Illinois
Wheel	THE MASCOT	Dec. 1894

Matheson Novelty and Manufacturing Company, Los Angeles, California
Target	ORIGINAL AUTOMATIC VENDER	1925
Target	KITZMILLER'S AUTOMATIC SALESMAN	1926

J. T. Mathews and Company, Cincinnati, Ohio
(Unknown)	(UNKNOWN)	1894

Charles May, Cincinnati, Ohio
(Unknown)	(UNKNOWN)	1895

S. May Company, Greensboro, North Carolina
Race Game	ELECTRIC DERBY	1927
Dice	AFRICAN DOODOO	1932

William McClellan, Danbury, Connecticut
Dice	BOARD OF CRAPS AND GUM MACHINE	1907

McCusker Supply Company, Philadelphia, Pennsylvania
(Unknown)	(UNKNOWN)	1906

William C. McDowell, Beaver, Pennsylvania
Coin Drop	(UNKNOWN)	Mar. 1906

George C. McGovern, Richmond, Virginia
Race Game	(UNKNOWN)	1895

Edward S. McLoughlin, New York, New York
Pointer	BANKER WHO PAYS	1876
Pointer	GUESSING BANK	1878
Pointer	DRINKS	1878
Pointer	PRETTY WAITER GIRL	1880

W. Nichols McManus, New York, New York
Race Game	RACE TRACK	June 1888
Race Game	COUNTER RACE TRACK	1888
Race Game	IMPROVED RACE TRACK	Nov. 1888
Race Game	RACE TRACK	Feb. 1889
Race Game	COUNTER RACE TRACK	Feb. 1889

Mead and Taylor, Detroit, Michigan
Wheel	THE DEWEY	1900

Menu Wheel Company, Cincinnati, Ohio
Pointer	MENU WHEEL	Oct. 1904

Merchants Advertising Company, Adrian, Michigan
Race Game	BELMONT-JR.	1933
Race Game	DELUXE BELMONT-JR.	1933
Race Game	MUTUAL 5	1933

Merkle and Cross Machine Company, Gary, Indiana
Card Machine	HIGH HANDS	1934

Metropolitan Coin Machines, Inc., Brooklyn, New York
Target	PITCH-TO-THE-LINE	1941

Michigan Metal and Wood Novelty Works, Detroit, Michigan
(Unknown)	COLUMBIAN	1893

Michigan Novelty Company, Detroit, Michigan
Roulette	DETROITER	1927

Michigan Sales Company, Detroit, Michigan
(Unknown)	(UNKNOWN)	1909

Mid-West Novelty Machine Manufacturing Company, Chicago, Illinois
Coin Drop	(UNKNOWN)	1921

Midwest Novelty (Manufacturing) Company, Chicago, Illinois
Baby Bell	LION PURITAN BABY BELL	Oct. 1928
Baby Bell	THE CHRYSLER	May 1929
Baby Bell	PERFECTED CHRYSLER	June 1929
Baby Bell	THE ACE	1929
Baby Bell	LION PURITAN BABY VENDOR	Nov. 1929
Cigarette	LION PURITAN BABY VENDOR	1929
Baby Bell	NATIONAL	1929
Dice	LION DICE MACHINE	1929
Specialty Reel	LION PURITAN BABY VENDOR (Fortune)	1930
Baby Bell	PURITAN BABY BELL	1930
Baby Bell	JACK POT PURITAN BABY BELL	1930
Baby Bell	TWO-IN-ONE	1930
Baby Bell	TWO-IN-ONE JACK POT	1930
Novelty/Skill	ZIP-O	Apr. 1931
Dice	CHUCK-LUCK	1931
Baby Bell	IMPROVED LION BABY VENDOR	1932
Dice	ARLINGTON	Feb. 1933
Specialty Reel	PURITAN BABY BELL (Fortune)	1933
Dice	(UNKNOWN)	1933

Mikro-Kall-It Inc., New York, New York
Dice	MIKRO-KALL-IT	Aug. 1938

Milark Manufacturing Company, Chicago, Illinois
Roulette	ROULETTE	1903

H. L. Miles Novelty Works, Denver, Colorado
(Unknown)	(UNKNOWN)	1904

Miller Novelty Company, Chicago, Illinois
Coin Drop	LITTLE DREAM	1904
Coin Drop	LITTLE DREAM PLAY BASEBALL	1907

F. W. Mills Manufacturing Company, Chicago, Illinois
Card Machine	JOCKEY	1917
Single Reel	PREMIUM TRADER	1917
(Unknown)	(UNKNOWN)	1923
Baby Bell	BABY VENDER	1927

F. W. Mills Manufacturing Company, Hoboken, New Jersey
Card Machine	AUTOMATIC CARD MACHINE	Oct. 1900
Wheel	LITTLE KLONDIKE	Apr. 1901
Coin Drop	THE BOOSTER	Jan. 1902

MBM (Mortimer B. Mills) Cigar Vending Machine Company, Chicago, Illinois
Coin Drop	HORSE SHOE	1897

Mills Novelty Company, Chicago, Illinois
Roulette	LITTLE MONTE CARLO	1898
Card Machine	PERFECTION CARD	1898
Card Machine	UPRIGHT PERFECTION CARD	1898
Card Machine	CHECK CARD	1898
Specialty Reel	CHECK FIGARO (Numbers)	1898
Specialty Reel	CHECK POLICY (Numbers)	1898
Card Machine	JUMBO SUCCESS	1898
Specialty Reel	JUMBO SUCCESS (Numbers)	1898
Card Machine	SUCCESS	1898
Specialty Reel	SUCCESS (Numbers)	1898
Card Machine	THE GIANT	1898
Card Machine	THE JUMBO	1898
Card Machine	COUNTER THE JUMBO	1898
Card Machine	THE LITTLE DUKE	1898
Novelty/Skill	THE MANILA	1899
Card Machine	THE CHECK JUMBO	1900
Card Machine	JUMBO SUCCESS No.2 ("Big Success")	1900
Card Machine	SUCCESS No.3 ("Little Success")	1900
Card Machine	JUMBO GIANT	1900
Dice	I WILL	Jan. 1900
Card Machine	JOCKEY	1900
Card Machine	JOCKEY CARD ("Cabinet Jockey")	1900
Card Machine	YOUR NEXT (5-Way)	1900
Card Machine	YOUR NEXT (4-Way)	1900
Card Machine	UPRIGHT CARD MACHINE	1901
Card Machine	CHECK UPRIGHT CARD MACHINE	1901
Card Machine	SUCCESS No.4 ("Little Success")	1901
Card Machine	JUMBO SUCCESS No.4 ("Big Success")	1901
Novelty/Skill	LITTLE KNOCKER	1902
Card Machine	LITTLE PERFECTION ("Round Top")	1902
Roulette	IMPROVED LITTLE MONTE CARLO	1902
Target	GAME O'SKILL	1902
Card Machine	BEN FRANKLIN	Dec. 1902
Card Machine	UPRIGHT PERFECTION	Dec. 1902
Card Machine	CHECK UPRIGHT PERFECTION	Dec. 1902
Card Machine	SUCCESS No.5 ("Little Success")	1902
Card Machine	JUMBO SUCCESS No.5 ("Big Success")	1902
Card Machine	SUCCESS No.6 ("Little Success")	1902
Card Machine	COUNTER SUCCESS No.6 ("Little Success")	1902
Card Machine	JUMBO SUCCESS No.6 ("Big Success")	1902
Card Machine	IMPROVED JOCKEY	1902
Card Machine	IMPROVED JOCKEY ("Cabinet Jockey")	1902
Wheel	BULL'S EYE (1-Way)	1902
Wheel	BULL'S EYE (5-Way)	1902
Wheel	NEW IDEA CIGAR MACHINE (non-coin)	Jan. 1903
Card Machine	IMPROVED LITTLE PERFECTION	1903
Roulette	1903 LITTLE MONTE CARLO (1-Way)	1903
Roulette	1903 LITTLE MONTE CARLO (5-Way)	1903
Dice	1903 I WILL	1903
Card Machine	JUMBO SUCCESS No.7 ("Big Success")	1903
Card Machine	JUMBO SUCCESS, JR.	1903
Card Machine	SUCCESS No.8 ("Little Success")	1903
Card Machine	SUCCESS JR.	1902
Card Machine	RELIABLE	1903
Card Machine	KING DODO (5-Way)	1903
Card Machine	KING DODO (3-Way)	1903
Card Machine	FLORADORA	1903

Type	Name	Date
Card Machine	1904 RELIABLE	1904
Card Machine	DRAW POKER	1904
Specialty Reel	MILLS PURITAN (Numbers)	1904
Card Machine	COMMERCIAL	1904
Card Machine	COMMERCIAL VENDER	1904
Card Machine	SPECIAL COMMERCIAL	1904
Card Machine	COMMERCIAL ("Turntable")	1904
Card Machine	SENTRY	1904
Card Machine	CALIFORNIA JACK	1904
Card Machine	PEERLESS	1904
Card Machine	PEERLESS CALIFORNIA JACK	1904
Pointer	ARROW (CIGAR SALESMAN)	1904
Card Machine	SUPERIOR	1904
Card Machine	SUPERIOR WITH STAND	1904
Card Machine	NATIVE SON	1904
Card Machine	THE TRADER	1904
Single Reel	ELK ("Card Reel")	Aug. 1905
Single Reel	ELK	Mar. 1906
Single Reel	ELK (Novelty)	1906
Single Reel	EXPORT ELK (Colors)	1906
Single Reel	SPECIAL (Poker)	1906
Single Reel	SPECIAL (Novelty)	1906
Single Reel	SPECIAL	1906
Single Reel	PILOT (Novelty)	1906
Single Reel	PILOT (Poker)	1906
Card Machine	NEW DRAW POKER	Apr. 1906
Single Reel	IMPROVED ELK	1907
Card Machine	VICTOR	1907
Card Machine	SPECIAL VICTOR	1907
Card Machine	VICTOR (MIRROR TOP)	1907
Card Machine	VICTOR CALIFORNIA JACK	1907
Coin Drop	LITTLE DREAM	1907
Pointer	SKILL-A-GALLE	June 1907
Single Reel	EAGLE ("Little Pilot")	June 1907
Single Reel	EAGLE ("Little Pilot") (Colors)	June 1907
Card Machine	NATIONAL	1907
Specialty Reel	IMPROVED PURITAN (Numbers)	1907
Card Machine	LITTLE GEM	1907
Dice	CRAP SHOOTER	1908
Dice	ON THE LEVEL	1909
Dice	MILLS PIPPIN	Mar. 1909
Card Machine	HY-LO	1909
Single Reel	IMPROVED SPECIAL	1909
Single Reel	IMPROVED SPECIAL (Novelty)	1909
Single Reel	IMPROVED SPECIAL (Poker)	1909
Single Reel	IMPERIAL MILLS PURITAN (Numbers)	1910
Card Machine	PILGRIM	1910
Card Machine	COUNTER HY-LO	1910
Single Reel	E'LAN (ELK) (Colors)	1910
Single Reel	SPECIAL EXPORT (Novelty)	1910
Single Reel	AEROPLANE	1910
Single Reel	L'AEROPLAN	1910
Single Reel	UMPIRE (Baseball)	1910
Pointer	PROFIT SHARING REGISTER STYLE A	1912
Pointer	PROFIT SHARING REGISTER STYLE B	1912
Pointer	PROFIT SHARING REGISTER STYLE C	1912
Coin Drop	(UNKNOWN)	Feb. 1912
Pointer	ARROW	1913
Single Reel	IMPROVED UMPIRE (Baseball)	1913
Dice	PIPPIN JACKPOT	1914
Roulette	LITTLE SCARAB	1914
Roulette	LITTLE SCARAB (GUM) VENDER	1914
Wheel	DANDY VENDER	1915
Pointer	NEW ARROW	1916
Coin Drop	THE PREMIUM	1916
Coin Drop	SILENT SALESMAN	1916
Target	TARGET PRACTICE	1918
Target	STAR TARGET PRACTICE	1919
Target	TARGET PRACTICE (Cast iron)	1920
Target	TARGET PRACTICE (Aluminum)	1922
Card Machine	LITTLE PERFECTION ("Flat Top")	Jan. 1926
Card Machine	JOCKEY ("Wooden Jockey")	Jan. 1926
Target	NEW TARGET PRACTICE	1926
Specialty Reel	PURITAN ("Aluminum Puritan") (Numbers)	1926
Baby Bell	PURITAN BELL ("Aluminum Puritan")	1926
Specialty Reel	PURITAN BELL ("Aluminum Puritan") (Numbers)	1926
Race Game	RACE HORSE	1926
Race Game	RACE HORSE PIANO	1926
Novelty/Skill	BASE BALL	1926
Novelty/Skill	AUTOMATIC PUNCH BOARD	1926
Novelty/Skill	(UNKNOWN)	1927
Target	TARGET PRACTICE VENDER	1928
Card Machine	NEW JOCKEY ("Aluminum Jockey")	1928
Baby Bell	THE BELL BOY	1931
Specialty Reel	NUMBERS (Numbers)	1932
Novelty/Skill	CATCH THE BALL	Feb. 1932
Novelty/Skill	TICKETTE	1935
Novelty/Skill	JACKPOT TICKETTE	1935
Specialty Reel	BLACK JACK (Numbers)	1935
Specialty Reel	DIAL (Novelty)	1935
Specialty Reel	DIAL VENDER (Novelty)	1935
Specialty Reel	KOUNTER KING (Numbers)	1938
Pointer	THE SPINNER	1938
Specialty Reel	WILD DEUCES (Numbers)	1938
Baby Bell	FRUIT KING	1938
Pointer	COUNTERETTE	1939
Novelty/Skill	FLIP SKILL	Feb. 1939
(Unknown)	(UNKNOWN)	1946

Mills Sales Company, Oakland, California

Type	Name	Date
Roulette	MIDGET ROULETTE	1926
Dice	36 LUCKY SPOT MIDGET	1926
Novelty/Skill	ARISTOCRAT POKER	1931
Novelty/Skill	ARISTOCRAT DOMINO	1931
Novelty/Skill	ARISTOCRAT GOLF	1931

Milwaukee Furniture and Show Case Repairing Company, Milwaukee, Wisconsin

Type	Name	Date
Coin Drop	(UNKNOWN)	1903

Modern Games and Novelty Company, Akron, Ohio

Type	Name	Date
Dice	(UNKNOWN)	Jan. 1932

Modern Novelty Company, Detroit, Michigan

Type	Name	Date
(Unknown)	(UNKNOWN)	1895

Lionel H. Moise, San Francisco, California

Type	Name	Date
Card Machine	(UNKNOWN)	1894

Charles Molitor Novelty Manufacturing Company, Chicago, illinois

Type	Name	Date
Card Machine	JOCKEY	Jan. 1900
Card Machine	FLOOR JOCKEY	Jan. 1900

Monarch Card Machine Company, San Francisco, California

Type	Name	Date
Card Machine	MONARCH CARD MACHINE	1895
Card Machine	IMPROVED MONARCH CARD MACHINE	1895
Card Machine	DRAW POKER	1896
Card Machine	BROWNIE	1897

Monarch Coin Machine Company, Chicago, Illinois

Type	Name	Date
Novelty/Skill	CHANGEMASTER	1938
Dice	INDIAN DICE	1939
Novelty/Skill	FLIP SKILL	1939
Novelty/Skill	FOOTBALL	1947

Monarch Sales Company, Indianapolis, Indiana

Type	Name	Date
Baby Bell	MONARCH	1927
Roulette	PEE-WEE ROULETTE	1927
Dice	36 LUCKY PLAY PEE-WEE	1927

Monarch Manufacturing and Sales Company, Indianapolis, Indiana

Type	Name	Date
Dice	BULL DOG	Sept. 1927
Roulette	BULL DOG GUM VENDER	Sept. 1927
Baby Bell	PENNY GUM VENDER	1927
Cigarette	PENNY GUM VENDER	1931

Monterey Woodcrafters

Type	Name	Date
Dice	COLOR CUBE	1942
Dice	MIAMI CRAP	1942
Dice	HI-LO FIELD	1942

Moon & Avera Barabo, Wisconsin

Type	Name	Date
Novelty/Skill	FLICKER	July 1933

Mooney and Goodwin Advertising Company, Glendale, California

Type	Name	Date
Pointer	TRADE STIMULATOR	1946

M. E. Moore, Chicago, Illinois

Type	Name	Date
Dice	DICE BOX (non-coin)	1887

Cornelius S. Morris, Moline, Illinois

Type	Name	Date
Pointer	(UNKNOWN)	Dec. 1890

J. D. Morris, Portland, Oregon

Type	Name	Date
(Unknown)	(UNKNOWN)	1906

Morris Novelty Company, St. Louis, Missouri

Type	Name	Date
Baby Bell	PURITAN BABY BELL VENDER	1932
Baby Bell	JACKPOT PURITAN BABY BELL VENDER	1932

M-R Advertising System

Type	Name	Date
Pointer	M-R	1936

Mueller and Bader Novelty Company, Cincinnati, Ohio

Type	Name	Date
(Unknown)	(UNKNOWN)	1895

Mueller Specialties, Wichita, Kansas

Type	Name	Date
Novelty/Skill	CHANGE MASTER	1938

Munves Corporation, New York, New York
| Target | TID-BIT | 1939 |

Mike Munves Corporation, New York, New York
| Novelty/Skill | SCOOTER | 1941 |

Munves Manufacturing Corporation, Chicago, Illinois
| Novelty/Skill | BAT-A-BALL JUNIOR | 1947 |

Murdock and Murdock, Washington, District of Columbia
| Race Game | (UNKNOWN) | June 1889 |
| Race Game | (UNKNOWN) | Aug. 1890 |

D. P. Murphy, New York
| Race Game | (UNKNOWN) | 1888 |

William H. Murphy, Brenham, Texas
| Race Game | (UNKNOWN) | Dec. 1886 |

Murray, Spink and Company, Providence, Rhode Island
| Pointer | (UNKNOWN) | June 1892 |
| Pointer | 3 DIAL FORTUNE | June 1892 |

Samuel Nafew Company, New York, New York
Coin Drop	TRADE VENDING MACHINE	1894
Card Machine	LITTLE MODEL CARD MACHINE	1894
Card Machine	MODEL CARD MACHINE	1895
Card Machine	MODEL	1896
Dice	AUTOMATIC DICE MACHINE	1898
Wheel	THE L.A.W. (BICYCLE)	1898

Nafew-Goldberg Company, New York, New York
| Dice | SQUARE DEAL | Apr. 1893 |
| Card Machine | CARD MACHINE | 1894 |

National Amusement Machine Company, Bronx, New York
| Target | BOXING SKILL | 1928 |
| Target | BIG GAME HUNT | 1928 |

National Automatic Device Company, Minneapolis, Minnesota
| Race Game | AUTOMATIC RACE COURSE | Feb. 1889 |

National Automatic Machines Company, St. Paul, Minnesota
Dice	POT LUCK (Number)	1933
Dice	POT LUCK (Novelty)	1933
Roulette	JITTERS	1933

National Automaton Company, Washington Court House, Ohio
| Specialty Reel | THE ELECTRICAL SHELL MAN (Dice) | Apr. 1893 |

National Cash Register Company, Dayton, Ohio
| Specialty Reel | NATIONAL (Numbers) | 1929 |

National Coin Machine Company, Chicago, Illinois
Target	NATIONAL TARGET PRACTICE	1926
Target	NATIONAL TARGET PRACTICE BALL GUM	1926
Baby Bell	PENNY FORTUNE	1926

National Coin Machine Exchange, Toledo, Ohio
Dice	HAZARD	1934
Dice	EVEN UP	1934
Dice	BIG SIX	1934
Cigarette	PENNY KING	1934
Card Machine	DRAW POKER	1934
Card Machine	DRAW POKER GUM VENDER	1935
Cigarette	SMOKES	1936
Dice	TIA JUANA	1936

National Institute For The Blind
| Novelty/Skill | A GOOD TURN | 1910 |

National Manufacturing Company, New York, New York
| Roulette | LITTLE MONTE CARLO | Nov. 1897 |
| Coin Drop | THE DEWEY | 1898 |

National Novelty Manufacturing Company, Chicago, Illinois
| Novelty/Skill | BASE BALL | 1928 |

National Sales Machine Company
| (Unknown) | (UNKNOWN) | 1932 |

National Table Company
| Race Game | NATIONAL TABLE | 1928 |
| Dice | NATIONAL DICE TABLE | 1928 |

Neff Novelty Company, Danville, Illinois
| Specialty Reel | KING ROW (Horses) | 1938 |

T. J. Nertney Manufacturing Company, Ottawa, Illinois
Wheel	COINOGRAPH SALESMAN	1898
Roulette	IMPROVED AUTOMATIC ROULETTE	1898
Card Machine	UPRIGHT PERFECTION	1898
Card Machine	JUMBO	1898
Card Machine	CHECK JUMBO	1898
Wheel	MASCOT	1898

L. Nessue, Portland, Oregon
| (Unknown) | (UNKNOWN) | 1907 |

Frank Netschert, New York, New York
| (Unknown) | SLOT MACHINE | Sept. 1904 |

New Era Manufacturing Company, Chicago, Illinois
Dice	JUMPING JACK	July 1933
Dice	IMPROVED JUMPING JACK	Oct. 1933
Dice	NEW ERA VENDER	1934
Dice	NEW ERA VENDER (Fruit)	1934
Dice	NEW ERA VENDER (Coin slide)	1934
Dice	CASINO	1934
Dice	BAZAAR	1934

New York Amusement Company, New York, New York
| Card Machine | (UNKNOWN) | 1890 |

Joseph Nichols, Chicago, Illinois
| Pointer | (UNKNOWN) | May 1897 |
| Pointer | DEWEY SALESMAN | 1898 |

Sidney T. Nimmo, Baltimore, Maryland
| Novelty/Skill | GAME-OF-CHANCE | May 1896 |

Milton Nitzberg, Chicago, Illinois
| Specialty Reel | (UNKNOWN) | 1938 |

Edward J. Noble, Hartford, Connecticut
| Race Game | (UNKNOWN) | July 1905 |

Nonpariel Novelty Company, New Haven, Connecticut
| Pointer | (UNKNOWN) | 1892 |

Norris Manufacturing Company, Columbus, Ohio
Card Machine	REEL AMUSEMENT VENDER	1935
Card Machine	(UNKNOWN)	1935
Card Machine	REEL AMUSEMENT	1935
Card Machine	REEL AMUSEMENT VENDOR	1937
Cigarette	LUCKY (Fortune)	1938
Dice	ROLL-EM	1939

Northland Vending Machine Manufacturing Company, Minneapolis, Minnesota
| Dice | THREE LITTLE BONES | 1934 |

Northwest Coin Machine Company, Chicago, Illinois
Coin Drop	CHURCH-HILL DOWNS	Nov. 1932
Coin Drop	CHURCHILL DOWNS	Feb. 1933
Roulette	BELL-SKILL DELUXE	1933
Roulette	JACK POT BELL-SKILL DELUXE	1933
Dice	HI-LO	1933

Northwestern Automatic Machine Company, Seattle, Washington
| (Unknown) | (UNKNOWN) | 1902 |

Northwestern Coin Machine Company, Chicago, Illinois
| Coin Drop | CHURCHILL DOWNS | Nov. 1932 |

Northwestern Novelty Company, Minneapolis, Minnesota
| Coin Drop | (UNKNOWN) | 1898 |

Northwestern Sales and Service Company, Brooklyn, New York
| Novelty/Skill | PINCH-HITTER | Dec. 1941 |

Norwood Manufacturing Company, Chicago, Illinois
| Novelty/Skill | PUNCH BALL | Mar. 1939 |

Novelty Cigar Company, Portland, Oregon
| (Unknown) | (UNKNOWN) | 1907 |

Novelty Iron Works, Allentown, Pennsylvania
| Coin Drop | (UNKNOWN) | Jan. 1912 |

Novelty Manufacturing Company, Chicago, Illinois
| (Unknown) | (UNKNOWN) | 1891 |

Novelty Manufacturing Company, Chicago, Illinois
| (Unknown) | (UNKNOWN) | 1938 |

Novelty Manufacturing Company, Cincinnati, Ohio
| Card Machine | (UNKNOWN) | 1895 |

Novelty Manufacturing Company, Los Angeles, California
| (Unknown) | (UNKNOWN) | 1893 |

Novelty Manufacturing Company, Oshkosh, Wisconsin
| (Unknown) | (UNKNOWN) | 1932 |

Novelty Manufacturing Company, South Grand Rapids, Michigan
| (Unknown) | (UNKNOWN) | 1895 |

Novelty Works Company, Detroit, Michigan
| (Unknown) | (UNKNOWN) | 1887 |

Novix Specialties, New York, New York
Target	INDOOR BASEBALL	Sept. 1927
Target	INDOOR BASEBALL MODEL B	Nov. 1927
Target	KOIN-KICK FOOTBALL	1928
Target	INDOOR AVIATOR	1928
Target	GEM	1928
Coin Drop	TRY-SKILL	1929
Novelty/Skill	POLO SKILL	Dec. 1931

Oakland Novelty Company, Oakland, California
| Card Machine | OAKLAND | 1902 |

James D. O'Donoghue, Brooklyn, New York
| Race Game | (UNKNOWN) | Oct. 1883 |

Ogden and Company, Chicago, Illinois
Card Machine	CARD MACHINE	1897
Coin Drop	HEAD OR TAIL	1897
Dice	BIG SIX	1897

Coin Drop	THE HOOSIER		1898
Card Machine	UPRIGHT-PERFECTION		1898
Pointer	DEWEY SALESMAN		1898

Ohio Automatic Machine Company, Cincinnati, Ohio
Dice	AUTO DOMINOES		1930
Target	GEM CONFECTION		1930
Target	CALVERT INDIAN SHOOTER		1930

Ohio Specialty Company, Cincinnati, Ohio
Specialty Reel	TAVERN STYLE A (Whiskey)		1935
Cigarette	TAVERN STYLE B		1935
Specialty Reel	THE BARTENDER (Beer) (Style A)		1935
Cigarette	THE BARTENDER (Style B)		1935

One-Cent Amusement Company, Camden, New Jersey
Coin Drop	KING OF THE DIAMOND		1929

R. H. Osbrink Manufacturing Company, Chicago, Illinois
Baby Bell	WILD CHERRIES		1933
Baby Bell	BELL FRUIT		1933
(Unknown)	KNOCK-KNOCK		1933
Card Machine	PENNY ANTE DRAW POKER		1934

Osbrink Games Company, Chicago, Illinois
(Unknown)	WAHOO		1934
(Unknown)	DOUBLE UP		1934

W. M. Ostrander and Company, Philadelphia, Pennsylvania
(Unknown)	(UNKNOWN)		1896

Overton Manufacturing Company, Topeka, Kansas
Wheel	THE WHEEL		1902

Pace Manufacturing Company, Chicago, Illinois
Target	TARGET ("All Aluminum Target")		1927
Target	TARGET PRACTICE		1927
Target	AUTO-TARGET		1927
Baby Bell	IMPROVED BALL GUM VENDER		1928
Baby Bell	BELL PURITAN		1929
Dice	DICE MACHINE		1930
Novelty/Skill	WHIZ-BALL		Jan. 1931
Novelty/Skill	BASE-BALL		Mar. 1931
Novelty/Skill	PEN-NEE GOLF		Mar. 1931
Novelty/Skill	BOOP-A-DOOP		Mar. 1931
Novelty/Skill	WHIZ-BALL (NO. 2)		June 1931
Novelty/Skill	DEUCES WILD		Oct. 1931
Baby Bell	DANDY VENDER		Oct. 1931
Novelty/Skill	SKIL-FLIP		Feb. 1932
Novelty/Skill	SPIRAL GOLF		Feb. 1932
Novelty/Skill	BAT-A-BALL		Feb. 1932
Cigarette	HOL-E-SMOKES		1933
Card Machine	NEW DEAL		1933
Card Machine	NEW DEAL JAK-POT		1935
Specialty Reel	NEW DEAL JAK-POT (Numbers)		1935
Card Machine	CARDINAL		1935
Card Machine	CARDINAL JAK-POT		1935
Card Machine	JAK-POT CARDINAL		1935
Baby Bell	CARDINAL BELL		1935
Baby Bell	CARDINAL MYSTERY BELL		1935
Cigarette	CARDINAL SMOKE		1935
Specialty Reel	CARDINAL BEER (Beer)		1935
Baby Bell	DANDY VENDER		1935
Baby Bell	JAK-POT DANDY VENDER		1935
Card Machine	HIT ME		1935
Card Machine	JAK-POT HIT ME		1935
Cigarette	HOL-E-SMOKES		1935
Specialty Reel	SUDS (Beer)		1935
Specialty Reel	ARMY 21 GAME (Numbers)		1935

Pacific Amusement Company, Los Angeles, California
Novelty/Skill	INDOOR STRIKER		July 1930

Pacific Amusement Manufacturing Company, Chicago, Illinois
Roulette	MARBLO		Dec. 1934
Dice	P.D.Q.		1935
Dice	ELECTRIC SPINNER		1935

Pacific Manufacturing Corporation, Chicago, Illinois
Dice	BROADWAY		1940

Pacific Electrical Works, San Francisco, California
Card Machine	PERFECTION		1895

Page Manufacturing Company, Chicago, Illinois
Pointer	SALES INCREASER ("Drinks On The House")		1909
Pointer	SALES INCREASER ("Profits Shared")		1909
Pointer	SALES INCREASER ("Free Merchandise")		1909

Palamedes Sales Company, Inc., Fort Wayne, Indiana
Dice	LOTTA-DOE		May 1933

Pana Enterprise Manufacturing Company, Pana, Illinois
Pointer	IMPROVED FAIREST WHEEL (THE FAIREST WHEEL)		1907

Pardue Novelty Company, Chicago, Illinois
Specialty Reel	FORTUNE TELLER (Fortune)		1936

Park Novelty Company, Kalamazoo, Michigan
Pointer	RED BIRD		1903

E. D. Parker and Company, Springfield, Ohio
Wheel	SPIRAL		1904

Frank T. Parritt, Bloomington, Illinois
Coin Drop	(CIGAR)		Apr. 1897

Parmly Engineering Company, Cleveland, Ohio
Specialty Reel	WORLD SERIES BASE BALL		1931
Novelty/Skill	THE LUNATIC		1931

C. Passow and Sons, Chicago, Illinois
Card Machine	CHICAGO (PERFECTION)		1909

Patent Purchase Company, New York, New York
Dice	HONEST DICE BOX (non-coin)		1892

Paupa and Hochriem, Chicago, Illinois
Novelty/Skill	THE MANILA		Dec. 1898
Coin Drop	NEW TRADE MACHINE		Nov. 1899
Single Reel	THE ELK (Model 7)		Feb. 1904
Single Reel	ELK		1905
Single Reel	PILOT (Model 8)		Feb. 1906
Single Reel	IMPROVED ELK (Model 9)		Mar. 1906
Single Reel	GOOSE		1906
Single Reel	DUCK		1906
Single Reel	EAGLE		1906
Dice	CRAP SHOOTER		1906
Single Reel	SPECIAL ELK		Jan. 1907
Single Reel	COMET (Model 10)		1910
Single Reel	LA COMETE (Model 10)		1910
Novelty/Skill	(UNKNOWN)		1917
Novelty/Skill	BASE BALL		1917

Pearsall and Finkbeiner, Syracuse, New York
Coin Drop	THE STAR		June 1898
Wheel	(FIRE ENGINE)		June 1898

Peerless Amusement Machine Company, Los Angeles, California
Specialty Reel	PEERLESS (Numbers)		1929
Specialty Reel	PEERLESS BALL GUM VENDER (Numbers)		1927
Specialty Reel	PEERLESS LITTLE WONDER (Numbers)		1928
Specialty Reel	PEERLESS LITTLE WONDER GUM VENDER (Numbers)		1928

Peerless Manufacturing Company, Chicago, Illinois
Dice	CRYSTAL GAZER		1933
Dice	TROJAN		1933
Coin Drop	ODD PENNY MAGNET		1934

Peerless Manufacturing Company, Fenton, Michigan
(Unknown)	(UNKNOWN)		1893

Peerless Products Company, North Kansas City, Missouri
Novelty/Skill	THE RACE OF HORSES AND MARBLES		1932
Coin Drop	ODD PENNY MAGNET		1933

Peerless Sales and Products Company, Kansas City, Missouri
Race Game	DERBY RACEHORSE		1932

Pennant Novelty Company, Chicago, Illinois
Dice	KENTUCKY DERBY		1934

Penny Ante Amusement Company, Inc., Riverside, California
Novelty/Skill	PETE'S PENNY ANTE		1934
Card Machine	PENNY ANTE		1934
Card Machine	PENNY ANTE VENDER		1934
Card Machine	5¢ PENNY ANTE PENNY DRAW		1934
Card Machine	PENNY AMUSE-U		1934

Peo Manufacturing Corporation, Rochester, New York
Novelty/Skill	VEST POCKET BASKET BALL		Apr. 1929
Novelty/Skill	LITTLE WHIRLWIND		Aug. 1930
Novelty/Skill	PLAY POKER		Oct. 1931
Novelty/Skill	BARN YARD GOLF		Feb. 1932
Novelty/Skill	(UNKNOWN)		Feb. 1933

Peo Sales Corporation, Rochester, New York
Coin Drop	WEE-GEE		1935

Peoria Slot Machine Company, Peoria, Illinois
(Unknown)	(UNKNOWN)		1916

Orin L. Percival, Champaign, Illinois
Wheel	CIGAR WHEEL		Aug. 1896

Perfection Manufacturing Company, Detroit, Michigan
(Unknown)	PERFECTION		1897

Perfection Novelty Company, New York, New York
Coin Drop	WALL STREET BANK		Jan. 1895
Card Machine	PERFECTION		1896

Perfection Novelty Company, Philadelphia, Pennsylvania
(Unknown)	PERFECTION		1914

Perpichnick, Chicago, Illinois

Card Machine	(UNKNOWN)	1897

Peter Manufacturing Company, St. Louis, Missouri
(Unknown)	(UNKNOWN)	1902

Peterson Brothers, Chicago, Illinois
Roulette	DICE ROULETTE	1935

Phillips Farm Supply, Hardware and Furniture, Carbondale, Illinois
Dice	CIGAR VENDER	Feb. 1904
Dice	PHILLIPS	July 1912

Albert Pick and Company, Chicago, Illinois
Dice	(UNKNOWN)	1896
Race Game	RACE TRACK	1903
Wheel	FAIREST WHEEL	1906

Pierce Tool and Manufacturing Company, Chicago, Illinois
Coin Drop	BUGG HOUSE	1932
Coin Drop	TARGETS	1932
Novelty/Skill	UNIVERSAL	Feb. 1933
Dice	BOUNCING BONES	1933
Coin Drop	WIN-A-PACK	1933
Dial	WHIRLWIND (Electric)	Feb. 1933
Dial	WHIRLWIND (Mechanical)	Feb. 1933
Baby Bell	WHIRLWIND	June 1933
Card Machine	THE NEW DEAL ("Regular") (MODEL 553)	1933
Card Machine	JACKPOT THE NEW DEAL (MODEL 555)	1933
Card Machine	THE NEW DEAL ("3 IN 1")	1933
Baby Bell	THE NEW DEAL ("Cherry Play")	1933
Baby Bell	MYSTERY AWARD THE NEW DEAL	1933
Card Machine	JACKPOT THE NEW DEAL("3 IN 1")	1933
Baby Bell	JACKPOT THE NEW DEAL ("Cherry Play")	1933
Baby Bell	JACKPOT MYSTERY AWARD THE NEW DEAL	1933
Novelty/Skill	CIG-O-ROL (Cigarette)	1934
Novelty/Skill	CIG-O-ROL (Card spot)	1934
Wheel	MATCH-A-PAK	1934
Novelty/Skill	GYPSY	Feb. 1934
Dial	4 LEAF CLOVER	1934
Dial	CIGARETTE DIAL 4 LEAF CLOVER	1934
Card Machine	1934 THE NEW DEAL	1934
Card Machine	1934 JACKPOT THE NEW DEAL	1934
Card Machine	HIT ME	1934
Card Machine	JACKPOT HIT ME	1934
Card Machine	ARMY 21 GAME	1934
Dice	BOTTOMS UP	1935
Dice	CHERRY JITTERS (Fruit)	1935
Wheel	PAIR IT	1935
Coin Drop	LEAPING LENA	1935
Dice	PIX-IT (Cigarette)	1935
Specialty Reel	HERE'S HOW (Beer)	1935
Cigarette	HERE'S HOW	1935
Specialty Reel	HERE'S HOW (Numbers)	1935
Cigarette	HOL-E-SMOKES	1935
Specialty Reel	BAR BOY (Beer)	1936
Cigarette	GEM	1936
Cigarette	PICK A PACK	1936
Pointer	PLAY-TEX	1936
Cigarette	PRINCE	1937

Pioneer Coin Machine Company, Chicago, Illinois
Coin Drop	SMILEY	Apr. 1946

Pioneer Games Company, Minneapolis, Minnesota
Dice	BIG BONES	1933

Irving L. Pitkin, Ravenna, Ohio
Dice	(UNKNOWN)	Nov. 1892

Pitton Novelty Company Limited, Detroit, Michigan
(Unknown)	(UNKNOWN)	1901

Plamor Novelty Company, Amarillo, Texas
Roulette	COIN-O-LUCK	1932
(Unknown)	(UNKNOWN)	1933

Planet Manufacturing Company, Detroit, Michigan
Novelty/Skill	"V"	Mar. 1942

Play-Write Corporation, Akron, Ohio
Number Reels	PLAY-WRITE (non-coin)	1950

D. S. Plumb Company, Inc., Newark, New Jersey
(Unknown)	(UNKNOWN)	1925

Pogue, Miller and Company, Richmond, Indiana
Novelty/Skill	WIZARD CLOCK	1910

Poole Brothers, Manufacturers, Chicago, Illinois
Wheel	THE BICYCLE	1903
Wheel	THE BICYCLE DISCOUNT WHEEL	1904

Portland Novelty Company, Portland, Oregon
(Unknown)	(UNKNOWN)	Mar. 1893

The Portland Novelty Company, Portland, Oregon
(Unknown)	(UNKNOWN)	Aug. 1901

Portland Novelty Works, Portland, Oregon
Card Machine	OREGON	Aug. 1901

Pratt and Letchworth, Buffalo, New York
(Unknown)	(UNKNOWN)	1890

Premier Novelty Company, Chicago, Illinois
(Unknown)	LITTLE EGYPTIAN FORTUNE TELLER	1919

Premium Novelty Works, Chicago, Illinois
Specialty Reel	LITTLE GYPSY VENDER (Fortune)	1927

Charles K. Probes, Elmira, New York
Coin Drop	(UNKNOWN)	Jan. 1904

Progressive Manufacturing Company, Pana, Illinois
Novelty/Skill	AUTOMATIC TRADE CLOCK ("2 column")	1906
Novelty/Skill	AUTOMATIC TRADE CLOCK ("4 column")	1906
Novelty/Skill	WIZARD CLOCK ("2 Column")	May 1906
Novelty/Skill	WIZARD CLOCK ("4 Column")	May 1906
Novelty/Skill	DIXON SPECIAL	1907

Progressive Novelty Company, Pana, Illinois
Wheel	THE FAIREST WHEEL	1899
Wheel	FAIREST WHEEL	1900
Wheel	OUR VERY BEST	1900

P. and S. Machine Company, Chicago, Illinois
Novelty/Skill	SHOOTING STAR COUNTER GAME	1947

P. and T. Club Equipment Company, Kansas City, Missouri
Dice	MAIN STREET MAGIC DICE	Oct. 1935

Puritan Machine Company, Detroit, Michigan
Specialty Reel	PURITAN (Numbers)	May 1904
Specialty Reel	PURITAN (Colors)	May 1904
Card Machine	HIAWATHA JR.	1904
Card Machine	MAYFLOWER	1905
Specialty Reel	CHECK-PAY PURITAN (Numbers)	1905
Card Machine	PILGRIM	1906
Card Machine	PURITAN WITH CIGAR CUTTER	June 1906

Quality Supply Company, Sioux Falls, South Dakota
Dice	HORSES	1949
Dice	HI-HAND	1949
Dice	FOUR OF A KIND	1949
Dice	ADD 'EM	1949
Dice	WIN YOUR SMOKES	1949
Dice	BEAT THE HORSE	1949
Dice	WIN A BEER	1949
Dice	HI-LOW SEVEN	1949
Dice	PRESTO	1949

Queen City Novelty Works, Cincinnati, Ohio
(Unknown)	(UNKNOWN)	1895

Reber and Rund, Seattle, Washington
(Unknown)	(UNKNOWN)	1908

Reel-O-Ball Company, Dayton, Ohio
Single Reel	REEL-O-BALL (Baseball)	1933

Reel Profits, Inc., Englewood, Colorado
Card Machine	YOUR DEAL	1967

ReFinders, Highland Park, Illinois
Specialty Reel	THE REPEATER (Novelty)	1987

George H. Reid Slot Machines, Cleveland, Ohio
Card Machine	CARD GRIP	Sept. 1893
Card Machine	COUNTER CARD GRIP	Sept. 1893

Reliable Coin Machine Exchange, Chicago, Illinois
Target	TARGET PRACTICE	1928
Target	GUM VENDER TARGET PRACTICE	1928
Baby Bell	4-WAY PURITAN BABY BELL	1928

Reliable Metal Engineering Company, Chicago, Illinois
Dice	IMP	1947
Roulette	POKO BALL	1947

Reliance Manufacturing Company, New Haven, Connecticut
(Unknown)	RELIANCE	1895

Reliance Novelty Company, San Francisco, California
Card Machine	RELIANCE	June 1896
Card Machine	VICTOR	1896
Card Machine	TROPHY	1897
Card Machine	STANDARD	1897
Pointer	THREE SPINDLE	1897
Card Machine	PEERLESS 5-SLOT	1897
Card Machine	ELITE	1898
Card Machine	IMPROVED RELIANCE	1898

Rennert Novelty Company, Detroit, Michigan
Wheel	(UNKNOWN)	Feb. 1902

Renouf Manufacturing Corporation, Chicago, Illinois
Coin Drop	LUCKY DUCKS	1938

Rex Novelty Company, Chicago, Illinois

Target	TARGET PRACTICE	1923	
Card Machine	LITTLE PERFECTION	1924	

Rialto Sales Company, Chicago, Illinois
Dice	IMPROVED SEVEN GRAND	1938	

Riggs Amusement Company, New York, New York
(Unknown)	(UNKNOWN)	Dec. 1902	

Ristaucrat Manufacturing Company, Kaukauna, Wisconsin
(Unknown)	RISTAURANT	1939	

D. Robbins and Company, Brooklyn, New York
Target	AUTOMATIC BASEBALL	1928	
Target	ALL ABOARD	1929	
Target	GEM CONFECTION VENDER	Feb. 1929	
Target	PENNY CONFECTION VENDER	1929	
Novelty/Skill	EVERLASTING AUTOMATIC SALESBOARD	1930	
Novelty/Skill	INDOOR STRIKER	Feb. 1931	
Target	MINIATURE GOLF	1931	
Novelty/Skill	FOOTBALL	1931	
Novelty/Skill	WHIRL-WIND SKILL CONTEST	1931	
(Unknown)	LITTLE GIANT	1938	
(Unknown)	BARON	1938	
Novelty/Skill	BINGO	Aug. 1938	
Novelty/Skill	CRISS-CROSS	Apr. 1939	
Coin Drop	PROFIT SHARING 5¢ PENCIL VENDOR	JULY 1939	
Novelty/Skill	MIDGET BASEBALL	Nov. 1939	
Target	BASEBALL GUM VENDOR	1940	

Roberts Novelty Company, Utica, New York
Pointer	5¢ SPIN THE ARROW	1929	

Earl A. Robinson Novelties, Providence, Rhode Island
Target	THE NEW PIANO GAME	1909	

Roche Novelty Company, Fort Wayne, Indiana
Dice	PENNY DICE	1932	

Rock-Ola Manufacturing Company, Chicago, Illinois
Baby Bell	DANDY VENDER	1930	

Rock-Ola Manufacturing Corporation, Chicago, Illinois
Baby Bell	DANDY VENDER	1932	
Baby Bell	DANDY JACK-POT VENDER	1932	
Race Game	OFFICIAL SWEEPSTAKES	1933	
Card Machine	POK-O-REEL	1933	
Baby Bell	4-IN-1 JACKPOT	1933	
Fruit reels	4-IN-1 VENDER	1933	
Novelty/Skill	BULLS EYE	1933	
Race Game	OFFICIAL SWEEPSTAKES BALL GUM VENDER	1933	
Wheel	RADIO WIZARD	1934	
(Unknown)	SKYSCRAPER	1934	
Card Machine	HOLD AND DRAW	1934	
Specialty Reel	HOLD AND DRAW (Numbers)	1934	
Baby Bell	HOLD AND DRAW	1934	
Cigarette	HOLD AND DRAW	1934	
Specialty Reel	HOLD AND DRAW (Dice)	1934	
Novelty/Skill	SHIP AHOY	1936	
Novelty/Skill	MAJOR SERIES	1936	

August F. Roesch, St. Louis, Missouri
(Unknown)	(UNKNOWN)	1912	

Julius Roever, Brooklyn, New York
Roulette	(UNKNOWN)	Feb. 1894	

Rogers Manufacturing Company, New York, New York
Wheel	PENNY CIGAR	July 1904	

Rola-Ball Vending Machine Company, St. Louis, Missouri
Novelty/Skill	ROLA-BALL	1939	
(Unknown)	(UNKNOWN)	1940	

Roll-Etto Novelty Company, Chicago, Illinois
Roulette	ROLL-ETTO	1933	

Roovers Brothers, Brooklyn, New York
(Unknown)	(UNKNOWN)	1908	

Rose City Importing Company, Portland, Oregon
Coin Drop	(UNKNOWN)	Nov. 1911	

David Rosen, Philadelphia, Pennsylvania
Baby Bell	ZENO (non-coin)	1947	

Rosenfield Manufacturing Company, New York, New York
Card Machine	CARD MACHINE	1894	
Card Machine	3-SLOT CARD MACHINE ("Jockey")	June 1900	

Joseph M. Roth, Brooklyn, New York
Race Game	(UNKNOWN)	1935	

R. Rothschild's and Sons, New York, New York
Race Game	AUTOMATIC RACE TRACK	1889	
Pointer	AMUSEMENT MACHINE	1889	

Roulette Clock Company, Chicago, Illinois
Pointer	ROULETTE CLOCK	1957	

Royal Card Machine Company, San Francisco, California
Card Machine	PERFECTION	1897	

Royal Machine Company, Kent, Ohio
(Unknown)	(UNKNOWN)	Oct. 1890	

Royal Manufacturing Company, Syracuse, New York
Novelty/Skill	(UNKNOWN)	1931	

Royal Manufacturing Company, Jersey City, New Jersey
Novelty/Skill	BOUNCING BALL	Feb. 1933	

Royal Novelty Company, Baltimore, Maryland
Novelty/Skill	BOUNCING BUSY BALL	Feb. 1933	

Royal Novelty Company, Indianapolis, Indiana
Target	TARGET	1921	
Baby Bell	BABY BELL	1930	
Target	PENNY BASEBALL	1930	

Royal Novelty Company, San Francisco, California
Card Machine	ROYAL TRADER	1902	
Dice	DICE	1912	

Royal Scale Company, Jersey City, New Jersey
Novelty/Skill	BOUNCING BALL	Feb. 1933	

R. and S. Company, New York, New York
Coin Drop	CHIP GOLF GAME	1929	

Rubini Cigar Company, Chicago, Illinois
Dice	TWENTY SIX	1936	

F. A. Ruff, Detroit, Michigan
Dice	CRAP SHOOTER'S DELIGHT	1900	
Wheel	THE DEWEY	1904	

Runyan Sales Company, Newark, New Jersey
Coin Drop	KEEP EM BOMBING	Apr. 1942	

Edward C. Russell, Seattle, Washington
Dial	(UNKNOWN)	1933	

Huntley Russell, Grand Rapids, Michigan
Coin Drop	WATCH YOUR MONEY	July 1893	

St. Louis Patent and Manufacturing Company, St. Louis, Missouri
Coin Drop	(UNKNOWN)	1895	

Sales Stimulator Company, Albany, New York
Specialty Reel	HONEST JOHN (Numbers)	July 1933	

J. Salm Manufacturing Company, Philadelphia, Pennsylvania
Card Machine	(UNKNOWN)	1899	

Sammis Manufacturing Company, Philadelphia, Pennsylvania
(Unknown)	(UNKNOWN)	1907	

J. M. Sanders Manufacturing Company, Chicago, Illinois
Baby Bell	BABY VENDER	1928	
Baby Bell	FORTUNE	1928	
Baby Bell	PURITAN BABY VENDOR	1931	
Baby Bell	BABY JACK POT VENDER	1933	
Roulette	ARISTOCRAT DELUXE VENDER	1933	
Roulette	DELUXE BABY VENDER	1933	
Card Machine	IDEAL	1933	
Baby Bell	IDEAL CHERRY	1934	
Baby Bell	IDEAL MYSTERY	1934	
Specialty Reel	IDEAL RACE HORSE (Horses)	1934	
Specialty Reel	GEM GUM VENDER (Numbers)	1935	
Cigarette	GEM GUM VENDER	1935	
Card Machine	LITTLE POKER FACE	1938	
Card Machine	DEUCES WILD	1939	
Card Machine	STREAMLINER	1939	
Cigarette	TOKETTE	1939	
Card Machine	LITTLE POKER FACE NO. 2	1941	
Cigarette	LUCKY PACK	1941	
Cigarette	ZIP	1941	

Sanderson and Son, Chicago, Illinois
Dice	(UNKNOWN)	Nov. 1892	

Sands Manufacturing Company
Novelty/Skill	SKILL KATCH	1936	

Edward A. Sanquinet, St. Louis, Missouri
Race Game	(UNKNOWN)	July 1902	

John Sassoe, San Francisco, California
(Unknown)	(UNKNOWN)	1904	

T. R. Savage and Company, Bangor, Maine
Dice	(UNKNOWN)	1893	

Savoy Vending Company, Brooklyn, New York
(Unknown)	BASEBALL GUM VENDER	1940	

Schaeffer, Redwood City, California
Pointer	MIDGET	1932	

D. N. Schall and Company, Chicago, Illinois
Card Machine	PERFECTION	1899	
Card Machine	SUCCESS	1901	
Card Machine	JUMBO SUCCESS	1901	
Card Machine	FANCY JUMBO	1901	

Schiemer-Yates Company, Detroit, Michigan

Card Machine	HI-LO	May 1904	
Card Machine	HI-LO VENDER	May 1904	

Schloss and Company, Boston, Massachusetts
Dice	(UNKNOWN)	1893	

Clarence Ray Schultz, Independence, Kansas
Card Machine	CARDS AND DICE	1938	

Schultze Novelty Company, San Francisco, California
Dice	MIDGET	Oct. 1910	
Roulette	MIDGET	Oct. 1910	

Scientific Machine Corporation, Brooklyn, New York
Novelty/Skill	TOTALIZER	Jan. 1940	
Novelty/Skill	SPITFIRE	Nov. 1940	
Novelty/Skill	HOLE IN ONE	Jan. 1941	

Scot Industries, Chicago, Illinois
Cigarette	CUB	1948
Baby Bell	CUB	1948
Specialty Reel	CUB (Numbers)	1948
Card Machine	ACE	1948
Cigarette	MARVEL	1948
Cigarette	NON-COIN MARVEL	1948
Baby Bell	AMERICAN EAGLE	1948
Baby Bell	NON-COIN AMERICAN EAGLE	1948
Specialty Reel	FREE PLAY (Numbers)	1948

Scott Sales Company, Detroit, Michigan
(Unknown)	(UNKNOWN)	1926

J. P. Seeburg Piano Company, Chicago, Illinois
Race Game	GRAYHOUND RACE PIANO	1926

See-Con, Inc., Detroit, Michigan
Dice	SEE-DICE	1937

Richard M. Shaffer, Baltimore, Maryland
Dice	(UNKNOWN)	Dec. 1892

Sheffler Brothers, Inc., Los Angeles, California
Dial	COMMERCE	1932

Shelby Agency, Inc., Lexington, Kentucky
Dice	LUCKY 7	1978
Dice	IMPROVED LUCKY 7	1980

Sherman and Fey Manufacturing Company, Chicago, Illinois
Dice	36 ROULETTE	1920
Dice	36 ROULETTE GUM VENDOR	1920
Roulette	SKILL ROLL	1920
Roulette	SKILL ROLL GUM VENDOR	1920
Novelty/Skill	JACK POT CHECK SALESBOARD	1920

Shipman Manufacturing Company, Los Angeles, California
Race Game	SPIN-IT (1¢ Peanut)	1947
Race Game	SPIN-IT (5¢ Almond)	1947

Shirley Novelty Manufacturing Company, Chicago, Illinois
Novelty/Skill	BASEBALL	Mar. 1923

Sicking Manufacturing Company, Cincinnati, Ohio
Card Machine	SUCCESS	1902
Card Machine	THE HAMILTON	1903
Card Machine	IRON CARD MACHINE NO.8	1905
Card Machine	COUNTER IRON CARD MACHINE NO.8	1905
Card Machine	CENTURY GRAND	1906
Card Machine	COUNTER CENTURY GRAND	1906
Cigarette	SPIN-A-PACK	1936

Sidway Manufacturing Company, Chicago, Illinois
(Unknown)	(UNKNOWN)	June 1898

M. Siersdorfer and Company, Cincinnati, Ohio
Target	HILLMAN COIN TARGET BANK	Mar. 1894
Target	COIN TARGET BANK	1894

Silver King Novelty Company, Indianapolis, Indiana
Specialty Reel	BASEBALL (Baseball)	Oct. 1914
Specialty Reel	SQUAW (PURITAN) (Numbers)	1917
Card Machine	OLD FORT (PILGRIM)	1917
Specialty Reel	INDIAN (MAYFLOWER) (Numbers)	1917
Card Machine	DRAW POKER	1917
Coin Drop	1918 INDIAN PIN POOL	1917
Target	1920 TARGET PRACTICE NO. 73	1918
Target	1922 TARGET PRACTICE NO. 75	1920
Coin Drop	SILENT SALESMAN	1920
Target	THE TARGET	1922
Baby Bell	PENNY BELL	1925
Baby Bell	BALL GUM VENDER	1926
Baby Bell	IMPROVED PENNY BELL	1926
Baby Bell	(IMPROVED) BALL GUM VENDER	1926
Target	PENNY BASE BALL	1932
Baby Bell	DOUBLE DOOR BALL VENDER (DANDY VENDER)	1932
Baby Bell	LITTLE PRINCE	1932
Baby Bell	DUVAL JACK POT (DAVAL JACKPOT)	1932

Cigarette	LITTLE PRINCE (WHIRLWIND)	1933

Sittman and Pitt, Brooklyn, New York
Card Machine	MODEL CARD MACHINE	Dec. 1891
Card Machine	MODEL (AUTOMATIC POKER PLAYER)	1893
Card Machine	LITTLE MODEL CARD MACHINE	1894
Card Machine	MODEL DRAW POKER	1895
Card Machine	BROWNIE	1897
Card Machine	DRAW POKER	1898

Skeen and Farmer, St. Louis, Missouri
(Unknown)	(UNKNOWN)	1893

Skipper Sales Company, Philadelphia, Pennsylvania
Coin Drop	SKIPPER	Mar. 1942

William J. Slattery, Beverly Hills, California
Novelty/Skill	ELECTRO-SKILL	Oct. 1931

James B. Slinn, San Francisco, California
Dice	DICE BOX	July 1891

T. H. Sloan Manufacturing and Sales Company, Charlevoi, Pennsylvania
Card Machine	THE POKEROLL GAME	1932

Sloan Novelty Company, Philadelphia, Pennsylvania
Coin Drop	THE LEADER	Mar. 1910

Slot Machine Company, New York, New York
Dice	SLOT MACHINE	1892

Samuel I. Smith, Philadelphia, Pennsylvania
Wheel	(UNKNOWN)	1894

Vernon K. Smith, Boise, Idaho
Novelty/Skill	GOOD LUCK	1950

Willard A. Smith, Providence, Rhode Island
Pointer	3 DIAL FORTUNE	June 1892
Pointer	LITTLE JOKER STYLE A	1893
Pointer	LITTLE JOKER STYLE B	1893

Smith, Winchester Manufacturing Company, South Windham, Connecticut
Pointer	GUESSING BANK	May 1877

Smokers Supply Company, Boston, Massachusetts
Wheel	CIGAR DICE	1918

Ora O. Snider Agency, Fort Wayne, Indiana
Dice	GIRO	1934

W. G. Souder, New York, New York
Card Machine	CARD MACHINE	Dec. 1896

South Side Machine Shop, Chicago, Illinois
Novelty/Skill	SPIRAL VENDOR	Feb. 1932

Southern Novelty Company, Atlanta, Georgia
Dice	CHUCK-O-LUCK	1926

Specialty Coin Machine Builders, Chicago, Illinois
Target	TARGET	1924

Specialty Machine Works, San Francisco, California
Dice	PORTOLA (non-coin)	1909
Dice	IMPROVED PORTOLA (non-coin)	1910

Specialty Manufacturing Company, Chicago, Illinois
Novelty/Skill	SLEEPIN' SAM	Dec. 1927
Novelty/Skill	PLAY POOL	1928
Novelty/Skill	AUTOMATIC CHAMPION	1927
Target	TARGET PRACTICE BALL GUM VENDER	Sept. 1931

Specialty Manufacturing Company, Detroit, Michigan
(Unknown)	(UNKNOWN)	1904

Specialty Sales Company, Madison, Wisconsin
Novelty/Skill	FLICKER	June 1933

Spin-O Sales Company, Minneapolis, Minnesota
Pointer	SPIN-O	1940

Square Deal Machine Company, New York, New York
Dice	SQUARE DEAL	1904

Standard
Pointer	THE STANDARD	1892

Standard Coin Machine Company, Chicago, Illinois
Novelty/Skill	SKILL-A-RETTE	Nov. 1941

Standard Games Company, Chicago, Illinois
Novelty/Skill	STEEPLECHASE	July 1932

Standard Games Company, Columbus, Ohio
Novelty/Skill	WINDMILL	1946
Novelty/Skill	WINDMILL JR.	1946

Standard Novelty Company, Cincinnati, Ohio
Novelty/Skill	SUNKEN TREASURE	Aug. 1936

Star Amusement Company, West Columbia, South Carolina
Dial	SPARKY	1952
Dice	JOKERS WILD	1953
Roulette	ROULETTE	1954
Novelty/Skill	GO-GO GIRL	1955

Star Machine Manufacturers, Inc., New York, New York

Novelty/Skill	GYRO		Feb. 1932
Novelty/Skill	IMPROVED GYRO		Apr. 1932

Star Manufacturing and Sales Company, Kansas City, Missouri

Roulette	SQUARE SHOOTER	1938
Coin Drop	FLIP FLOP FLUZZEE	1939
Dice	STRUTTIN KUBES	1939
Dice	PASTIME	1939
Coin Drop	SINK A JAP SHIP	1943

Star Novelty Company, Toledo, Ohio

Dice	WINNER DICE	1925
Novelty/Skill	MINIATURE BASEBALL GAME	Apr. 1931

Star Sales Company, Kansas City, Missouri

Pointer	SPINAROUND	1936
Coin Drop	DUCK SOUP	1936
Dice	DANCING DOMINOES	1937
Roulette	SQUARE SHOOTER	1937
(Unknown)	MOUSIE MOUSIE	1937
(Unknown)	BUCK-A-DAY	1938

Star Specialty Company, Philadelphia, Pennsylvania

Coin Drop	SCENIC RAILROAD	1929

Star Trade Register Company, Montpelier, Vermont

Wheel	STAR TRADE REGISTER	Oct. 1901

Star Vending Machine Company, Philadelphia, Pennsylvania

Card Machine	LARK	May 1931
Baby Bell	LARK	May 1931
Novelty/Skill	FOOT BALL	May 1931
Novelty/Skill	DWARF GOLF	1931

Stark, Buffalo, New York

Card Machine	JUMBO SUCCESS	1894

Starved Rock Novelty Manufacturing Company, Ottawa, Illinois

Wheel	ROLL-O-DICE	1935

State Coin Machine Corporation, Columbus, Ohio

Roulette	BEAT-IT	1936

Henry J. Steinberg, Worchester, Massachusetts

Wheel	KLONDIKE	1932

Clarence M. Steiner, New York, New York

Novelty/Skill	(UNKNOWN)	Apr. 1892

The Steinmetz Manufacturing Company, Rochester, New York

Novelty/Skill	BATAPENNY	May 1925

A. J. Stephens and Company, Kansas City, Missouri

Coin Drop	ODD PENNY MAGNET	1932
(Unknown)	NEW DEAL	1933
Pointer	SILENT SALESMAN	1933
(Unknown)	SHARPSHOOTER DELUXE	1933
(Unknown)	NATIONAL BASEBALL	1933
Novelty/Skill	LUCKY COIN TOSSER	1934
Novelty/Skill	PITCH A PENNY FOR A PACK	1934
Dice	DIZZY	1934
Target	ANNIE-OAKLEY	1934
Specialty Reel	MAGIC BEER BARREL (Pretzels)	1934
Cigarette	CIGARETTE KEG	1934
Card Machine	PENNY DRAW	1934
Dice	FLIP	1934
Coin Drop	LUCKY STAR	1938

Stephen's Novelty Company, Milwaukee, Wisconsin

Novelty/Skill	BABE RUTH BASEBALL	1931

B. A. Stevens Company, Toledo, Ohio

Dice	SLOT DICE SHAKER	1893
Dice	RIVAL (non-coin)	1894
Wheel	STAR	1894

J. and E. Stevens Company, Cromwell, Connecticut

Race Game	RACE COURSE	Aug. 1871
Race Game	BIG RACE COURSE	1871
Race Game	RACE AGAINST TIME	1871

Stewart and McQuire, New York, New York

Novelty/Skill	POK-O-MAT	1934

J. W. Stirrup Manufacturing Company, New York, New York

Roulette	FAIREST ROULETTE	1896
Roulette	WINNER ROULETTE	1897
Roulette	AUTOMATIC CIGAR SELLER	Nov. 1897

H. J. Stock, Milwaukee, Wisconsin

Wheel	FLYING HEELS	1934
Wheel	FLYING THRILLS	1934
Wheel	CORK TIP	1934

Nathan M. Stone Company, Chicago, Illinois

Dice	KENTUCKY DERBY (non-coin)	1932

Stone Brothers, Inc., Chicago, Illinois

Dice	KENTUCKY DERBY (non-coin)	1934
Dice	KENTUCKY DERBY	1934

Stone and Jones, Chicago, Illinois

Coin Drop	NEW SPIRAL	Sept. 1933

Stratford Games, Chicago, Illinois

Coin Drop	(UNKNOWN)	1939

Streater, Streater, Illinois

Wheel	STAR (non-coin)	1904

W. E. Struthers, Providence, Rhode Island

Coin Drop	(UNKNOWN)	1932

Stuckey Cigar Company, Lancaster, Ohio

Wheel	STUCKEY CIGAR	1900

Success Manufacturing Corporation, Chicago, Illinois

(Unknown)	(UNKNOWN)	1940

Sun Manufacturing Company, Columbus, Ohio

Wheel	BICYCLE	1898
Wheel	IMPROVED BICYCLE	1903

Sundwall Company, Seattle, Washington

Card Machine	ELITE	1905
Card Machine	TUXEDO	1906
Single Reel	THE EAGLE	1907

King and Sundwall, Seattle, Washington

Card Machine	ELITE	1903

Superior Confection Company, Columbus, Ohio

Baby Bell	FORTUNE BALL GUM VENDER	1927
Baby Bell	THREE WAY SELECTIVE MINT VENDER	1928
Specialty Reel	SUPERIOR CONFECTION VENDOR	1929
Specialty Reel	BASEBALL BALL GUM VENDER	1930
Dice	AUTO-DICE AMUSEMENT TABLE	1930
Card Machine	AUTOMATIC CARD TABLE	1930
Card Machine	AUTOMATIC BRIDGE TABLE	1930
Cigarette	CIGARETTE BALL GUM VENDER	1931
Cigarette	CIGARETTE	1935
Cigarette	CIGARETTE GUM VENDER	1935
Cigarette	CIGARETTE SALESMAN BALL GUM VENDER	1936
Cigarette	CIGARETTE SALESMAN	1936
Cigarette	CIGARETTE SALESMAN MINT VENDER	1936
Cigarette	PENNY PAK	1937

Superior Games Manufacturing Company, New York, New York

Target	ROLO	1938

Superior Products, Chicago, Illinois

Pointer	TORNADO	1941

William Suydam, New York, New York

Race Game	NEW STYLE RACE TRACK	1880
Pointer	DIAMOND WHEEL	1883

Harry W. Swanson and Company, Chicago, Illinois

Novelty/Skill	TOPSY TURVY DERBY	Dec. 1931

Sweeney, California

Card Machine	CARD MACHINE	1907

Tap-It Manufacturing Company, Alhambra, California

Coin Drop	TAP-A-PENNY	May 1928
Coin Drop	ANOTHER TAP-A-PENNY	1928
Coin Drop	TAP-A-PENNY FORTUNE TELLER	1928

James P. Taylor, Fort Worth, Texas

Novelty/Skill	DREAM REVEALER	Apr. 1893

The Tecumseh Sales Company, Cincinnati, Ohio

Novelty/Skill	MINIATURE BASEBALL	Mar. 1931

Edmund A. Thompson, Amherst, Massachusetts

Race Game	RACE COURSE	July 1874

Tibbils Manufacturing Company, Rochester, New York

Card Machine	4 CARD	Feb. 1891
Card Machine	POKER SOLITAIRE	1891
Card Machine	CARD MACHINE	1891
Card Machine	JUMBO CARD MACHINE	1892
Card Machine	SUCCESS	1893
Card Machine	JUMBO SUCCESS	1893

Tillitson Specialty Company, Aurora, Illinois

Roulette	MIDGET ROULETTE	1925

Charles Timroth, Philadelphia, Pennsylvania

(Unknown)	(UNKNOWN)	Jan. 1900

The Tip Slot Machine Company, Brooklyn, New York

Coin Drop	THE TIP	May 1925

Tippecanoe Manufacturing Company, Benton Harbor, Michigan

Cigarette	GET A PACK	1939
Baby Bell	WONDER BELL	1939

Tivoli Automatic Machine and Amusement Company, Seattle, Washington

Card Machine	TIVOLI	1906

Roy Torr, Philadelphia, Pennsylvania

Race Game	DERBY RACES	1940
Novelty/Skill	HU-LA	Feb. 1942

Transcontinental Machine Company, Portland, Oregon
    Card Machine   (UNKNOWN)   1901

Tratsch and Wayman, Chicago, Illinois
    (Unknown)   (UNKNOWN)   1926

Troxler Novelty Introduction Company, Newark, New Jersey
    Target   10 TO 1   Dec. 1893

Try Me Manufacturing Company
    (Unknown)   TRY ME   1939

Twentieth Century Novelty Company, Springfield, Ohio
    Wheel   SPIRAL   Feb. 1903

Twico Corporation, Chicago, Illinois
    Dice   TWICO   1957

Twin City Novelty Company, Minneapolis, Minnesota
    Target   BASEBALL TARGET   1927
    Dice   IMPROVED EAGLE   1928

Unit Sales Company, Lincoln, Nebraska
    Dice   WINNER DICE   1925

United Amusement Company, Kansas City, Missouri
    Coin Drop   PENNY SKILLO   1937
    Target   PITCH-A-PENNY   1938

United Automatic Machine Company, Kansas City, Missouri
    Dice   6-WAY DICE   1905
    Dice   ELECTRIC DICE   1931

United Enterprises, Ltd., Staten Island, New York
    Card Machine   JOKER'S WILD   1971

United Specialty Amusement Company, Chicago, Illinois
    Specialty Reel   CANDY SPINNER   1933
    Specialty Reel   BEER SPINNER (Beer)   1933
    Cigarette   CIGARETTE SPINNER   1933

United States Bulletin Company, Kittery, Maine
    Novelty/Skill   UNIVERSAL ADVERTISING MACHINE   Mar. 1898

United States Music Company, Chicago, Illinois
    (Unknown)   (UNKNOWN)   1909

United States Novelty Company, Chicago, Illinois
    Dice   WINNER   Nov. 1893
    Dice   LUCKY   1894
    Pointer   JOKER   Feb. 1894

U. S. Electric Manufacturing Company, New York, New York
    Novelty/Skill   HOTCHA   Feb. 1933

U. S. Novelty Company, Kansas City, Missouri
    Dice   (UNKNOWN)   Nov. 1893

Universal
    Race Game   AUTOMATIC RACE TRACK   1889
    Race Game   IMPROVED HORSERACE GAME   1891

Universal Advertising Company, Buffalo, New York
    Novelty/Skill   UNIVERSAL ADVERTISER   1897

Universal Advertising Machine Company, Kittery, Maine
    (Unknown)   UNIVERSAL ADVERTISING MACHINE   Jan. 1894

UPI
    Novelty/Skill   GOLD MINE   1965

Valley City Novelty Company, Grand Rapids, Michigan
    Pointer   (UNKNOWN)   1894

Valley City Slot Machine Company, Grand Rapids, Michigan
    Pointer   (UNKNOWN)   May 1893

Valley Sales Service, Aurora, Illinois
    Dice   HI-LO-FIELD   1941
    Dice   NUMBER ROLL   1942

Van-White Novelty Company, Syracuse, New York
    Coin Drop   TRY SKILL   1929

Vending Machine Company, Fayetteville, North Carolina
    Baby Bell   MINT VENDER   1930

Vendor Manufacturing Company, Scranton, Pennsylvania
    Target   VEMCO CONFECTION VENDER   1932

Veteran
    Novelty/Skill   LITE-A-PACK   (1946)

Victor Games, Chicago, Illinois
    (Unknown)   (UNKNOWN)   1946

Victor Novelty Works, Chicago, Illinois
    Pointer   ARROW   1901

Victor Vending Corporation, Chicago, Illinois
    Dice   ROLL A PACK (Cigarette)   1941

C. C. Vogolsong, Emeryville, California
    (Unknown)   (UNKNOWN)   1894

John M. Waddell Manufacturing Company, Greenfield, Ohio
    Novelty/Skill   ROOLO (non-coin)   1896
    Wheel   THE BICYCLE WHEEL   1896
    Wheel   THE BICYCLE DISCOUNT WHEEL   1896
    Wheel   THE BICYCLE DISCOUNT WHEEL
        ("Square Wheel")   May 1897
    Wheel   THE BICYCLE DISCOUNT WHEEL
        ("Large Square Wheel")   July 1897
    Novelty/Skill   PLAY BALL (non-coin)   1897
    Coin Drop   DANCING DOLLS   1897
    Wheel   THE BICYCLE   July 1897
    Pointer   THE BOOMER   Oct. 1897

Waddell Wooden Ware Works Company, Greenfield, Ohio
    Wheel   THE BICYCLE WHEEL   Jan. 1901
    Wheel   THE BICYCLE DISCOUNT WHEEL   Jan. 1901
    Wheel   THE BICYCLE   Jan. 1901
    Wheel   REVOLUTION CIGAR WHEEL   Mar. 1917

Wagaer Amusement Company, St. Louis, Missouri
    (Unknown)   (UNKNOWN)   1902

Charles A. Wagner Manufacturing Company, Chicago, Illinois
    Wheel   COMMODORE   1899

Wagner Manufacturing Company, Chicago, Illinois
    Card Machine   JUMBO   1902

Wain and Bryant Company, Detroit, Michigan
    Wheel   ZODIAC   Feb. 1902

Wakeley Novelty Works, Pasadena, California
    (Unknown)   (UNKNOWN)   1892

Wales Manufacturing Company, Syracuse, New York
    (Unknown)   (UNKNOWN)   1889

Harry O. Walker, Phoenix, Arizona
    Race Game   (UNKNOWN)   1937

Walker Sales Company, Fort Worth, Texas
    Specialty Reel   MUTUAL (Horses)   1935

Wall Novelty Company, Huntington Park, California
    Specialty Reel   HUSTLER (Numbers)   1927

R. C. Walters Manufacturing Company, St. Louis, Missouri
    Novelty/Skill   5¢ JACKPOT CHARLEY   1944
    Novelty/Skill   10¢ JACKPOT CHARLEY   1945
    Novelty/Skill   25¢ JACKPOT CHARLEY   1948
    Novelty/Skill   JACKPOT SUE   1949
    Novelty/Skill   PROFESSOR CHARLEY   1949
    Novelty/Skill   BINGO   1949
    Novelty/Skill   STRIKE IT RICH   1949
    Novelty/Skill   DOLLARS FOR DIMES   1949
    Novelty/Skill   THE BIG FIFTY   1950
    Novelty/Skill   PLENTY OF CASH   1950

A. M. Walzer and Company, Minneapolis, Minnesota
    Coin Drop   HORSE SHOE   1935
    Novelty/Skill   TIPSY TUMBLERS   Feb. 1933

Watling Manufacturing Company, Chicago, Illinois
    Card Machine   PERFECTION   Jan. 1902
    Card Machine   SUCCESS   Jan. 1902
    Card Machine   JUMBO SUCCESS   Jan. 1902
    Card Machine   FANCY JUMBO   Jan. 1902
    Card Machine   FANCY JUMBO SUCCESS   1903
    Card Machine   THE CLOVER (3-WAY)   1903
    Card Machine   THE CLOVER PINOCHLE   1903
    Card Machine   No.9 CARD MACHINE   1903
    Card Machine   THE JOCKEY   1903
    Card Machine   THE JOCKEY CABINET   1903
    Pointer   THREE ARROW   1904
    Pointer   THE FULL DECK   Feb. 1905
    Wheel   THE PURITANA   1905
    Wheel   MECCA   1905
    Wheel   BUFFALO, JR.   1905
    Single Reel   THE MOOSE   1905
    Card Machine   LITTLE DUKE   1905
    Card Machine   DRAW POKER   1907
    Single Reel   IMPROVED ELK   1906
    Single Reel   SPECIAL ELK   1906
    Single Reel   PILOT   1907
    Card Machine   NATIONAL   1908
    Dice   WINNER DICE   1909
    Baby Bell   THE MERCHANT   1910
    Card Machine   PILGRIM   1910
    Specialty Reel   PURITAN (Numbers)   1910
    Card Machine   MAYFLOWER STYLE A   1911
    Baby Bell   MAYFLOWER STYLE B   1911
    Specialty Reel   MAYFLOWER STYLE C (Numbers)   1911
    Specialty Reel   MAYFLOWER STYLE D (Numbers)   1911
    Wheel   CHECK TRADER (6-Way)   1914
    Wheel   CHECK TRADER (5-Way)   1914
    Single Reel   BASE-BALL   1915
    Dial   GOOD LUCK RACE HORSE   1931
    Dial   GOOD LUCK GYPSY   1931

Watling Scale Company, Chicago, Illinois
Wheel	(UNKNOWN)	1923

John J. Watson, Buffalo, New York
Card Machine	COMBINATION CARD AND DICE MACHINE	Jan. 1894

W. B. Specialty Company, St. Louis, Missouri
Specialty Reel	SALESBOARD (Numbers)	1934
Baby Bell	SALESBOARD	1934

W. C. Steel Ball Table Company, Fort Worth, Texas
Pointer	PUT AND TAKE	May 1931

Webb Distributing Company, Chicago, Illinois
(Unknown)	(UNKNOWN)	1946

Webb Novelty Company, Chicago, Illinois
Novelty/Skill	MIRROR OF FORTUNE	1927

Webster Manufacturing Company, Bay City, Michigan
(Unknown)	(UNKNOWN)	1895

Jacob Wedesweiler, Chicago, Illinois
Dice	DICE BOX (non-coin)	Dec. 1876

Samuel Welsh Cigars, Philadelphia, Pennsylvania
Dice	(UNKNOWN)	Oct. 1892

West Coast Novelty Company, Seattle, Washington
Novelty/Skill	HONEST JOHN	Aug. 1931

Western Automatic Machine Company, Cincinnati, Ohio
Coin Drop	NICKEL TICKLER	Aug. 1893
Pointer	ECLIPSE	1894
Roulette	IMPROVED ROULETTE	1894

Western Automatic Machine Company, San Francisco, California
Dial	FAIR-N-SQUARE	1933
Dial	DELUXE FAIR-N-SQUARE	1933
(Unknown)	BELL BOY	1936
Novelty/Skill	VETERAN	1937

Western Company, Chicago, Illinois
Dice	CIGAR AND BLOWING MACHINE	1898
Wheel	WESTERN WHEEL	1898

Western Distributors, Inc., Seattle, Washington
Dice	TOM MACK	1933
Dice	ROLETTA	1933

Western Electric Piano Company, Chicago, Illinois
Race Game	DERBY	1930
Race Game	SWEEP STAKES	Apr. 1932

Western Equipment and Supply Company, Chicago, Illinois
Race Game	KING'S HORSES	Dec. 1933
Target	GEE-WHIZ	1936
Specialty Reel	REEL RACES (Horses)	1936
Specialty Reel	DRAW 21 (Numbers)	1937
Cigarette	MATCH 'EM	1937
Dice	BABY TRACK (Numbers)	1938
Dice	CIGARETTES (Cigarettes)	1938

Western Novelty Company, Kalamazoo, Michigan
(Unknown)	(UNKNOWN)	1899

Western Products, Inc., Chicago, Illinois
Dice	MONTE CARLO	1939
Cigarette	EMPIRE	1939
Card Machine	REEL POKER	1939
Cigarette	WHOOPERDOO	1939
(Unknown)	HORSE-SHOES	1939
Novelty/Skill	JITTERBUG BALL	1939
Novelty/Skill	OOMPH	1940
Cigarette	TOT	1940
Baby Bell	TOT	1940
Cigarette	TOT GUM VENDOR	1940
Baby Bell	TOT GUM VENDOR	1940

Western Weighing Machine Company, Cincinnati, Ohio
Coin Drop	NICKEL TICKLER	1893
Coin Drop	IMPROVED NICKEL TICKLER	1894
Pointer	ECLIPSE (TWO ARROW MACHINE)	1894

Weston and Smith, Syracuse, New York
(Unknown)	(UNKNOWN)	1884

Weston Slot Machine Company, Syracuse, New York
Novelty/Skill	SLOT MACHINE	1892

William H. Wheaton, Chicago, Illinois
(Unknown)	(UNKNOWN)	1909

Charles E. Wheeland and Company, Salt Lake City, Utah
Coin Drop	(Unknown)	Mar. 1894
Coin Drop	(Unknown)	Mar. 1896
Coin Drop	(Unknown)	Sept. 1897

Wheeland Novelty Company, Seattle, Washington
Coin Drop	(UNKNOWN)	1900
Card Machine	PERFECTION	Apr. 1901
Card Machine	CALIFORNIA	1901
Card Machine	FLORADORA	1902

Charles Wheeler, Denver, Colorado
(Unknown)	(UNKNOWN)	1909

Wheeler Novelty Company, Chicago, Illinois
(Unknown)	(UNKNOWN)	Mar. 1891

(R. J.) White Manufacturing Company, Chicago, Illinois
Card Machine	THE SUCCESS	Oct. 1902
Card Machine	JUMBO SUCCESS	Oct. 1902
Card Machine	PERFECTION	Oct. 1902
Card Machine	LITTLE PERFECTION	1902
Target	GAME O'SKILL	1902
Card Machine	JOCKEY 3-WAY	1902
Card Machine	COUNTER JOCKEY 3-WAY	1902
Card Machine	THE TRADER No.1 (6-WAY)	1902
Card Machine	THE TRADER No.2 (3-WAY)	1902
Card Machine	THE TRADER No.3 (non-coin)	1902
Card Machine	THE TRADER No.4 (1-WAY)	1902
Card Machine	THE TRADER No.5 (3-WAY)	1902
Card Machine	THE TRADER No.6 (1-WAY)	1902

William M. White Company, St. Louis, Missouri
Novelty/Skill	WIZARD CLOCK	1907

Whiteside Specialty Company, Los Angeles, California
Novelty/Skill	BASE BALL GAME OF SKILL	1925

J. W. Whitlock Company, Rising Sun, Indiana
Race Game	THE DARBY	1931

Percy G. Williams, Brooklyn, New York
Dice	MONKEY DICE	Dec. 1889

Williams Manufacturing Company, Minneapolis, Minnesota
Roulette	THE ADDER (SKILL ROLL)	1925

O. A. Williams Manufacturing Company, Indianapolis, Indiana
Baby Bell	(UNKNOWN)	1926

Willoughby Company, Grand Rapids, Michigan
Coin Drop	SLOT MACHINE	Jan. 1901

Winchell Novelty Works. Syracuse, New York
(Unknown)	(UNKNOWN)	1909

Winner Novelty Company, David City, Nebraska
Dice	WINNER DICE	1923

Winner Sales Company, Chicago, Illinois
Dice	ALL ELECTRIC DICE MACHINE	1939
Novelty/Skill	PULL-A-BALL	1939

Wire and Metal Manufacturing Company, Glendale, California
Pointer	(UNKNOWN)	1938

Wisconsin Deluxe Corporation, Milwaukee, Wisconsin
Novelty/Skill	TOSS A COIN-HIT HITLER	Mar. 1942
Novelty/Skill	TOSS A COIN-MUSS UP MUSSOLINI	Mar. 1942

Wisconsin Novelty (Manufacturing) Company, Kaukauna, Wisconsin
Roulette	TRIPLEX	1923
Roulette	TRIPLE ROULETTE (ON THE LEVEL)	1923
Roulette	ON THE LEVEL	1923

Wisconsin Novelty Company, Fond du Lac, Wisconsin
Roulette	ON THE LEVEL	1925
Roulette	TRIPLEX	1925

B. A. Withey Company, Chicago, Illinois
Dice	THE SIZZLER	1932
Novelty/Skill	AUTOMATIC SALESBOARD	1932
Dice	CIGARETTE DICE (Cigarette)	1933
Dice	DIXIE ROULETTE	1933
Dice	KING SIX	1934
Dice	KING SIX JR.	1935
Dice	IMPROVED KING SIX	1936
Dice	SHAKE AND DRAW	1936
Dice	FIVE OF A KIND	1937
Dice	SEVEN GRAND	1938
Dice	IMPROVED SEVEN GRAND	1938
Roulette	ROLL-A-WAY	1939

Withey Manufacturing Company, Chicago, Illinois
Dice	SEVEN GRAND	1940
Coin Drop	ROLL-A-CENT	1940

Wolf Manufacturing Company, Inc., Seattle, Washington
(Unknown)	(UNKNOWN)	1915

Wolverine Novelty Company, Detroit, Michigan
(Unknown)	(UNKNOWN)	1891

Wolverine Supply and Manufacturing Company, Pittsburgh, Pennsylvania
Race Game	GEE-WIZ	1938

Wonder Novelty Company
Specialty Reel	SQUARE SPIN (Colors)	1933

A. R. Wood Manufacturing Company, Santa Cruz, California

Wheel	MARGARET BURNHAM'S	1938

World's Fair Slot Machine Company, Bridgeport, Connecticut
Pointer	COLUMBIAN FORTUNE TELLER	Dec. 1892

Lindley A. Wright, Champaign, Illinois
Wheel	RED, WHITE AND BLUE	Aug. 1896

Frank Wuotila, Floodwood, Minnesota
Card Machine	HONESTY	June 1939

Levi W. Yaggy, Lake Forest, Illinois
Dial	FUNNY FACES	Jan. 1898

Yale Wonder Clock Company, Burlington, Vermont
Novelty/Skill	YALE WONDER CLOCK	July 1899
Novelty/Skill	ADVERTISING AND DISCOUNT MACHINE	Apr. 1900
Novelty/Skill	1901 ADVERTISING AND DISCOUNT MACHINE	Feb. 1901
Novelty/Skill	AUTOMATIC CASHIER AND DISCOUNT MACHINE	Oct. 1905

Yendes Manufacturing and Sales Company, Dayton, Ohio
(Unknown)	DOG HOUSE RUN	Feb. 1932
Single Reel	REEL-O-BALL	Feb. 1932

Yendes Service, Inc., Dayton, Ohio
Single Reel	REEL-O-BALL (Baseball)	1934

Charles P. Young, York, Pennsylvania
Roulette	AUTOMATIC ROULETTE	Nov. 1893
Roulette	PREMIUM	1894

W. J. Young and Company, San Francisco, California
Dice	THE LARK	1907

Willard B. Young, New York, New York
Dice	(UNKNOWN)	Mar. 1893

Ypsilanti Novelty Works, Ypsilanti, Michigan
(Unknown)	(UNKNOWN)	1891

Zeno Manufacturing Company, Chicago, Illinois
Coin Drop	YUCCA	1908

Unidentified Manufacturers
Coin Drop	JUMBO (THE ELEPHANT)	Circa 1890
Dice	GOOD LUCK	Circa 1892
Dice	(6-WAY)	Circa 1894
Dice	AUTOMATIC DICE	1894
Dice	"THE SPITTOON DICER"	Circa 1894
Dice	(UNKNOWN)	1894
Wheel	CIGAR WHEEL	Circa 1894
Wheel	CIGAR ONE WHEEL	Circa 1894
Coin Drop	THE COMBINATION	Circa 1894
Card Machine	I.X.L. JR.	Circa 1895
Dice	(Triangular Base)	Circa 1896
Dice	(Hexagonal Base)	Circa 1896
Wheel	THE UMBRELLA	Circa 1897
Coin Drop	THE KLONDIKE GUM MACHINE	Circa 1900
Pointer	FULL WEIGHT	Circa 1902
Target	SKILLIARD	Nov. 1903
Novelty/Skill	GAME-O-SCIENCE	Circa 1903
Target	LITTLE HELPER	1908
Card Machine	ROYAL CARD MACHINE	Circa 1920
Dial	(FORTUNE COIN DROP)	1922
Coin Drop	OUR LEADER	1922
Coin Drop	DIVIDEND GUM MACHINE	1922
Coin Drop	BASE BALL	1926
Novelty/Skill	BASE BALL	1927
Coin Drop	THE BOUNCER	1928
Novelty/Skill	THE CHAMPION SPEED TESTER	1930
Coin Drop	FOOTBALL	1930
Novelty/Skill	GREAT AMERICAN GAME	1931
(Unknown)	SKILL-O-METER	1931
Dice	BOARD OF TRADE	1932
Coin Drop	THREE PENNY	1934
Novelty/Skill	PLACE THE COPPERS	1936
Specialty Reel	THE MIDGET SMOKER (Numbers)	1936
Dice	ROL-LUCK	1936
Novelty/Skill	THE MIDGET BARTENDER	1938
Wheel	PLACE YOUR BETS	1938
Wheel	HERSHEY'S	1939
(Unknown)	JESTERS	1939
Dice	LUCKY SEVEN	1939
Coin Drop	RINGER	1940
Coin Drop	(MUSSOLINI)	1943
Coin Drop	LUCKY COINS	1947
Coin Drop	FREE BEER	1947
Pointer	MAKE TWENTY ONE	1948
Coin Drop	BOUNCERINO	1952

Australian

Manufacturer and Location
Format	Name	Date

A. O. Buchanan, Sydney
Card Machine	THE AUSTRALIA	1937

Majestic Scale Company
Novelty/Skill	CIRCLE SKILL	1930

George McMullen, Perth, Western Australia
Specialty Reel	McMULLEN'S NUMERATOR	July 1901

Modern Amusements Company
Novelty/Skill	PENNY SKILL	1945

Charles Shelley Pty., Ltd., Sydney
Single Reel	SHELSPESCHEL	1948

Unidentified Manufacturers
Novelty/Skill	CARLOS SKILL	1950

Canadian

Manufacturer and Location
Format	Name	Date

Canada (Novelty Company)
Card Machine	LITTLE PERFECT	1916

Enterprise Novelty Company, Calgary, Alberta
(Unknown)	(UNKNOWN)	1917

Fowler and Wheeler, Calgary, Alberta
(Unknown)	(UNKNOWN)	1910

Northwestern Novelty Company Limited, Calgary, Alberta
(Unknown)	(UNKNOWN)	1910

Charles P. Potter, Toronto, Ontario
Pointer	(UNKNOWN)	1936

Totem Manufacturing Company, Calgary, Alberta
Coin Drop	TOTEM	1914

West Coast Novelty Company, Vancouver, British Columbia
Coin Drop	HONEST JOHN	1929

British

Manufacturer and Location
Format	Name	Date

Amusement Equipment Company, Wembley, England
Novelty/Skill	LITTLE MICKEY	1938
Novelty/Skill	CITY WONDER	1938
Novelty/Skill	LUCKY CIRCLE	1938
Novelty/Skill	CRESTA RUN	1938
Novelty/Skill	LUCKY STAR	1938
Novelty/Skill	FLYER	1938
Novelty/Skill	PLAYBALL	1938

Archibald
Race Game	HORSES	Mar. 1899

Automatic Machines (Haydon and Urry's Patents) Limited, Islington, England
Novelty/Skill	FAIRPLAY SWITCHBACK	Sept. 1900
Novelty/Skill	THE COLONIAL SHOOTING RANGE	1901

Automatic Skill Machines Company, London, England
Coin Drop	PAVILLION	1901
Coin Drop	FORTUNA	1901

H. Aylward
Novelty/Skill	(UNKNOWN)	June 1923

L. Bradley
Coin Drop	CHALLENGER	1929

William H. Britain, London, England
Race Game	FOUR-HORSE RACE	1886
Race Game	THE MECHANICAL WALKING RACE	1888
Race Game	PENNY-FARTHING BICYCLE RACE	1888

Bryans Automatic Works, Derby, England
Coin Drop	THE TRICKLER	1933

Bucknell
Novelty/Skill	(UNKNOWN)	Jan. 1899

Cocozza and Jannece
Novelty/Skill	(UNKNOWN)	Feb. 1915

Coin Operating Company, Birmingham
Novelty/Skill	CLOWN	1922

Cook and Bauer
Race Game	(UNKNOWN)	Jan. 1898

Cresset Automatic Machine Company, Kent, England
Target	(UNKNOWN)	June 1914

G. W. De Melven
Novelty/Skill	THE POSTMAN	Sept. 1900

Demelius

Coin Drop	(UNKNOWN)	Jan. 1906	

Eisner
Novelty/Skill	(UNKNOWN)	Apr. 1910	

F. E. Fensom
Novelty/Skill	(UNKNOWN)	Oct. 1897	

Foster and Foster
Novelty/Skill	TIVOLI	Oct. 1912	

Gallery
Target	(UNKNOWN)	July 1898	

Gamages of Holborn, England
Coin Drop	TIVOLI CIGAR MACHINE	1905	

Haigh and Pickles
Race Game	SCULL RACE	July 1899	
Race Game	HORSE RACE	Nov. 1899	

Hardyman
Wheel	(UNKNOWN)	Mar. 1914	

Walter Hart, Kent, England
Pointer	(UNKNOWN)	Mar. 1899	

Harvey
Coin Drop	(UNKNOWN)	Apr. 1914	

Hawtins
Novelty/Skill	ALL-WIN	1934	

Haydon and Urry Limited, Islington, England
Coin Drop	TIVOLI	1892	
Target	SHOOTING FOR GOAL	May 1895	
Novelty/Skill	(UNKNOWN)	Oct. 1899	
Target	GAME OF BARRELS	Oct. 1900	
Novelty/Skill	COLONIAL SHOOTING RANGE	1900	
Target	AUTOMATIC SHOOTING RANGE	1900	
Novelty/Skill	FAIRPLAY SKILL TELLER	1900	

Humphris and Forster
Novelty/Skill	THE ANVIL	Nov. 1899	

Interchangeable Syndicate Ltd., Islington, England
Dice	(UNKNOWN)	1896	

Jackson
Pointer	(UNKNOWN)	Dec. 1919	

Jaconelli
Novelty/Skill	(UNKNOWN)	Dec. 1914	

John Jaques and Son Limited, London, England
Race Game	ELECTROLETTE	1888	
Race Game	THE NEW RACING GAME	1890	

J. D. Equipment Sales Ltd., Wiltshire
Novelty/Skill	AUTO-SKILL	Feb. 1972	

E. J. Jofeh, England
Novelty/Skill	(UNKNOWN)	Sept. 1915	
Novelty/Skill	(UNKNOWN)	Sept. 1916	

Kaiser and Cushion
Novelty/Skill	THE CLOWN	June 1914	

London Automatic Machine Company, Ltd., London, England
Novelty/Skill	SWITCHBACK	1924	

Mancini
Novelty/Skill	TIVOLI	Jan. 1909	

Man
Race Game	(UNKNOWN)	Mar. 1900	

W. Margot
Novelty/Skill	(UNKNOWN)	June 1898	
Novelty/Skill	(UNKNOWN)	Apr. 1900	

Maxfield and Company, London, England
Race Game	PARLOR RACE GAME	Mar. 1882	

Maynard
Novelty/Skill	(UNKNOWN)	June 1919	

Charles Middlebrook, London, England
Race Game	(UNKNOWN)	Mar. 1900	

Mocogni
Novelty/Skill	CLOWN	Apr. 1915	

Muller
Novelty/Skill	(UNKNOWN)	Oct. 1899	

Newton, Smith and Rhodes
Target	(UNKNOWN)	Apr. 1911	

William S. Oliver, London, England
Race Game	HORSE RACE	Feb. 1887	

Palmer and Hartley, Birmingham, England
Race Game	CYCLE RACE	Sept. 1899	
Target	(UNKNOWN)	Apr. 1901	

Persichini
Novelty/Skill	FLAGS	Dec. 1915	

Pessers and Moody, London, England
Target	(UNKNOWN)	Mar. 1898	
Novelty/Skill	(UNKNOWN)	Sept. 1899	

Pessers, Moody, Wraith and Gurr, London, England
Novelty/Skill	(UNKNOWN)	Mar. 1916	

George Pinder
Novelty/Skill	BALL RACE	Apr. 1927	

Price and Castell, London, England
Novelty/Skill	GAME OF SKILL	Oct. 1900	

Reppmann
Dial	(UNKNOWN)	Mar. 1908	

Rhodes
Novelty/Skill	(UNKNOWN)	Apr. 1911	

R. C. Richards
Novelty/Skill	(UNKNOWN)	June 1913	

Robbins
Novelty/Skill	(UNKNOWN)	Jan. 1913	

Robertson
Target	(UNKNOWN)	Dec. 1898	

Sandoz
Race Game	HORSE RACE	1901	
Race Game	CYCLE RACE	1901	
Race Game	LOCOMOTIVE RACE	1901	

Scottish Automatic Machine Company, Glasgow, Scotland
Novelty/Skill	CONQUERED FLAGS	1918	

Gordon H. Smith
Roulette	(UNKNOWN)	Nov. 1900	

Solomon
Novelty/Skill	BALANCING ACT	Sept. 1928	

Southern Amusements and General Trading Company
Novelty/Skill	(UNKNOWN)	June 1923	

Stapleton
Novelty/Skill	(UNKNOWN)	Apr. 1910	

Verrecchia
Novelty/Skill	(UNKNOWN)	Dec. 1914	
Novelty/Skill	(UNKNOWN)	Mar. 1916	
Novelty/Skill	(UNKNOWN)	Nov. 1916	

Walter
Wheel	(UNKNOWN)	Jan. 1913	

Wegg
Novelty/Skill	(UNKNOWN)	Nov. 1912	

R. E. Wickes
Novelty/Skill	(UNKNOWN)	Nov. 1899	

Unidentified Manufacturers
Novelty/Skill	HALF PENNY	1900	
Pointer	DUNSTAN'S CIGAR SELLER	1905	
Novelty/Skill	SPORTS	1910	
Novelty/Skill	SKILLO	1930	
Novelty/Skill	HI-BALL	1948	
Novelty/Skill	(NEW) HI-BALL	1960	

French

Manufacturer and Location		
Format	Name	Date

Barme
Dial	MEPHISTO (THE DEVIL)	1905

Beraud, Paris
Novelty/Skill	LE DIABOLIQUE	1900
Novelty/Skill	CINQ GAGNANT	1901
Novelty/Skill	LE MAGIC (MAGIC)	1910

Bidard
Dial	LA GRENOUILLERE (FROG'S POND)	1900
Dial	LE PERE BIDARD (OLD MAN BIDARD)	1910

Tolerie de Boulogne, Boulogne
(Unknown)	TOL BOUL (COLOR BALLS)	1930
(Unknown)	LE CAMELEON (THE CHAMELEON)	1935

Boux and Company
Novelty/Skill	LE FRANCAIS	1914

Brevete S.G.D.G.
Dice	SUPER-POKER	1910

La Compagnie Caille (Cie Caille), Paris
Single Reel	L'AEROPLAN	1905
Single Reel	LE TIGRE (3-Way)	1905
Single Reel	LE TIGRE (5-Way)	1905
Single Reel	LA COMETE (3-Way)	1906
Single Reel	LA COMETE (5-Way)	1906
Specialty Reel	MATADOR (Dice)	1910

Cie Jost, Paris
Race Game	DE COURSE (non-coin)	1889
Race Game	RACE TRACK (4 Horse) (non-coin)	1902
Race Game	RACE TRACK (6 Horse) (non-coin)	1902

Race Game	RACE TRACK (8 Horse) (non-coin)	1902
Jost and Cie, Paris		
Race Game	GRANDS JEUX DE PETITS CHAVAUX (Non-coin)	1930
Race Game	PETIT JEU A TABLEAU NO. 2 (Non-coin)	1930
Race Game	LE KLONDIKE (Non-coin)	1930
Race Game	PETITS CHAVAUX DE SALON (Non-coin)	1930
Race Game	PETITS CHAVAUX POUR FAMILLE (Non-coin)	1930
Race Game	DERBY (Non-coin)	1930
Pierre-Abel Nau, Paris		
Single Reel	LES PETITES CHEVAUX (THE PONY)	1905
Single Reel	L'ELAN FRANCAIS (ELK)	1905
Single Reel	LES 3 COULEURS (TRI-COLOR)	1910
Single Reel	LES TROIS COULEURS (TRI-COLOR)	1910
Single Reel	LES TROIS COULEURS (Marquee Coin Head)	1910
Single Reel	L'ELAN (LES 3 COULEURS)	1910
Single Reel	SPECIAL (NOUVEAU LES PETITS CHEVAUX)	1911
Single Reel	SPECIAL (NOUVEAU LES TROIS COULEURS)	1911
Single Reel	L'AIGLE (PILOT)	1911
S.G.D.G. Roguez, Paris		
Specialty Reel	L'ARBITRE (THE JUDGE) (Dice)	1900
Specialty Reel	LES DOMINOS (Dice)	1915
Unidentified Manufacturers		
Novelty/Skill	L'UNIE	1895
Dial	LE SOURIRE (THE SMILE)	1900
Dice	SUPER POKER	1910
Dice	APPAREIL ELECTRIQUE MUSICAL	1912
Dial	QUEEN TOP (Style 1)	1937
Dial	QUEEN TOP (Style 2)	1937
Dial	LE' 21	1937

German

Manufacturer and Location

Format	Name	Date
Auffangschalen		
Novelty/Skill	RING-SCHLEUDER-SPIEL	1928
Max Conde, Berlin		
Race Game	(UNKNOWN)	1920
Otto Eisner		
Target	(UNKNOWN)	Apr. 1910
Gartmann, Hamburg		
Dial	ALLOTRIA	1934
Hoehne, Heidelberg		
Novelty/Skill	GLOBETROTTER	1953
Jentzsch and Meerz, Leipzig		
Novelty/Skill	I CALCIATORI (KICKER AND CATCHER)	1900
Novelty/Skill	BAJAZZO (CLOWN)	1910
Novelty/Skill	BAJAZZO (NEW CLOWN)	1913
Novelty/Skill	(UNKNOWN)	1922
Novelty/Skill	AIRSHIP PROFIT SHARER	1929
Novelty/Skill	1930 BAJAZZO (CLOWN)	1930
Lowen Spielautomaten, Bingen		
Novelty/Skill	ZEPPELIN	1910
Race Game	HANDICAP	1922
Metallindustrie Schönebeck A. G.		
Coin Drop	DUPLEX	1902
Reppman and Schade		
Novelty/Skill	(UNKNOWN)	Mar. 1908
Plagwitzer Musikwerke, Leipzig		
Dice	FORTUNA	1895
Roulette	JAPANISCHES KUGELSPIEL	1895
Friedrich Ernst Thomas, Steinigwoolmsdorf		
Dice	(UNKNOWN)	Jan. 1897
West Deutsche Automaten-Grosshandlung		
Novelty/Skill	BAJAZZO (THE CLOWN)	1950
Unidentified Manufacturers		
Novelty/Skill	BRAVO	1898
Dice	DICE AUTOMATON	1898
Coin Drop	FORTUNA	1904
Dice	MONACO	1905
Coin Drop	ONKEL THEODOR	1910
Dice	MONACO	1920
Race Game	HANDICAP	1930

Irish Trade Stimulators

Manufacturer and Location

Format	Name	Date
John Dundon, Cork		
Race Game	COIN-CONTROLLED RACE	Feb. 1899
John Stewart Wallace, Belfast		
Target	CANNON SHOOT	Dec. 1890

Venezuelan Trade Stimulators

Manufacturer and Location

Format	Name	Date
Jacob M. Henriquez, Coro		
Coin Drop	(UNKNOWN)	Jan. 1892

COLLECTIBLE TRADE STIMULATORS (1870-1919) AND COUNTER GAMES (1920-1996)

(by name alphabetically)

Many trade stimulators and counter games do not reveal their maker or date. In the early years games were quite often produced with this purposeful level of ambiguity in order to mask the name of the maker in order to allow the sales agents to take credit for the manufacture and distribution. This multiplication of origin makes it exceedingly difficult to determine the prime manufacturer in many cases, with identical machines of the same name often being made and sold by a number of suppliers, while other suppliers changed the names of the same machines for their catalog, mail and advertising promotions. The collectible trade stimulators and counter game lists reveal a number of such examples. To add to this area of confusion, many makers copied the names of other games. At least a dozen different machines carried the name PERFECTION, yet were radically different in design and operation. The SUCCESS and JUMBO SUCCESS cast iron card machines exhibit much the same level of differentiation as well a high level of copycat sameness.

In the counter game era the practice was maintained by a number of manufacturers to make it possible for the machine distributors to apply their decals or nameplates. Having the name of a machine and an identifying decal is not always an accurate representation of the game and maker. The collectible trade stimulators and counter game lists have been carefully created and culled to reduce this error of identification as much as possible, although multiple examples of the some machines are still listed by the name of the selling agent in addition to the manufacturer.

Then there is the other identification gap, that of knowing a machine name and not the manufacturer. Either the maker identification has worn off, or in many cases was never applied in the first place. Such a machine can only be identified by game name, leaving the facts of its origins and time frame to conjecture.

In order to combat this lack, and aid in machine identification, the following name interchange listing has been created for the machines made in the United States and Canada over the years. It is not foolproof, as numerous makers gave their machines names already in use by others. The enormous variety of BABY BELL and PURITAN machines, all with different cabinet castings, points up the difficulty of pinning down absolute answers from a list. Some study of the machines and their cabinets will also be needed to draw a finite conclusion, if indeed it can be done in some cases.

All of the known North American producers are included to help pin down the game and maker name. Once that has been indicated, you can look up the makers in the corresponding collectible trade stimulators and counter game lists to identify the maker and location. This will often reveal the date of introduction, with some of these machines having both a long production and service life. The trade machines made in the U.K., Europe and elsewhere have not been included as too few are known to have survived the years. North America was the primary producing area in any event, with the list covering most of the machines made.

If you know the name of a machine, check it out in the alphabetical listing, and then check the makers in the larger collectible trade stimulators and counter game lists. If luck is with you, you will be able to identify your machine by maker, location, name and date, the basic building blocks of machine identification and evaluation.

A

A.B.C.	A. B. T.
ACE	ABCO Novelty
ACE	Comet Industries
ACE	Daval
ACE	Scot Industries
ACE, THE	Edmund Fey
ACE, THE	Midwest Novelty
ACME, THE	Acme Sales
ACME	Leo Canda
ADD 'EM	Quality Supply
ADDER, THE	Williams Manufacturing
AD-LEE VENDER	Ad-Lee Company
ADD YOUR SCORE	Consolidated Automatic
ADVERTISER, THE	Jarbo
ADVERTISING AND DISCOUNT MACHINE	Yale Wonder Clock Company
ADVERTISING REGISTER	Drobisch Brothers
AEROPLANE	Mills Novelty
AFRICAN DOODOO	Sam May
AFRICAN GOLF	H. C. Evans
AFRICAN GOLF	Frey
AFRICAN GOLF	Hollingsworth
ALL ABOARD	Calvert
ALL ABOARD	D. Robbins & Company
ALL ELECTRIC DICE MACHINE	Winner Sales
ALL-IN-ONE	All-In-One Company
ALL-WAYS CONFECTION VENDER	Chicago Mint
ALWIN	Buckley
AMERICAN EAGLE	ABCO Novelty
AMERICAN EAGLE	Comet Industries
AMERICAN EAGLE	Daval
AMERICAN EAGLE	Scot Industries
AMERICAN FLAGS	Daval
AMERICAN SPORT	Elm City Novelty
AMERICAN SPORT	Care Leonardo Novelty
AMERICA'S SMOKE	Majestic
AMUSEMENT MACHINE	R. Rothschild's & Sons
ANOTHER TAP-A-PENNY	Tap-It Manufacturing
ARCADE DICE FORTUNE TELLER	Exhibit Supply Company
ARISTOCRAT	Douglis Machine
ANNIE-OAKLEY	A. J. Stephens
ARISTOCRAT DELUXE VENDER	J. M. Sanders
ARISTOCRAT DOMINO	Mills Sales
ARISTOCRAT GOLF	Mills Sales
ARISTOCRAT POKER	Mills Sales
ARLINGTON	Midwest Novelty
ARMY 21 GAME	Garden City Novelty
ARMY 21 GAME	Pace
ARMY 21 GAME	Pierce Tool
ARROW	Amusement Machine Company
ARROW	Jennings
ARROW	Mills Novelty
ARROW (CIGAR SALESMAN)	Mills Novelty
ARROW	Victor Novelty
ATOM	Groetchen Tool
AUTO-DICE AMUSEMENT TABLE	Auto-Dice
AUTO-DICE AMUSEMENT TABLE	Superior Confection
AUTO DOMINOES	Ohio Automatic
AUTOMAT	Bally
AUTO-PUNCH	Daval
AUTO-TARGET	Pace
AUTOMATIC ADVERTISER	Coyle & Rogers
AUTOMATIC BASEBALL	D. Robbins & Company
AUTOMATIC BRIDGE TABLE	Superior Confection
AUTOMATIC CARD MACHINE	Leo Canda Manufacturing
AUTOMATIC CARD MACHINE	Columbian Automatic Card Machine
AUTOMATIC CARD MACHINE	Maley
AUTOMATIC CARD MACHINE	F. W. Mills
AUTOMATIC CARD TABLE	Superior Confection
AUTOMATIC CASHIER AND DISCOUNT MACHINE	Almy Manufacturing
AUTOMATIC CASHIER AND DISCOUNT MACHINE	Yale Wonder Clock Company
AUTOMATIC CHAMP, THE	A. C. Bindner
AUTOMATIC CHAMPION	Specialty Manufacturing
AUTOMATIC CIGAR SELLER	Stirrup
AUTOMATIC DICE	American Automatic Machine
AUTOMATIC DICE	Automatic Machine
AUTOMATIC DICE	Automatic Novelty Machine
AUTOMATIC DICE (SHAKER)	Clawson Slot Machine
AUTOMATIC DICE	Coyle & Rogers
AUTOMATIC DICE BOX	Fey
AUTOMATIC DICE MACHINE	Samuel Nafew
AUTOMATIC DICE SHAKER	Bucyrus
AUTOMATIC DICE SHAKER	Clawson Slot Machine
AUTOMATIC DICE SHAKING MACHINE	American Automatic Machine
AUTOMATIC DICE SHAKING MACHINE	Automatic Manufacturing
AUTOMATIC DICE SHAKING MACHINE	Hudson Moore
AUTOMATIC DICE SHAKING SLOT MACHINE	Maley
AUTOMATIC DICE VENDING MACHINE	Coyle & Rogers
AUTOMATIC FORTUNE TELLER	Clawson Slot Machine
AUTOMATIC GUM TARGET	Blue Bird Sales
AUTOMATIC POKER MACHINE	Columbian Automatic Machine
AUTOMATIC POKER PLAYER	Automatic Machine Company
AUTOMATIC POKER PLAYER	Hudson Moore
AUTOMATIC PUNCH BOARD	Mills Novelty
AUTOMATIC RACE COURSE	Coyle & Rogers
AUTOMATIC RACE COURSE	Flour City
AUTOMATIC RACE COURSE	National Automatic Device
AUTOMATIC RACE TRACK	Chicago Nickel
AUTOMATIC RACE TRACK	Kennedy & Diss
AUTOMATIC RACE TRACK	R. Rothschild's & Sons
AUTOMATIC RACE TRACK	Universal
AUTOMATIC REGISTERING BANK	Cooley
AUTOMATIC ROULETTE	American Automatic Machine
AUTOMATIC ROULETTE	Hudson Moore
AUTOMATIC ROULETTE	C. P. Young
AUTOMATIC SALESBOARD	B. A. Withey
AUTOMATIC SALESMAN	Clawson Slot Machine
AUTOMATIC SALESMAN AND PHRENOLOGIST	Lichty
AUTOMATIC SHOW CASE	Baltimore Vending Machine
AUTOMATIC TRADE CLOCK	Jones Machine Shop
AUTOMATIC TRADER	Automatic Trading Company
AUTOMATIC VOTE RECORDER AND CIGAR SELLER	Brunhoff
AUTOMATIC WIZARD CLOCK	Jones Machine Shop

B

BABE RUTH BASEBALL	Stephen's Novelty
BABY BELL	Arlington Heights Machine
BABY BELL	Blue Bird Sales
BABY BELL	Field
BABY BELL	Royal Novelty (Indianapolis)
BABY CARD MACHINE	Amusement Machine Company
BABY JACK	Garden City Novelty
BABY JACK-POT VENDER	Field Paper Products
BABY JACK POT VENDER	J. M. Sanders
BABY RESERVE	Bally
BABY SHOES	Buckley
BABY SKILL VENDER	Field Paper Products
BABY TRACK	Western Equipment & Supply
BABY VENDER	Blue Bird Sales
BABY VENDER	B. and M. Products
BABY VENDER	Burnham Gum
BABY VENDER	Burnham & Mills
BABY VENDER	Field Paper Products
BABY VENDER	Keeney and Sons
BABY VENDER	F. W. Mills
BABY VENDER	J. M. Sanders
BALL GUM VENDER	Silver King Novelty
BALL-QUET	M. T. Daniels
BALL/WALK	Cointronics
BALLY	Bally
BALLY BABY	Bally
BALLY BOOSTER	Bally
BALRICKY	
BANKER	Caille Bros.
BANKER WHO PAYS	McLoughlin
BANNER PERFECTION	Banner Specialty
BANNER PURITAN	Banner Specialty
BANNER TARGET PRACTICE	Banner Specialty
BAR BOY	Garden City Novelty
BAR BOY	Pierce Tool
BARN YARD GOLF	Peo Manufacturing
BARON	D. Robbins
BARTENDER, THE	Ohio Specialty
BASE BALL	Atlas Indicator
BASE BALL	Exhibit Supply Company
BASE BALL	Industry Novelty
BASE BALL	Mills Novelty
BASE BALL	National Novelty
BASE-BALL	Pace

Title	Manufacturer
BASE BALL	Paupa And Hochriem
BASE-BALL	Watling
BASE BALL	(Unknown)
BASE BALL GAME OF SKILL	Whiteside Specialty
BASE BALL TARGET	Blue Bird Sales
BASEBALL	Amusement Machine Company
BASEBALL	Bonus Sales
BASEBALL	J. W. Calvert
BASEBALL	J. F. Frantz
BASEBALL	Shirley Novelty
BASEBALL	Silver King Novelty
BASEBALL ("The Tiger")	Caille Bros.
BASEBALL (1918)	Industry Novelty
BASEBALL	(Unknown)
BASEBALL BABY VENDER	Field
BASEBALL BALL GUM VENDER	Superior Confection
BASEBALL GUM VENDOR	D. Robbins & Company
BASEBALL GUM VENDOR	Savoy Vending
BASEBALL MUTUEL	Four Jacks
BASEBALL ROLL-ET	Gottlieb
BASEBALL TARGET	Twin City Novelty
BASKEE BALL	B. Madorsky
BASKET BALL	Genco
BASKETBALL	Champion Manufacturing
BASKETBALL STARS	Associated Amusements
BAT-A-BALL	C. D. Fairchild
BAT-A-BALL	Pace
BAT-A-BALL JUNIOR	Munves Manufacturing
BAT-A-PENY	Advance Machine
BAT-A-PENY	Bat-A-Peny Corporation
BATAPENY	Steinmetz Manufacturing
BATTER-UP	Exhibit Supply
BAZAAR	New Era
BEAT IT	Exhibit Supply Company
BEAT IT	State Coin
BEAT THE DEALER	Frey
BEAT THE HORSE	Quality Supply
BEER DANDY VENDER	Groetchen Tool
BEER MARVEL	Daval
BEER SPINNER	United Specialty
BEER TARGET	Great States
BELL BOY	Western Automatic Machine
BELL BOY, THE	Mills Novelty
BELL FRUIT	Osbrink
BELL PURITAN	Pace
BELL SKILL	A.B.C. Coin Machine
BELL SKILL	K. & S. Sales
BELL-SKILL DELUXE	Northwest Coin
BELL SLIDE	Daval
BELMONT-JR.	Merchants Advertising
BEN FRANKLIN	Mills Novelty
BEST HAND	Daval
BICYCLE	Kelley
BICYCLE	Sun Manufacturing
BICYCLE, THE	Poole Brothers
BICYCLE, THE	John M. Waddell
BICYCLE, THE	Waddell Wooden Ware Works
BICYCLE DISCOUNT WHEEL, THE	Poole Brothers
BICYCLE DISCOUNT WHEEL, THE	John M. Waddell
BICYCLE DISCOUNT WHEEL, THE	Waddell Wooden Ware Works
BICYCLE RIDER	David W. Dunn
BICYCLE WHEEL, THE	John M. Waddell
BICYCLE WHEEL, THE	Waddell Wooden Ware Works
BIG BONES	Pioneer Games
BIG FIFTY, THE	R. C. Walters
BIG GAME HUNT	National Amusement
BIG RACE COURSE	J. & E. Stevens
BIG STAR SIX	Caille Bros.
BIG THREE, THE	M. O. Griswold
BIG TOP	J. F. Frantz
BIG SIX	Harlich
BIG SIX	National Coin Machine Exchange
BIG SIX	Ogden & Company
BING	Frey
BINGO	Bingo Sales
BINGO	Buckley
BINGO	D. Robbins
BINGO	R. C. Walters
BLACK BEAUTY	Field
BLACK CAT, THE	M. O. Griswold
BLACK JACK	Mills Novelty
BLUE BONNET	Groetchen Tool
BLUE STREAK	Hutchison Engineering
BOARD OF CRAPS AND GUM MACHINE	McClellan
BOARD OF TRADE	(Unknown)
BOMB HIT	Baker Novelty
BOMB HITLER	Coin Machine Company of America
BOMBER	Erie Machine
BON TON	Caille Bros.
BONANZA	Leo Canda
BONES	Gamemasters
BONUS WHEEL	Monroe Barnes
BONUS WHEEL	Drobisch Brothers
BOOMER, THE	John M. Waddell
BOOP-A-DOOP	Pace
BOOSTER	G. F. Hochriem
BOOSTER, THE	Exhibit Supply Company
BOOSTER, THE	F. W. Mills
BOSCO	Bally
BOTTOMS UP	Buckley
BOTTOMS UP	Globe Novelty
BOTTOMS UP	Pierce Tool
BOUNCER	(Unknown)
BOUNCER, THE	Exhibit Supply
BOUNCERINO	(Unknown)
BOUNCING BALL	Royal Manufacturing
BOUNCING BALL	Royal Scale
BOUNCING BONES	K. And S. Sales
BOUNCING BONES	Pierce Tool
BOUNCING BUSY BALL	Royal Novelty (Baltimore)
BOXING SKILL	National Amusement
BROADWAY	Pacific Manufacturing
BRODIE (STANDARD)	Howard Sales
BRODIE DELUXE	Howard Sales
BRODIE LIBERTY	Howard Sales
BROWNIE	T. F. Holtz & Company
BROWNIE	Monarch Card Machine
BROWNIE	Sittman & Pitt
BUCK-A-DAY	Great States
BUCK-A-DAY	Star Sales
BUDDY	ABCO Novelty
BUDDY	Comet Industries
BUDDY	Daval
BUDDY BALL GUM VENDER	Buddy Sales
BUDDY GUM VENDER	Budin's Specialties
BUGG HOUSE	Pierce Tool
BUFFALO, JR.	Watling
BULL DOG	Monarch Manufacturing
BULLS EYE	J. F. Frantz
BULL'S EYE	Mills Novelty
BULLS EYE	Rock-Ola
BULLS-EYE BALL GUM VENDER	Exhibit Supply Company
BUSY BEE	Caille Bros.
BUSY BEE, LITTLE	Caille-Schiemer
BUY AMERICAN	Hawkeye Novelty
BY-A-BLADE	H. C. Evans

C

Title	Manufacturer
CALIFORNIA	Wheeland Novelty
CALIFORNIA BEAR	Caille Bros.
CALIFORNIA JACK	Mills Novelty
CALVERT INDIAN SHOOTER	Ohio Automatic
CANDA CARD MACHINE	Leo Canda
CANDY	Jennings
CANDY RACE TRACK	H. C. Evans
CANDY SPINNER	United Specialty
CANDY TARGET	Great States
CAN'T LOSE	Jackson
CARD EXHIBITING MACHINE	Ideal Toy
CARD GRIP	George H. Reid
CARD MACHINE	Amusement Machine Company
CARD MACHINE	Clawson Machine
CARD MACHINE	M. O. Griswold
CARD MACHINE	L. G. Grund
CARD MACHINE	T. F. Holtz & Company
CARD MACHINE	Jennings
CARD MACHINE	Kennedy & Diss
CARD MACHINE	LaBuff
CARD MACHINE	George E. Maple

CARD MACHINE	Nafew-Goldberg
CARD MACHINE	Ogden & Company
CARD MACHINE	Rosenfield
CARD MACHINE	W. G. Souder
CARD MACHINE	Sweeney
CARD MACHINE	Tibbils
CARD MACHINE, NO. 9	Watling
CARD MACHINE, 3-SLOT ("Jockey")	Rosenfield
CARDINAL	Pace
CARDINAL BEER	Pace
CARDINAL BELL	Pace
CARDINAL MYSTERY BELL	Pace
CARDINAL SMOKE	Pace
CARDS AND DICE	C. R. Schultz
CASHIER	Maley
CASINO	New Era
CATCH-N-MATCH	Berger Manufacturing
CATCH THE BALL	Mills Novelty
CENT-A-PACK	Buckley
CENTA-SMOKE	Daval
CENTURY BELL	A.B.C. Coin Machine
CENTURY GRAND	Sicking
CHAMPION	Buckley
CHAMPION SPEED TESTER, THE	(Unknown)
CHANGE MASTER	Mueller Specialties
CHANGEMASTER	Monarch Coin Machine
CHECK CARD	Mills Novelty
CHECK FIGARO	Mills Novelty
CHECK POLICY	Mills Novelty
CHECK TRADER	Watling
CHECK-PAY PURITAN	Puritan Machine
CHECKER MATCH	Checkerboard
CHERRY JITTERS	Garden City Novelty
CHERRY JITTERS	Pierce Tool
CHERRY ROLL	G. & M. Machinery
CHICAGO (PERFECTION)	C. Passow
CHICAGO CLUB HOUSE	Daval
CHICAGO DERBY	H. C. Evans
CHICAGO EXPRESS	Daval
CHICLE-CHOCLE	Chicle Chocle Products
CHIEF	Field Paper Products
CHIEF, THE	Indian Head Novelty
CHIP GOLF GAME	R. & S. Company
CHRYSLER, THE	Midwest Novelty
CHUCK-A-ROLL	Frey
CHUCK-LUCK	Midwest Novelty
CHUCK-O-LUCK	Southern Novelty
CHUCK-O-LUCK	Keystone Novelty
CHUCK-O-LUCK DICE	Gottlieb
CHUCK-O-LUCK DICE	Keeney And Sons
CHUCK-O-LUCK FOOTBALL	Gottlieb
CHUCK-O-LUCK FOOTBALL	Keeney And Sons
CHUCK-O-LUCK HORSE RACING	Gottlieb
CHUCK-O-LUCK HORSE RACING	Keeney And Sons
CHUCK-O-LUCK POKER	Gottlieb
CHUCK-O-LUCK POKER	Keeney And Sons
CHUCK-O-LUCK 21 BLACK JACK	Gottlieb
CHUCK-O-LUCK 21 BLACK JACK	Keeney And Sons
CHURCH-HILL DOWNS	Northwest Coin
CHURCHILL DOWNS	Field Manufacturing
CHURCHILL DOWNS	Ideal Amusement
CHURCHILL DOWNS	Northwest Coin
CHURCHILL DOWNS	Northwestern Coin
CIGA ROLA	A. B. T.
CIGAR AND BLOWING MACHINE	Western Company
CIGAR WHEEL	Percival
CIGAR CUTER	Keane Novelty
CIGAR CUTTER	Brunhoff
CIGAR DICE	Smokers Supply
CIGAR ONE WHEEL	(Unknown)
CIGAR TARGET	Great States
CIGAR VENDER	Phillips Farm Supply
CIGAR WHEEL	(Unknown)
CIGARETTE	Superior Confection
CIGARETTE DICE	B. A. Withey
CIGARETTE GUM VENDER	Superior Confection
CIGARETTE SALESMAN	Superior Confection
CIGARETTE SALESMAN BALL GUM VENDER	Superior Confection
CIGARETTE SALESMAN MINT VENDER	Superior Confection
CIGARETTE TARGET	Great States
CIGARETTES	Western Equipment & Supply
CIGGY	Comet Industries
CIG-O-ROL	Pierce Tool
CIRCUS	Atlas Games
CIRCUS	Holly Manufacturing
CIVILIAN DEFENSE	Atlas Games
CLEARING HOUSE	Daval
CLEARING HOUSE	Field
CLIPPER	Caille Bros.
CLOVER	Leo Canda Manufacturing
CLOVER, THE (3 WAY)	Watling
CLOVER INOCHLE, THE	Watling
CLOVER EAF	Leo Canda
CLOVERLEAF (PINOCHLE SUCCESS)	Clawson Machine
CLOWN	Caille Bros.
CLOWN, THE	Arcade Supply
CLUB JACK	Four Jacks
CLUB VENDER	Jennings
COCO-NUTS	Amusement Coin
COIN GETTER	Bell Fruit Vending
COINOGRAPH SALESMAN	T. J. Nertney
COIN-O-LUCK	Plamor Novelty
COIN OPERATED WHEEL	Ad-Lee Company
COINSKILL	Advance Machine
COIN TARGET BANK	M. Siersdorfer
COLOR CUBE	Monterey Woodcrafters
COLOR-ROLL	Frey
COLUMBIAN	Michigan Metal & Wood
COLUMBIAN FORTUNE TELLER	World's Fair Slot Machine
COMBINATION, THE	(Unknown)
COMBINATION CARD AND DICE MACHINE	Watson
COMBINATION JACK POT	Amusement Machine Company
COMBINATION LUNG TESTER	Colby Specialty Supply
COMET	Caille Bros.
COMET	Daval
COMET	Paupa & Hochriem
LE COMETE	Caille Bros.
LE COMETE	Paupa & Hochriem
COMMERCE	Sheffler
COMMERCIAL	W. H. Clune
COMMERCIAL	Mills Novelty
COMMODORE	C. A. Wagner
COMMON SENSE DICE MACHINE	Maley
COMSTOCK	Huestis
CONGO	Bally
COPY CAT	Fun Industries
CORK TIP	Hub Manufacturing
CORK TIP	H. J. Stock
COUNTER CARD	Amusement Machine Company
COUNTER CENTURY GRAND	Sicking
COUNTER HAMILTON	Hamilton
COUNTER IRON CARD	Amusement Machine Company
COUNTER IRON CARD MACHINE (NO.8)	Sicking
COUNTER JUMBO	Caille Bros.
COUNTER JUMBO	Leo Canda
COUNTER JUMBO	Leo Canda Manufacturing
COUNTER JUMBO SUCCESS	Leo Canda
COUNTER PERFECTION	Leo Canda
COUNTER PERFECTION	Kelley
COUNTER RACE TRACK	McManus
COUNTER SUCCESS	Caille Bros.
COUNTER SUCCESS	Leo Canda
COUNTER THE JUMBO	Mills Novelty
COUNTERETTE	Mills Novelty
COURT HOUSE	Bennett and Company
CRAPS	Four Jacks
CRAP SHOOTER	Mills Novelty
CRAP SHOOTER	Paupa & Hochriem
CRAP SHOOTER'S DELIGHT	F. A. Ruff
CRAZY	Brunhoff
CRESCENT	Monroe Barnes
CRISS-CROSS	D. Robbins
CRYSTAL DESTINY	Ad-Lee Company
CRYSTAL GAZER	Ad-Lee Company
CRYSTAL GAZER	Automatic Coin Machine
CRYSTAL GAZER	Peerless Manufacturing
CUB	ABCO Novelty
CUB	Bally

CUB	Comet Industries
CUB	Daval
CUBA	Ad-Lee Company
CUB	Scot Industries
CUBE ROULETTE	H. C. Evans

D

DAILY RACES JR.	Gottlieb
DAISY ("Hump Back")	Brunhoff
DAISY	Ebersole
DAISY	Hamilton
DAISY NO BLANK	Hamilton
DANCING DOLLS	John M. Waddell
DANCING DOMINOES	Star Sales
DANDY GUM VENDER	Caille Bros.
DANDY VENDER	Auto-Vender Company
DANDY VENDER	A. S. Douglis
DANDY VENDER	Groetchen Tool
DANDY VENDER	Mills Novelty
DANDY VENDER	Pace
DANDY VENDER	Rock-Ola
DARBY	J. W. Whitlock
DART, THE	Maley
DAVAL DERBY	Daval
DAVAL GUM VENDER	Daval
DAVAL GUM VENDOR	Douglis Machine
DEALER'S CHOICE	Associated Distributors
DEALER'S CHOICE	Dealer's Choice, Inc.
DEFENSE AMERICAN EAGLE	Daval
DELUXE BABY VENDER	J. M. Sanders
DELUXE BELMONT-JR.	Merchants Advertising
DELUXE CENT-A-PACK	Buckley
DELUXE FAIR-N-SQUARE	Western Automatic Machine
DELUXE VENDER	Garden City
DERBY	The Alb Company
DERBY	Daval
DERBY	Western Electric Piano Company
DERBY RACEHORSE	Peerless Sales & Products
DERBY RACES	Roy Torr
DERBY VENDER	Keeney & Sons
DETROITER	Michigan Novelty
DEUCES WILD	Paul Bennett & Company
DEUCES WILD	Genco
DEUCES WILD	Pace
DEUCES WILD	J. M. Sanders
DEWEY	Jonas D. Bell & Company
DEWEY, THE	Mead & Taylor
DEWEY, THE	National Manufacturing
DEWEY, THE	F. A. Ruff
DEWEY SALESMAN	Nichols
DEWEY SALESMAN	Ogden & Company
DEWEY SALESMAN, THE	John Henry Davis
DIAL	Maley
DIAL	Mills Novelty
DIAL-IT	Groetchen Tool
DIALIT	Kimsey
DIAMOND	Field
DIAMOND	Marvel
DIAMOND EYE	Brunhoff
DIAMOND WHEEL	William Suydam
DICE	Genco
DICE	Royal Novelty (San Francisco)
DICE BOX	Bucyrus
DICE BOX	Moore
DICE BOX	Slinn
DICE BOX	Wedesweiler
DICE FORTUNE TELLER	Exhibit Supply Company
DICE MACHINE	Leo Canda
DICE MACHINE	Leo Canda Manufacturing
DICE MACHINE	Clawson Slot Machine
DICE MACHINE	Cowper
DICE MACHINE	M. O. Griswold
DICE MACHINE	Maley
DICE MACHINE	Pace
DICE-MAT	Dependable Enterprises
DICE-O-MATIC	Groetchen Tool
DICE ROULETTE	Peterson
DICE SHAKER	Amusement Enterprises
DICE SHAKER	Lighton
DICE SLOT MACHINE	D. Kernan
DICE TOSSER	Clawson Slot Machine
DICELESS DICE	Automatic Corporation
DICETTE	Bally
DICETTE	Jennings
DICTA-CARD	Dicta-Card, Inc.
DICTA-RACE	Dicta-Card, Inc.
DING THE DINGER	Charle
DING THE DINGER	Field
DISCOUNT WHEEL	Decatur Fairest Wheel Company
DIVIDEND GUM MACHINE	(Unknown)
DIXIE	B. & F. Sales & Manufacturing
DIXIE	Bowman Specialty
DIXIE	Dixie Music
DIXIE BELL	Dixie Manufacturing
DIXIE DICE	Bowman Specialty
DIXIE DOMINOS	Groetchen Tool
DIXIE ROULETTE	B. A. Withey
DIXIE SPELLING BEE	Groetchen Tool
DIXON SPECIAL	Progressive Manufacturing
DIZZY	A. J. Stephens
DIZZY DIALS	Electra Corporation
DIZZY DISKS	Acme Game
DOG HOUSE	Bally
DOG HOUSE FUN	Yendes Manufacturing
DOLL PITCH	Issac T. Bomar
DOLLARS FOR DIMES	R. C. Walters
DOMINO JR.	H. C. Evans
DONKEY	Cowper
DOUBLE DECK	Daval
DOUBLE DICE	Exhibit Supply Company
DOUBLE DOOR BALL GUM VENDER	Silver King Novelty
DOUBLE HEADER	J. F. Frantz
DOUBLE TARGET	Barr Novelty
DOUBLE UP	Osbrink Games
DOUGH BOY	Paul Bennett & Company
DRAW FIVE	Lyon Novelty
DRAW POKER	Buckley
DRAW POKER	Caille Bros.
DRAW POKER	Cowper
DRAW POKER	Fey
DRAW POKER	J. L. Foley
DRAW POKER	Leonhardt
DRAW POKER	McClatchie Manufacturing
DRAW POKER	Mills Novelty
DRAW POKER	Monarch Card Machine
DRAW POKER	National Coin Machine Exchange
DRAW POKER	Sittman & Pitt
DRAW POKER	Silver King Novelty
DRAW POKER	Watling
DRAW 21	Western Equipment & Supply
DREAM REVEALER	Taylor
DRINKMASTER	Dutch
DRINKS	McLoughlin
DROP IT	Great States
DUCK	Paupa & Hochriem
DUCK SOUP	Five Boro Machine
DUCK SOUP	Star Sales
DUKE, THE	Fey
DUVAL JACK POT (DAVAL JACKPOT)	Silver King Novelty
DWARF GOLF	Star Vending

E

EAGLE	Eagle Amusement
EAGLE	Eagle Manufacturing
EAGLE	Mills Novelty
EAGLE	Paupa & Hochriem
EAGLE, THE	Leo Canda
EAGLE, THE	Sundwall Company
EAGLE EYE	Bally
ECLIPSE	Anthony (Cigar) Company
ECLIPSE	Anthony & Smith
ECLIPSE	Bucyrus
ECLIPSE	Grand Rapids Slot Machine
ECLIPSE	Maley
ECLIPSE	Western Automatic
ECLIPSE (TWO ARROW MACHINE)	Western Weighing Machine
E'LAN (ELK)	Mills Novelty
ELECTRIC AMERICAN EAGLE	Daval
ELECTRIC BALL/WALK	Cointronics
ELECTRIC BASEBALL	Keeney AND SONS

ELECTRIC DERBY	Sam May
ELECTRIC DICE	Bucyrus
ELECTRIC DICE	United Automatic
ELECTRIC SPINNER	Pacific Amusement
ELECTRICAL DICE	Coyle & Rogers
ELECTRICAL SHELL MAN, THE	National Automaton
ELECTRO-SKILL	Slattery
ELITE	Reliance Novelty
ELITE	Sundwall
ELITE	King & Sundwall
ELK	Caille Bros.
ELK	Fey
ELK	Mills Novelty
ELK	Paupa & Hochriem
ELK, IMPROVED	Watling
ELK, THE	Cowper
ELK, THE	Paupa & Hochriem
EMCO, THE (NERVE EXERCISE)	Erie Manufacturing
EMPIRE	Automatic Machine Company
EMPIRE	Western Products
EPSOM DOWNS	In And Out-Door Games
EUREKA	Eureka Novelty Sales
EVAN'S BABY BELL	H. C. Evans
EVEN UP	National Coin Machine Exchange
EVER READY TRADE BOARD	Dean Novelty
EVERLASTING AUTOMATIC SALESBOARD	D. Robbins & Company
EXCELSIOR	Leo Canda
EXCELSIOR	Excelsior Race Track Company
EXCELSIOR	Maley
EXCELSIOR (AUTOMATIC)	Richard K. Fox

F

FAIR-N-SQUARE	Fey
FAIR-N-SQUARE	Western Automatic Machine
FAIR-SELLING MACHINE	Clawson Slot Machine
FAIREST ROULETTE	Stirrup
FAIREST WHEEL	Monroe Barnes
FAIREST WHEEL	Clawson Machine
FAIREST WHEEL	Decatur Fairest Wheel Company
FAIREST WHEEL	Decatur Fairest Wheel Works
FAIREST WHEEL	Albert Pick
FAIREST WHEEL	Progressive Novelty
FAIREST WHEEL, THE	Progressive Novelty
FAIREST WHEEL, IMPROVED	Pana Enterprise
FANCY JUMBO	D. N. Schall
FANCY JUMBO	Watling
FANCY JUMBO SUCCESS	Watling
FAVORITE	Jennings
FAVORITE, THE	Douglass Specialties
FIGARO	Leo Canda
FIGARO	Leo Canda Manufacturing
FINGER TEST	A. B. T.
FIRE EAGLE	Cowper
"FIRE ENGINE"	Pearsall And Finkbeiner
5¢ A PUSH 5¢	Amusement Novelty Company
5¢ SPIN THE ARROW	Roberts Novelty
FIVE CIGARS	Brunhoff
FIVE OF A KIND	B. A. Withey
FIVE ON	Atlas Manufacturing
FIVE STAR	Adams Gum
5 STAR FINAL	Colonial
FLICKER	Bally
FLICKER	Moon & Avera
FLIP-A-KOPPER	Great States
FLIP FLAP (LOOP-THE-LOOP)	Kelley
FLIP FLOP FLUZZEE	Great States
FLIP FLOP FLUZZEE	Star Manufacturing
FLIP SKILL	Mills Novelty
FLIPPER	Markepp
FLIP SKILL	Monarch Coin Machine
FLIP TARGET	Chicago Slot Machine Exchange
FLORADORA	Mills Novelty
FLORADORA	Wheeland Novelty
FLTEL	American Manufacturing & Sales
FLYING FUN	Flying Fun Company
FLYING HEELS	Hub Manufacturing
FLYING HEELS	H. J. Stock
FLYING SKILL	Flying Skill Machine
FLYING THRILLS	H. J. Stock
FOOT BALL	B. Madorsky
FOOT BALL	Star Vending
FOOT BALL PRACTICE	Great States
FOOTBALL	Monarch Coin Machine
FOOTBALL	D. Robbins & Company
FOOTBALL	(Unknown)
FOOTBALL MUTUEL	Four Jacks
FORTUNE	J. M. Sanders
FORTUNE BALL GUM VENDER	Caille Bros.
FORTUNE BALL GUM VENDER	Superior Confection
FORTUNE COIN DROP	(Unknown)
FORTUNE MATCHES	G. & M. Manufacturing
FORTUNE TELLER	Fortune Machine
FORTUNE TELLER	Pardue Novelty
FORTUNE TELLER, THE	Groetchen Tool
4 CARD	Lewis & Strobel
4 CARD	Tibbils
4-11-44	Fey
4-IN-1	Rock-Ola
4 LEAF CLOVER	Pierce Tool
FOUR OF A KIND	Quality Supply
4-WAY FROLIC	Ad-Lee Company
4-WAY PURITAN BABY BELL	Reliable Coin
FREE BEER	(Unknown)
FREE PLAY	ABCO Novelty
FREE PLAY	Daval
FREE PLAY	Exhibit Supply Company
FREE PLAY	Scot Industries
FRENCH RACE GAME	Richard K. Fox
FRUIT KING	Mills Novelty
FULL DECK, THE	Watling
FULL WEIGHT	(Unknown)
FUNNY FACES	Yaggy

G

GALLOPING CUBES	Globe Novelty
GAME-OF-CHANCE	Nimmo
GAME-O-SCIENCE	Loss
GAME O'SKILL	Conrad Jackson Desk
GAME O'SKILL	Latimer & Company
GAME O'SKILL	Mills Novelty
GAME O'SKILL	R. J. White
GAME TABLE	Ganns, Clarke And Dengler
GAME WHEEL	William Dennings
GEE-WHIZ	Western Equipment & Supply
GEE-WIZ	Wolverine Supply
GEM	Calvert
GEM	Central City
GEM	Garden City Novelty
GEM	Novix Specialties
GEM	Pierce Tool
GEM CONFECTION	Ohio Automatic
GEM CONFECTION VENDER	Calvert
GEM CONFECTION VENDER	D. Robbins & Company
GEM GUM VENDER	J. M. Sanders
GET-A-PACK	Exhibit Supply Company
GET A PACK	Tippecanoe Manufacturing
GIANT (CARD)	Leo Canda
GIANT, THE	Mills Novelty
GIANT ARROW	Leo Canda
GIANT COUNTER CARD	Leo Canda
GIANT DICE	Leo Canda
GIANT POLICY	Leo Canda
GINGER	ABCO Novelty
GINGER	Groetchen Tool
GINGER BASEBALL	Groetchen Tool
GINGER PUNCHBOARD	Groetchen Tool
GIRO	Snider
GLOBE	Caille Bros.
GO-GO GIRL	Star Amusement
GOAL LINE	Exhibit Supply Company
GOLD MINE	UPI
GOLD MINE, THE	Coleman Novelty
GOLD NUT	Little Nut Vender
GOLD RUSH	Bally
GOLD RUSH	Groetchen Tool
GOLD RUSH CALENDER	Groetchen Tool
GOLDEN BELL	Exhibit Supply
GOLDEN HORSES	Buckley
GOOD LUCK	Caille Bros.
GOOD LUCK	Vernon K. Smith

GOOD LUCK	(Unknown)
GOOD LUCK GYPSY	Watling
GOOD LUCK RACE HORSE	Watling
GOOD LUCK SPECIAL	Caille Bros.
GOOD TURN, A	National Institute
GOOSE	Paupa & Hochriem
GRAND PRIZE	A. B. T.
GRANDSTAND	Jennings
GRAYHOUND RACE PIANO	Seeburg
GREAT AMERICAN GAME	(Unknown)
GUESSING BANK	McLoughlin
GUESSING BANK	Smith, Winchester Manufacturing
GUSHER	Daval
GYPSY	Pierce Tool
GYPSY FORTUNE TELLER	Field Paper Products
GYRO	Star Machine

H

HALF MILE	A. B. T.
HAMILTON, THE	Hamilton
HAMILTON, THE	Sicking
HAMILTON SUCCESS	Leo Canda Manufacturing
HAPPY DAY	ABCO Novelty
HAPPY DAYS	Great States
HAPPY JACK	Happy Jack Company
HAPPY THOUGHT	Clawson Slot Machine
HAV-A-SMOKE	American Manufacturing & Sales
HAVE SOME FUN	Jackson
HAZARD	National Coin Machine Exchange
HEADS AND TAILS	Clawson Slot Machine
HEAD OR TAIL	Ogden & Company
HEADS OR TAILS	Coin Machine Service
HEADS OR TAILS	Daval
HEARTS	Genco
HERE'S HOW	Pierce Tool
HERSHEY'S	(Unknown)
HI FLY	Central Manufacturing
HIGH DICE	Mabey Electric
HI-LO	Camco Products
HI-LO	Northwest Coin
HI-LO	Schiemer-Yates
HI-LO FIELD	Monterey Woodcrafters
HI-LO-FIELD	Valley Sales
HI-LO SEVEN	Quality Supply
HIAWATHA	Caille Bros.
HIAWATHA	Hiawatha Manufacturing
HIAWATHA CARD	Hiawatha Manufacturing
HIAWATHA JR.	Caille Bros.
HIAWATHA JR.	Puritan Machine
HIGH HANDS	Merkle & Cross
HIGH STAKES	Groetchen Tool
HIGH TENSION	Groetchen Tool
HI-HAND	Quality Supply
HILLMAN COIN TARGET BANK	M. Siersdorfer
HILLSBORO, THE	Hillsboro Wooden Ware
HIT A HOMER	ABCO Novelty
HIT-A-HOMER	Auto-Bell
HIT A HOMER	Central Manufacturing
HIT ME	Pace
HIT ME	Pierce Tool
HIT THE COON	Exhibit Supply
HOLD-A-BALL	King And Company
HOLD AND DRAW	Rock-Ola
HOL-E-SMOKES	Pace
HOL-E-SMOKES	Pierce Tool
HOLE IN ONE	Scientific Machine
HOLLYWOOD	Giles & Simpkins
HONEST DICE BOX	Patent Purchase
HONEST JOHN	Sales Stimulator Company
HONEST JOHN	West Coast Novelty (Seattle)
HONEST JOHN	West Coast Novelty (Canada)
HONESTY	Frank Wuotila
HONEY	Exhibit Supply Company
HOO DOO, THE	Aug. Grocery
HOO DOO CIGAR CUTTER	Clawson Slot Machine
HOOPS	Genco
HOOSIER, THE	Ogden & Company
HORSE PLAY	Exhibit Supply Company
HORSE RACE	Hammond & Jones
HORSE RACE	Lumley
HORSES	Buckley
HORSE SHOE	MBM (Mortimer B. Mills) Cigar
HORSE SHOE	A. M. Walzer
HORSE SHOES	Exhibit Supply Company
HORSE-SHOES	Western Products
HORSES	Quality Supply
HOT-SHOT	Hot Shot Sales
HOTCHA	U. S. Electric
HOW IS YOUR LUCK	Jonas D. Bell & Company
HOWARD'S FAVORITE	S. A. Cook
HULLABALOO	Field
HUMMER	Belk, Schafer & Company
HU-LA	Roy Torr
HUNCH	Bally
HUSTLER	Wall Novelty
HY-LO	Caille Bros.
HY-LO	Mills Novelty

I

DEAL	J. M. Sanders
IDEAL, THE	Cowper
IMP	ABCO Novelty
IMP	Bally
IMP	Engel Manufacturing
IMP	Groetchen Tool
IMP	Reliable Metal
IMP GOLF	A. B. T.
IMPERIAL	Imperial
IMPERIAL PURITAN	Mills Novelty
IMPROVED AUTOMATIC CHECK PAYING CARD MACHINE	Caille Bros.
IMPROVED AUTOMATIC ROULETTE	T. J. Nertney
IMPROVED BALL GUM VENDER	Pace
IMPROVED BICYCLE	Sun Manufacturing
IMPROVED DICE GAME	Richard K. Fox
IMPROVED DICE MACHINE	Keystone Novelty & Sales
IMPROVED EAGLE	Twin City Novelty
IMPROVED FAIREST WHEEL	Pana Enterprise
IMPROVED HORSERACE GAME	Universal
IMPROVED KELLEY	Kelley
IMPROVED MONARCH CARD MACHINE	Monarch Card Machine
IMPROVED NICKEL TICKLER	Western Weighing
IMPROVED PURITAN VENDOR	Buckley
IMPROVED ROULETTE	Leo Canda
IMPROVED ROULETTE	D. Kernan
IMPROVED ROULETTE	Western Automatic
IMPROVED SPECIAL	Mills Novelty
INDIAN (MAYFLOWER)	Silver King Novelty
INDIAN DICE	Exhibit Supply Company
INDIAN DICE	Gottlieb
INDIAN DICE	Monarch Coin Machine
INDIAN PIN POOL	Caille Bros.
INDIAN PIN POOL, 1918	Silver King Novelty
INDIAN SHOOTER	Calvert
INDOOR AVIATOR	Novix Specialties
INDOOR BASEBALL	Novix Specialties
INDOOR JUNIOR	Indian Head Novelty
INDOOR SPORTS	Field
INDOOR STRIKER	Consolidated Automatic
INDOOR STRIKER	Edmund Fey
INDOOR STRIKER	Indian Head Novelty
INDOOR STRIKER	Pacific Amusement
INDOOR STRIKER	D. Robbins
I OWE YOU	Exhibit Supply Company
IRON CARD MACHINE (NO.8)	Sicking
IS IT ANY OF YOUR BUSINESS?	Imperial
I WILL	I Will Novelty
I WILL	Mills Novelty
I.X.L. JR.	(Unknown)

J

JACK-POT BONUS	Bonus Advertising
JACK-POT-GAME	Buffalo Toy
JACKPOT CHARLEY	R. C. Walters
JACK POT CHECK SALESBOARD	Sherman & Fey
JACK POT IMPROVED PURITAN VENDOR	Buckley
JACK POT PURITAN BABY BELL	Buckley
JACK POT PURITAN BELL	Buckley
JACKPOT SUE	R. C. Walters
JACK POT TARGET PRACTICE	Buckley
JESTER	Caille Bros.
JESTERS	(Unknown)

JEWEL	Caille Bros.
JIFFY	Daval
JIGGER	Blake Manufacturing
JINGLE	Automatic Industries, Inc.
JINGLES	Checker Sales
JITTERBUG	Frey
JITTER-BONES	Keeney
JITTER CUBE	American Automatics
JITTER-ROLL	Frey
JITTERBUG BALL	Western Products
JITTERS	National Automatic Machines
JOCKEY	Automatic Machine & Tool
JOCKEY	Caille Bros.
JOCKEY	F. W. Mills
JOCKEY	Mills Novelty
JOCKEY	Molitor
JOCKEY, THE	Watling
JOCKEY CABINET, THE	Watling
JOCKEY CLUB	A.B.C. Coin Machine
JOCKEY 3-WAY	R. J. White
JOKER	United States Novelty
JOKER (JOKER WILD)	Daval
JOKER GUM VENDER	Daval
JOKERS WILD	Star Amusement
JOKER'S WILD	United Enterprises
JONES CARD MACHINE	Jones Novelty
JUGGLER	Boyce Coin
JUGGLING CLOWN	International Mutoscope
JUMBO	Leo Canda
JUMBO	Leo Canda Manufacturing
JUMBO	Clawson Machine
JUMBO	T. J. Nertney
JUMBO	Wagner Manufacturing
JUMBO, THE	Mills Novelty
JUMBO CARD MACHINE	Tibbils
JUMBO (THE ELEPHANT)	(Unknown)
JUMBO GIANT	Leo Canda
JUMBO GIANT	Leo Canda Manufacturing
JUMBO GIANT	Clawson Machine
JUMBO GIANT	Mills Novelty
JUMBO PURITAN	Caille Bros.
JUMBO SUCCESS	Automatic Machine & Tool
JUMBO SUCCESS	Paul E. Berger Manufacturing
JUMBO SUCCESS	Caille Bros.
JUMBO SUCCESS	Leo Canda
JUMBO SUCCESS	Leo Canda Manufacturing
JUMBO SUCCESS	Clawson Machine
JUMBO SUCCESS	D. Kernan
JUMBO SUCCESS	Mills Novelty
JUMBO SUCCESS	D. N. Schall
JUMBO SUCCESS	Stark
JUMBO SUCCESS	Tibbils
JUMBO SUCCESS	Watling
JUMBO SUCCESS	R. J. White
JUMPER	Blake Manufacturing
JUMPING JACK	New Era
JUNIOR	Exhibit Supply Company
JUNIOR BELL	Caille Bros.

K

KALAMAZOO	Kalamazoo Automatic Music
KAZOO	Kalamazoo Automatic Music
KAZOO-ZOO	Kalamazoo Automatic Music
KEEP 'EM BOMBING	Runyan Sales
KELLEY, THE	Kelley
KENTUCKY DERBY	Consolidated Service Bureau
KENTUCKY DERBY	Filmascope
KENTUCKY DERBY	General Novelty
KENTUCKY DERBY	Pennant Novelty
KENTUCKY DERBY	Stone
KENTUCKY DERBY	Stone Brothers
KEY HOLE	Exhibit Supply Company
KEYSTONE	Keystone Sales
KEYSTONE PURITAN BELL	Keystone Novelty & Sales
KICK 'N CATCH	Fun Industries
KICK 'N CATCH	Johnston Products
KICKER AND CATCHER	Baker Novelty
KICKER AND CATCHER	J. F. Frantz
KICKO ARISTOCRAT, THE	Keystone Engineering
KILL THE JAP	Groetchen Tool
KING	Comet Industries
KING BASE BALL MACHINE	Flatbush Gum
KING DODO	Mills Novelty
KING OF THE DIAMOND	One-Cent Amusement
KING OF THE DIAMOND	One-Cent Amusement
KING ROW	Neff Novelty
KING SIX	B. A. Withey
KING SIX, IMPROVED	B. A. Withey
KING SIX JR.	B. A. Withey
KING'S HORSES	Ad-Lee Company
KING'S HORSES	Western Equipment And Supply
KITZMILLER'S AUTOMATIC SALESMAN	Matheson Novelty
KLIX	ABCO Novelty
KLIX	Groetchen Tool
KLONDIKE	Fey
KLONDIKE	Steinberg
KLONDIKE, THE	Klondike Slot Machine
KLONDIKE GUM MACHINE	(Unknown)
KLONDYKE PROSPECTOR	Klondyke Prospector
KNOCK KNOCK	Osbrink
KNOCKOUT, THE	Levitt Manufacturing
KOIN-KICK FOOTBALL	Novix Specialties
KONTEST BOMBER	Gillespie Games
KONTEST POKER	Gillespie Games
KOUNTER KING	Mills Novelty
K'RAZY	Field
KROMO KOLORED KUBE	Exhibit Supply Company

L

L'AEROPLAN	Mills Novelty
LA WA-WO-NA	Caille Bros.
LARK	Star Vending
LARK, THE	Bradford Novelty Machine
LARK, THE	Lark Distributing
LARK, THE	W. J. Young
L. A. W., THE (BICYCLE)	Samuel Nafew
LEADER	Banner Specialty
LEADER	Knight Novelty
LEADER, THE	Drobisch Brothers
LEADER, THE	Sloan Novelty
LEAPING LENA	Garden City Novelty
LEGAL, THE	A. J. Fisher
LEAP FROG	Ad-Lee Company
LEAPING LENA	Pierce Tool
'LEVEN COME SEVEN	Kalamazoo Automatic Music
LIBERTY	ABCO Novelty
LIBERTY	Groetchen Tool
LIBERTY BELL	ABCO Novelty
LIBERTY SPORTS PARADE	Groetchen Tool
LILLIPUT GOLF	Lilliput Manufacturing
LINCOLN	Caille Bros.
LINE-A-BASKET	Bally
LION JR.	Caille-Richards
LITE-A-PACK	Veteran
LITE-A-PAX	Bally
LION DICE MACHINE	Midwest Novelty
LION PURITAN BABY BELL	Midwest Novelty
LION PURITAN BABY VENDOR	Lion Manufacturing
LION PURITAN BABY VENDOR	Midwest Novelty
LINE EM UP	A.B.C. Coin Machine
LITTLE BILLIKIN	H. C. Evans
LITTLE DREAM	Caille Bros.
LITTLE DREAM	Miller Novelty
LITTLE DREAM	Mills Novelty
LITTLE DREAM PLAY BASEBALL	Miller Novelty
LITTLE DUKE	Cowper
LITTLE DUKE	Watling
LITTLE DUKE, THE	Mills Novelty
LITTLE EGYPTIAN FORTUNE TELLER	Premier Novelty
LITTLE GEM	Bradford Novelty
LITTLE GEM	Mills Novelty
LITTLE GEM FORTUNE TELLER	Bradford Novelty
LITTLE GIANT	Little Giant
LITTLE GIANT	D. Robbins & Company
LITTLE GYPSY VENDER	Premium Novelty Works
LITTLE HELPER	(Unknown)
LITTLE JOE	Fort Wayne Novelty
LITTLE JOKER	Willard A. Smith
LITTLE KLONDIKE	F. W. Mills
LITTLE KNOCKER	Mills Novelty

LITTLE LEAGUE	J. F. Frantz
LITTLE LEAGUE	Fun Industries
LITTLE LEAGUE	Johnston Products
LITTLE MERCHANT	Jennings
LITTLE MODEL CARD MACHINE	Samuel Nafew
LITTLE MODEL CARD MACHINE	Sittman & Pitt
LITTLE MONTE CARLO	Mills Novelty
LITTLE MONTE CARLO	National Manufacturing
LITTLE MYSTERY	Jennings
LITTLE PERFECT	Canada Novelty
LITTLE PERFECTION	Leo Canda Manufacturing
LITTLE PERFECTION	Davis Novelty
LITTLE PERFECTION	Iowa Novelty
LITTLE PERFECTION	Mills Novelty
LITTLE PERFECTION	Rex Novelty
LITTLE PERFECTION	R. J. White
LITTLE POKER FACE	J. M. Sanders
LITTLE PRINCE	Silver King Novelty
LITTLE PRINCE (WHIRLWIND)	Silver King Novelty
LITTLE SCARAB	Mills Novelty
LITTLE SILENT POKER	Kalamazoo Automatic Music
LITTLE VENDER	Fey
LITTLE WHIRLWIND	Peo Manufacturing
LITTLE WINNER	Acme Scale
LITTLE WONDER	Caille-Richards
LITTLE WONDER, THE	Gatter Novelty
LIVELY CIGAR SELLER	Clawson Machine
LIVELY CIGAR SELLER	Clawson Slot Machine
LOG CABIN	Erickson
LONE STAR	Bally
LONG SHOT	J. F. Frantz
LONG SHOT	Fun Industries
LONG SHOT	Johnston Products
LOTTA-DOE	Palamedes Sales
LUCKY	Norris
LUCKY	United States Novelty
LUCKY BOY	Kentucky Gum
LUCKY COIN TOSSER	M. T. Daniels
LUCKY COIN TOSSER	A. J. Stephens
LUCKY COINS	(Unknown)
LUCKY DICE	Automatic Manufacturing
LUCKY DUCKS	Renouf Manufacturing
LUCKY EYES	Chicago Coin
LUCKY HOROSCOPE	Auto-Bell Novelty
LUCKY HOROSCOPE	Marvel
LUCKY PACK	Paul Bennett & Company
LUCKY PACK	Buckley
LUCKY PACK	J. M. Sanders
LUCKY ROLL	M. Brodie Company
LUCKY ROLL	Fortune Sales
LUCKY SEVEN	Shelby Agency
LUCKY SMOKES	Daval
LUCKY STAR	Camco Products
LUCKY STAR	Gottlieb
LUCKY STAR	A. J. Stephens
LUCKY STRIKE	Baker Novelty
LUCAT	Lu-Kat Novelty

M

MAGIC BEER BARREL	A. J. Stephens
MAGIC CLOCK	Keeney & Sons
MAIN STREET MAGIC DICE	P. & T. Club Equipment
MAJOR SERIES	Rock-Ola
MAKE TWENTY ONE	Exhibit Supply Company
MAKE TWENTY ONE	(Unknown)
MANILA, THE	Mills Novelty
MANILA, THE	Paupa And Hochriem
MARBLO	Pacific Amusement
MARGARET BURNHAM'S	A.R. Wood
MARVEL	ABCO Novelty
MARVEL	Advance Machine
MARVEL	Comet Industries
MARVEL	Daval
MARVEL	Scot Industries
MARK I TIC TAC TOE	Mark Gerard
MASCOT	Caille Bros.
MASCOT	Cowper
MASCOT	T. J. Nertney
MASCOT, THE	Mascot Machine
MASTER	Crown Machine
MATADOR	Caille Bros.
MATCH-A-BALL	Daval
MATCH-A-PAK	Garden City
MATCH-A-PAK	Pierce Tool
MATCH COLOR	Acme Game
MATCH-EM	Bally
MATCH 'EM	Western Equipment & Supply
MATCH IT	Bally
MATCH IT	Gottlieb
MAYFLOWER	Caille Bros.
MAYFLOWER	Puritan Machine
MAYFLOWER	Watling
MAE AND HER PALS	Carney
MECCA	Watling
MELODY GUM	Gumatic
MENU WHEEL	Menu Wheel Company
MERCHANT	Caille Bros.
MERCHANT, THE	Watling
MERCURY	ABCO Novelty
MERCURY	Groetchen Tool
MERCURY PUNCHBOARD	Groetchen Tool
MERRY-GO-ROUND	Genco
METEOR	Atlas Manufacturing
MEXICAN BASEBALL	Daval
MIAMI CRAP	Monterey Woodcrafters
MIDGET	Fey
MIDGET	L. C. Graham
MIDGET	D. Kernan
MIDGET	Schaeffer
MIDGET	Schultze Novelty
MIDGET, THE	Cowper
MIDGET BARTENDER, THE	(Unknown)
MIDGET BASEBALL	Hercules Novelty
MIDGET BASEBALL	D. Robbins
MIDGET RACES	Gottlieb
MIDGET ROULETTE	Tillitson Specialty
MIDGET SMOKER, THE	(Unknown)
MIDGET 36	Fey
MIDGET WITH SALESBOARD	Fey
MIDGET ROULETTE	Atlas Manufacturing
MIDGET ROULETTE	L. C. Graham
MIDGET ROULETTE	Mills Sales
MIKRO-KALL-IT	Mikro-Kall-It Inc.
MILL RACE	Electra Corporation
MILLARD'S MINIATURE BASEBALL	Coin Sales Corporation
MILLWHEEL	Bally
MINIATURE BASEBALL	Gottlieb
MINIATURE BASEBALL	Tecumseh Sales
MINIATURE BASEBALL GAME	Star Novelty
MINIATURE BASEBALL WORLD CHAMPION	Advance Machine
MINIATURE BELL	F. F. Sales
MINIATURE GOLF	D. Robbins & Company
MINIATURE MUTUEL RACE TRACK	H. C. Evans
MINIATURE RACE COURSE	H. C. Evans
MINIATURE RACE COURSE MACHINE	George Mason & Company
MINIATURE RACE TRACK	Cowper
MINT VENDER	Vending Machine Company
MIRROR OF FORTUNE	A. B. T.
MIRROR OF FORTUNE	Webb Novelty
MISTIC DERBY	Maitland
MITE	Comet Industries
MODEL	Samuel Nafew
MODEL (AUTOMATIC POKER PLAYER)	Sittman & Pitt
MODEL CARD MACHINE	Leo Canda
MODEL CARD MACHINE	Maley
MODEL CARD MACHINE	Samuel Nafew
MODEL CARD MACHINE	Sittman & Pitt
MODEL DRAW POKER	Sittman & Pitt
MONARCH	Monarch Sales
MONARCH, THE	Clawson Machine
MONARCH CARD MACHINE	Leonhardt
MONARCH CARD MACHINE	Monarch Card Machine
MONARCH DICE MACHINE (NO. 5)	Drobisch Brothers
MONKEY DICE	Percy G. Williams
MONTE CARLO	Paul E. Berger Manufacturing
MONTE CARLO	Burton Machine
MONTE CARLO	Frey
MONTE CARLO	Western Products
MOOSE, THE	Watling

247

Name	Manufacturer
MOUSIE MOUSIE	Star Sales
M-R	M-R Advertising System
(MUSSOLINI)	(Unknown)
MUTUAL	Walker Sales
MUTUEL	Hawkeye Manufacturing
MUTUEL 5	Merchants Advertising
MUTUEL HORSES	Buckley

N

Name	Manufacturer
NATIONAL	Midwest Novelty
NATIONAL	Mills Novelty
NATIONAL	National Cash Register
NATIONAL	Watling
NATIONAL BASEBALL	A. J. Stephens
NATIONAL DICE TABLE	National Table
NATIONAL TABLE	National Table
NATIONAL TARGET PRACTICE	National Coin Machine
NATIVE SON	Mills Novelty
NATURAL	Bally
NAVAJO	Caudle & McCrary
NECK AND NECK	Kemo Novelty
NERVE SCALE	Craft Engineering
NEVER LOSE	Jackson
NEW CARD MACHINE	Leo Canda
NEW DEAL	Pace
NEW DEAL	A. J. Stephens
NEW DEAL, THE	Pierce Tool
NEW DROP CASE	Cowper
NEW ERA VENDER	New Era
NEW IDEA	M. O. Griswold
NEW IDEA CIGAR MACHINE	Mills Novelty
NEW LARK BABY BELL	Automatic Games
NEW LARK POKER	Automatic Games
NEW MOON	Hawkeye Novelty
NEW PIANO GAME, THE	Robinson Novelties
NEW PURITAN	Caille Bros.
NEW SPIRAL	Stone & Jones
NEW STAR	Griswold Manufacturing
NEW TARGET PRACTICE	Mills Novelty
NEW TRADE MACHINE	Paupa & Hochriem
NEWARK RAINBOW, THE	Clawson Slot Machine
NEXT, THE	B. B. Novelty Company
NICKELSCOPE	Jonas D. Bell & Company
NICKEL TICKLER	Maley
NICKEL TICKLER	Western Automatic
NICKEL TICKLER	Western Weighing
NINE POCKET	Frederick W. Bishop
NON-COIN AMERICAN EAGLE	Comet Industries
NON-COIN COMET	Comet Industries
NON-COIN MARVEL	Comet Industries
NON-COIN METEOR	Comet Industries
NOVELTY GEM	Gatter Novelty
NEW 36 GAME	Fey
NUGGET, THE	Bally
NUMBER-ROLL	Frey
NUMBER ROLL	Valley Sales
NUMBERS	Buckley
NUMBERS	Mills Novelty

O

Name	Manufacturer
O-TO-GO	Aspin & Furry
OAKLAND	Oakland Novelty
ODD PENNY MAGNET	M. T. Daniels
ODD PENNY MAGNET	Peerless Manufacturing
ODD PENNY MAGNET	Peerless Products
ODD PENNY MAGNET	A. J. Stephens
OFFICIAL SWEEPSTAKES	Rock-Ola
OLD AGE PENSION	Exhibit Supply Company
OLD FORT (PILGRIM)	Silver King Novelty
ONE BALL	Clawson Slot Machine
ON THE LEVEL	Fey
ON THE LEVEL	Mills Novelty
ON THE LEVEL	Wisconsin Novelty
ON THE SQUARE	Fey
OOMPH	Daval
OOMPH	Western Products
OPERATORS DICE MACHINE	Exhibit Supply Company
OREGON	Portland Novelty
ORIGINAL AUTOMATIC VENDER	Matheson Novelty
OUR VERY BEST	Progressive Novelty
OVER THE TOP	Boyce Coin

P

Name	Manufacturer
PAIR IT	Pierce Tool
PAK-O-CIGS	Electro Ball
PAK-O-CIGS	Exhibit Supply Company
PAIR-IT	Garden City Novelty
PANAMA CANAL	Cawood Novelty
PANSY	Gatter Manufacturing
PAR-KET	Buddick Engineering
PASTIME	Star Manufacturing
PATIENCE DEVELOPER, THE	Boyce Coin
PAY DAY	Brunswick
PAYING TELLER	Fey
P. D. Q.	Pacific Amusement
PEE WEE	Fey
PEE-WEE ROULETTE	Monarch Sales
PEERLESS	Mills Novelty
PEERLESS	Peerless Amusement Machine
PEERLESS ADVERTISER (THE CODE)	Merriam Collins
PEERLESS BALL GUM VENDER	Peerless Amusement Machine
PEERLESS CALIFORNIA JACK	Mills Novelty
PEERLESS 5-SLOT	Reliance Novelty
PEERLESS LITTLE WONDER	Peerless Amusement Machine
PENCIL VENDER	Bally
PEN-NEE GOLF	Pace
PENN-E-WIZE	Harkness-Miller
PENNY AMUSE-U	Penny Ante Amusement
PENNY ANDY	Marathon Specialty
PENNY ANTE	Dean Novelty
PENNY ANTE	Penny Ante Amusement
PENNY ANTE DRAW POKER	Osbrink
PENNY ANTE PENNY DRAW, 5¢	Penny Ante Amusement
PENNY BACK	Boyce Coin Machine Amusement
PENNY BALL GUM VENDER	Caille Bros.
PENNY BASE BALL	Silver King Novelty
PENNY BASEBALL	A. S. Douglis
PENNY BASEBALL	Royal Novelty (Indianapolis)
PENNY BELL	Silver King Novelty
PENNY CIGAR	Rogers
PENNY CIGARETTE MACHINE	Groetchen Tool
PENNY CLUB	Jennings
PENNY CONFECTION VENDOR	D. Robbins
PENNY DICE	Roche Novelty
PENNY DRAW	Dean Novelty
PENNY DRAW	A. J. Stephens
PENNY FORTUNE	National Coin Machine
PENNY GUM VENDER	Monarch Manufacturing
PENNY JITTER BALL	W. H. Kelly
PENNY KING	National Coin Machine Exchange
PENNY PACK	ABCO Novelty
PENNY PACK	Daval
PENNY PAK	Superior Confection
PENNY PITCH	Barco Products
PENNY PLAY	Froom Laboratories
PENNY SKILLO	American Sales
PENNY SKILLO	Century Manufacturing
PENNY SKILLO	United Amusement
PENNY SLOT MACHINE	Joseph J. Lane
PENNY SMOKE	Groetchen Tool
PENNYSCOPE	Jonas D. Bell & Company
PENNY TICKLER	Maley
PEP	Brunswick
PERFECT SELLING MACHINE	Clawson Slot Machine
PERFECTION	Caille Bros.
PERFECTION	Dunn Brothers
PERFECTION	Ennis & Carr
PERFECTION	Hudson Moore
PERFECTION	Pacific Electrical Works
PERFECTION	Perfection Manufacturing
PERFECTION	Perfection Novelty
PERFECTION	Royal Card Machine
PERFECTION	D. N. Schall
PERFECTION	Watling
PERFECTION	Wheeland Novelty
PERFECTION	R. J. White
PERFECTION, THE	Comstock Novelty Works
PERFECTION CARD	Paul E. Berger Manufacturing
PERFECTION CARD	Leo Canda
PERFECTION CARD	Clawson Machine
PERFECTION CARD	Cowper

Name	Manufacturer
PERFECTION CARD	Diamond Novelty
PERFECTION CARD	Maley
PERFECTION CARD	Mills Novelty
PERFECTION FIGARO	Leo Canda
PERFECTION WHEEL	Comstock Novelty Works
PETE'S PENNY ANTE	Penny Ante Amusement
PHILLIPS	Phillips Farm Supply
PICK-A-PACK	Baker Novelty
PICK A PACK	Garden City Novelty
PICK A PACK	Pierce Tool
PICK-A-PLUM	Majestic
PICK A WINNER	Atlas Indicator
PIKES PEAK	Groetchen Tool
PILGRIM	Buckley
PILGRIM	Caille Bros.
PILGRIM	Industry Novelty
PILGRIM	Mills Novelty
PILGRIM	Puritan Machine
PILGRIM	Watling
PILGRIM VENDOR	Buckley
PILOT	J. Edward Cowles & Company
PILOT	Mills Novelty
PILOT	Paupa & Hochriem
PILOT	Watling
PIN BOARD TARGET PRACTICE	Buckley
PIN FORTUNE	Advance Machine
PINCH HITTER	Northwestern Sales
PIPE-EYE	Chicago Coin
PIPPIN, MILLS	Mills Novelty
PITCH-A-PENNY	Kenyon
PITCH-A-PENNY	United Amusement
PITCHEM	Amusement Enterprises
PITCH-TO-THE-LINE	Metropolitan Coin
PIX-IT	Pierce Tool
PLACE THE COPPERS	(Unknown)
PLACE YOUR BETS	(Unknown)
PLAY A HAND	Majestic
PLAY AND DRAW	Fey
PLAY BALL	A. B. T.
PLAY BALL	Auto-Bell
PLAY BALL	Burnham And Mills
PLAY BALL	Cawood Novelty
PLAY BALL	Exhibit Supply Company
PLAY BALL	John M. Waddell
PLAY BASKETBALL	E. E. Junior
PLAY FOOTBALL	E. E. Junior
PLAY GOLF	Lowe Vending
PLAY HORSESHOES	Booth Games
PLAY-PAX	Garden City Novelty
PLAY POKER	Peo Manufacturing
PLAY POOL	E. F. Driver
PLAY POOL	Specialty Manufacturing
PLAY-TEX	Pierce Tool
PLAY THE FIELD	C. & F. Manufacturing
PLAY-WRITE	Play-Write Corporation
PLEASURE ISLAND	Kenyon
PLENTY OF CASH	R. C. Walters
PLUS OR MINUS	Bally
POCKET BALL	Arcade Supply
POCKET BALL	A. L. Holt
POCKET EDITION GALLOPING DOMINOS	H. C. Evans
POCKET POOL	Automatic Amusement
POCKET POOL	Kramer
POISON THIS RAT	ASCO
POISON THIS RAT (HITLER)	Groetchen Tool
POKER	Buckley
POKER	Genco
POKER CARD MACHINE	Columbian Machine
POKER DICE	Grove Brothers
POKER DICE	Hudson Moore
POKER GAME	Jennings
POKER JR.	Amusement Enterprises
POKER PLAY	Field Manufacturing
POKER SOLITAIRE	Lewis & Strobel
POKER SOLITAIRE	Tibbils
POKEROLL GAME, THE	Sloan
POKERENO	Field Manufacturing
POK-O-MAT	Stewart And McQuire
POK-O-REEL	ABCO Novelty
POK-O-REEL	Groetchen Tool
POK-O-REEL	Rock-Ola
POK-O-REEL TRIPLEX	Groetchen Tool
POKO BALL	Reliable Metal
POLICY	Amusement Machine Company
POLICY	Fey
POLICY MACHINE	Leo Canda
POLO SKILL	Novix Specialties
PONIES	Bally
POPE DICE MACHINE	Automatic Manufacturing
POPPY	Bally
PORTOLA	Specialty Machine
POT OF GOLD	J. F. Frantz
POT-O-GOLD	Hawkeye Novelty
POP-UP	Marvel
POT LUCK	National Automatic Machines
PREMIUM	A. J. Fisher
PREMIUM	C. P. Young
PREMIUM, THE	Mills Novelty
PREMIUM TRADER	Industry Novelty
PREMIUM TRADER	F. W. Mills
PREMO MERCHANDISER	Flatbush Gum
PRESTO	Quality Supply
PRETTY WAITER GIRL	McLoughlin
PRINCE	Garden City Novelty
PRINCE	Pierce Tool
PRIZE PEPSIN GUM VENDING MACHINE	Lehigh Novelty
PROFESSOR CHARLEY	R. C. Walters
PROFIT SHARER	Automatic Cash Discount Register
PROFIT SHARING 5¢ PENCIL VENDOR	D. Robbins & Company
PROFIT SHARING REGISTER	Mills Novelty
PROMOTION SALES REGISTER	Boduwil Company
PROSIT	A. B. T.
PULL-A-BALL	Winner Sales
PUNCH A BALL	Genco
PUNCH BALL	Norwood Manufacturing
PUNCHBOARD	Bally
PUNCHETTE	Groetchen Tool
PURITAN	Caille Bros.
PURITAN	Cowper
PURITAN	Detroit Coin Machine
PURITAN	Industry Novelty
PURITAN	Mills Novelty
PURITAN	Puritan Machine
PURITAN	Watling
PURITAN, MILLS	Mills Novelty
PURITAN BABY	Field Paper Products
PURITAN BABY BELL	Buckley
PURITAN BABY BELL	Del Norte Specialty
PURITAN BABY BELL	Lion Manufacturing
PURITAN BABY BELL	Midwest Novelty
PURITAN BABY BELL VENDER	Field Manufacturing
PURITAN BABY BELL VENDER	Morris Novelty
PURITAN BABY VENDOR	Buckley
PURITAN BABY VENDOR	Devices Manufacturing Sales
PURITAN BABY VENDOR	Great States
PURITAN BABY VENDER	J. M. Sanders
PURITAN BELL	Buckley
PURITAN BELL	Caille Bros.
PURITAN BELL	J. F. Gleason
PURITAN BELL	Mills Novelty
PURITAN BELL FORTUNE	Keystone Novelty & Sales
PURITAN BELL GUM VENDER	Keystone Novelty & Sales
PURITAN CONFECTION VENDER	Chicago Mint
PURITAN GIRL	Jennings
PURITAN VENDOR	Buckley
PURITAN VENDOR	Caille Bros.
PURITAN VENDOR	A. S. Douglis
PURITAN VENDOR	Douglis Machine
PURITANA, THE	Watling
PUT AND TAKE	Four Jacks
PUT AND TAKE	A. R. Kiser
PUT AND TAKE	W. C. Steel Ball
PYRAMID, THE	Buchanan

Q

Name	Manufacturer
QUINTETTE	Ad-Lee
QUINTETTE	Caille Bros.
QUINTETTE	Leo Canda Manufacturing

R

RACE	Hudson Moore
RACE AGAINST TIME	J. & E. Stevens
RACE COURSE	Marean
RACE COURSE	J. & E. Stevens
RACE COURSE	Thompson
RACE HORSE	Genco
RACE HORSE	Mills Novelty
RACE HORSE PIANO	Mills Novelty
RACE OF HORSES AND MARBLES, THE	Peerless Products
RACES	Daval
RACE TRACK	H. C. Evans
RACE TRACK	McManus
RACE TRACK	Albert Pick & Company
RACE TRACK, NEW STYLE	William Suydam
RADIO	Ad-Lee Company
RADIO WIZARD	Rock-Ola
RAILBIRD	Elston Sales
RAINBOW	Bally
REBATER	Jennings
RED BIRD	Park Novelty
RED DOG	Exhibit Supply Company
RED 'N' BLUE	Daval
RED STAR	(Unknown)
RED, WHITE AND BLUE	L. A. Wright
RED, WHITE AND BLUE TARGET	Blue Bird Sales
REDWOOD	Bally
REEL AMUSEMENT	Norris
REEL DICE	Daval
REEL-O-BALL	Daycom
REEL-O-BALL	Reel-O-Ball Company
REEL-O-BALL	Yendes Manufacturing
REEL-O-BALL	Yendes Service
REEL POKER	Western Products
REEL RACES	Western Equipment & Supply
REEL SPOT	Daval
REEL 21	Daval
REGISTER	Caille Bros.
RELIABLE	Mills Novelty
RELIABLE NERVE AND EYE TESTER	Gent
RELIABLE PURITAN BABY VENDER	Field
RELIANCE	Caille Bros.
RELIANCE	Reliance Manufacturing
RELIANCE	Reliance Novelty
REPEATER	ReFinders
RESERVE JACK-POT PURITAN VENDOR	Buckley
REVOLUTION CIGAR WHEEL	Waddell Wooden Ware Works
REX	Daval
RINGER	(Unknown)
RISTAURANT	Ristaucrat Manufacturing
'RITHMATIC	Daval
RITZ	Genco
RIVAL	Bucyrus
RIVAL	B. A. Stevens
ROCK-IT	Doraldina Corporation
ROLA-BALL	Rola-Ball Vending
ROLL-A-CENT	Grand National Sales
ROLL-A-CENT	Koplo Sales
ROLL-A-CENT	Withey Manufacturing
ROLL-A-PACK	Buckley
ROLL A PACK	Victor Vending
ROLL A POINT JUNIOR	Biscayne Manufacturing
ROLL A POINT SENIOR	Biscayne Manufacturing
ROLL-A-WAY	B. A. Withey
ROLL 'EM	Cardinal
ROLL 'EM	Fey
ROLL-EM	Norris
ROLL-ET	Gottlieb
ROLETTA	Western Distributors
ROLL'ETTO	Game of Games Company
ROLL-ETTO	Roll-Etto Novelty
ROLLETTO	Century Games
ROLLETTO JR.	Century Games
ROLLING POKER	Blanchard
ROLL-O-DICE	Starved Rock Novelty
ROLL SKILL	Dixon
ROLL SKILL	J. D. Drushell
ROL-LUCK	(Unknown)
ROLO	Superior Games
ROOLO	John M. Waddel
ROTO-MATIC	Groetchen Tool
ROULETTE	Acme Novelty Works
ROULETTE	Atlas Manufacturing
ROULETTE	Leo Canda Manufacturing
ROULETTE	M. O. Griswold
ROULETTE	Mansfield Brass
ROULETTE	Milark Manufacturing
ROULETTE	Star Amusement
ROULETTE CLOCK	Roulette Clock
ROULETTE POKER	Keystone Novelty
ROYAL CARD MACHINE	Leo Canda
ROYAL CARD MACHINE	(Unknown)
ROYAL FLUSH	Groetchen Tool
ROYAL FLUSH	Hunts Club
ROYAL JUMBO	Caille Bros.
ROYAL REELS	A. B. T.
ROYAL SPORT DICE GAMER	Ad-Lee Company
ROYAL SPORT MARBLE GAME	Ad-Lee Company
ROYAL TRADER	Royal Novelty (San Francisco)
RUNABOUT	Boyce Coin Machine Amusement
RUMBA WHEEL, THE	Gist Cabinet

S

SAFE HIT	Daval
SALESBOARD	W. B. Specialty
SALES INCREASER	Page Manufacturing
SAMPLE EXHIBITOR	Ideal Toy
SANDY'S HORSES	Great States
SARATOGA SWEEPSTAKES	H. C. Evans
SAVE OUR BUSINESS	J. F. Frantz
SCARAB, LITTLE	Mills Novelty
SCENIC RAILROAD	Star Specialty
SCOTCH GOLF	Marcus
SCOOTER	Mike Munves Corporation
SCRAMBALL	J. H. Keeney
SEARCHLIGHT	Caille Bros.
SEE-DICE	See-Con, Inc.
SELECT-EM	Exhibit Supply Company
SELF PAY	Griswold Manufacturing
SENSATIONAL	Caille Bros.
SENTRY	Mills Novelty
SEVEN-AND-ONE-HALF	Buckley
SEVEN COME ELEVEN	Daval
SEVEN COME ELEVEN	Kalamazoo Automatic Music
SEVEN ELEVEN	Genco
7-GRAND	Bradley Industries
SEVEN GRAND	Koplo Sales
SEVEN GRAND	B. A. Withey
SEVEN GRAND	Withey Manufacturing
SEVEN GRAND, IMPROVED	Rialto Sales
SEVEN GRAND, IMPROVED	B. A. Withey
SHAKE AND DRAW	B. A. Withey
SHAKE RATTLE 'N' ROLL	Jess Manufacturing
SHARPSHOOTER DELUXE	A. J. Stephens
SHENANDOAH UP-TO-DATE	Bennett and Company
SHIMMER DICE	Liberty
SHIP AHOY	Rock-Ola
SHOOT-A-PAK	A. B. T.
SHOOT-A-PAK	Fey Automatic Machine
SHOOTEM	Chicago Coin
SHOOT-THE-MOON	Majestic
SHOOTING STAR COUNTER GAME	P. & S. Machine
SHUFFLER, THE	Leo Canda
SHUFFLER, THE	D. Kernan
SILENT SALESMAN	Industry Novelty
SILENT SALESMAN	Keystone Novelty & Manufacturing
SILENT SALESMAN	Mills Novelty
SILENT SALESMAN	Silver King Novelty
SILENT SALESMAN	A. J. Stephens
SINK A JAP SHIP	Star Manufacturing
SINK OR SWIM	Acme Sales
SIX HORSEMEN	A.B.T.
SIX IN ONE	Blue Bird Sales
6-WAY	(Unknown)
6-WAY DICE	United Automatic
SIZZLER, THE	Field
SIZZLER, THE	B. A. Withey
SKE-BAL-ETE	Buddick Engineering
SKEE-SHOT	Arcade Supply

Name	Manufacturer
SKIL-FLIP	Pace
SKILL-A-GALLE	Mills Novelty
SKILL-A-RETTE	Baker Novelty
SKILL-A-RETTE	Standard Coin
SKILL DRAW	A. B. T.
SKILL DRAW	Exhibit Supply Company
SKILL DRAW	Fey
SKILL KATCH	Sands Manufacturing
SKILL MACHINE	Fey
SKILL-O-METER	(Unknown)
SKILL ROLL	Fey
SKILL ROLL	Sherman & Fey
SKILL SHOT	Gottlieb
SKILL SHOT	Groetchen Tool
SKILL TEST	Paul Bennett & Company
SKILLiARD	Loss
SKILO	Bonus Sales
SKILSHOT	Advance Machine
SKIPPER	Skipper Sales
SKYSCRAPER	Leo Canda
SKYSCRAPER	Rock-Ola
SLEEPIN' SAM	Specialty Manufacturing
SLOT DICE	Boardman Rubber Stamp Works
SLOT DICE SHAKER	B. A. Stevens
SLOT MACHINE	Netschert
SLOT MACHINE	Slot Machine Company
SLOT MACHINE	Weston Slot Machine
SLOT MACHINE	Willoughby Company
SLOTLESS (CIGAR CUTTER)	Brunhoff
SLUGGER	Bally
SLUGGER	Marvel
SMILEY	Pioneer Coin
SMILING JOE	Exhibit Supply
SMOKE HOUSE	Groetchen Tool
SMOKES	National Coin Machine Exchange
SMOKE UP	A. B. T.
SMOKE REELS	Daval
SMOKEMASTER	Dutch
SNAKE-EYES	Chicago Coin
SNAP-A-BALL	Jayess Novelty
SNOOKY	Camco Products
SOCCER	Genco
SOLITAIRE	Groetchen Tool
SONG DICK	Friedman
SPARK-O-LITE	Bally
SPARKS	ABCO Novelty
SPARKS CHAMPION	ABCO Novelty
SPARKS CHAMPION	Groetchen Tool
SPARKS DELUX	Groetchen Tool
SPARKS PUNCHBOARD	Groetchen Tool
SPARKY	Star Amusement
SPECIAL (AUTOMATIC CHECK PAYING CARD MACHINE)	Caille Bros.
SPECIAL	Mills Novelty
SPECIAL COMMERCIAL	Mills Novelty
SPECIAL ELK	Paupa & Hochriem
SPECIAL ELK	Watling
SPECIAL EXPORT	Mills Novelty
SPECIAL TIGER	Caille Bros.
SPECIAL VICTOR	Mills Novelty
SPEEDWAY	International Amusement
SPIN-A-PACK	Buckley
SPIN-A-PACK	Daval
SPIN-A-PACK	Grand National Sales
SPIN-A-PACK	Sicking
SPINAROUND	Star Sales
SPIN-EM	Ad-Lee Company
SPIN-IT	Shipman
SPINNER	Bally
SPINNER, THE	Mills Novelty
SPINNER WINNER	J. H. Keeney
SPINNERINO	Bally
SPINNING TOP	Brunhoff
SPIN-O	Chase Vending
SPIN-O	Garden City Novelty
SPIN-O	Spin-O Sales
SPIRAL	Bally
SPIRAL	E. D. Parker
SPIRAL	Twentieth Century Novelty
SPIRAL GOLF	A. S. Douglis
SPIRAL GOLF	Genco
SPIRAL GOLF	Pace
SPIRAL VENDOR	South Side Machine Shop
SPITFIRE	Scientific Machine
SPITTOON DICER, THE	(Unknown)
SPOOK HOUSE	Calvert
SPORTLAND	A. B. T.
SPOT POKER	Keystone Novelty
SQUARE DEAL	Bernard Abel
SQUARE DEAL	Keane Novelty
SQUARE DEAL	Nafew-Goldberg
SQUARE DEAL	Square Deal Machine
SQUARE SHOOTER	Star Manufacturing
SQUARE SHOOTER	Star Sales
SQUARE SPIN	Wonder Novelty
SQUAW (PURITAN)	Silver King Novelty
STANDARD	Amusement Machine Company
STANDARD	Clawson Machine
STANDARD	Reliance Novelty
STANDARD, THE	Standard
STAR	M. O. Griswold
STAR	Griswold Manufacturing
STAR	Jennings
STAR	B. A. Stevens
STAR	Streater
STAR, THE	Pearsall And Finkbeiner
STAR ADVERTISER	Drobisch Brothers
STAR AMERICAN EAGLE	Daval
STAR CIGARETTE MERCHANDISER	Exhibit Supply Company
STAR GREEN	Bennett & Company
STAR POINTER	Cowper
STAR POINTER	H. C. Evans
STAR TARGET PRACTICE	Mills Novelty
STAR TRADE REGISTER	Star Trade Register Company
STAR VENDER	Jennings
STEEPLE-CHASE	Automatic Novelty
STEEPLE CHASE	Exhibit Supply
STEEPLECHASE	Brunswick
STEEPLECHASE	Keeney And Sons
STEEPLECHASE	Standard Games
STEEPLE RACES	Gottlieb
STEP UP	Daval
STEVE BRODIE	M. Brodie Company
STOP AND SOCK	Gottlieb
STREAMLINER	J. M. Sanders
STRIKE IT RICH	R. C. Walters
STRUTTIN KUBES	Star Manufacturing
STUCKEY CIGAR	Stuckey Cigar
SUCCESS	American Novelty
SUCCESS	American Specialty
SUCCESS	Amusement Machine Company
SUCCESS	Automatic Machine & Tool
SUCCESS	W. R. Bartley
SUCCESS	Bartley & McFarland
SUCCESS	Paul E. Berger Manufacturing
SUCCESS	Caille Bros.
SUCCESS	Leo Canda
SUCCESS	Leo Canda Manufacturing
SUCCESS	Clawson Machine
SUCCESS	Clawson Slot Machine
SUCCESS	Columbia Manufacturing
SUCCESS	A. Feinberg Company
SUCCESS	Hamilton
SUCCESS	D. Kernan
SUCCESS	Little Casino
SUCCESS	Mills Novelty
SUCCESS	D. N. Schall
SUCCESS	Sicking
SUCCESS	Tibbils
SUCCESS	Watling
SUCCESS, THE	R. J. White
SUCCESS CARD MACHINE	Leo Canda
SUDS	Pace
SUM FUN	Bally
SUNBURST	Caille Bros.
SUNKEN TREASURE	Standard Novelty
SUPERIOR	Mills Novelty
SUPERIOR CONFECTION VENDOR	Superior Confection
SUSIE Q	Exhibit Supply Company

SWEETHEART	Bally
SWEET SALLY	Exhibit Supply Company
SWEET SIXTEEN	Genco
SWEEP STAKES	Western Electric Piano Company
SWEEPSTAKES	Amusement Products
SWEET MUSIC	Jennings

T

TALLY	Daval
TANGO	Exhibit Supply Company
TAP-A-PENNY	Tap-It Manufacturing
TAP-A-PENNY FORTUNE TELLER	Tap-It Manufacturing
TARGET	Blue Bird Sales
TARGET	Pace
TARGET	Royal Novelty (Indianapolis)
TARGET	Specialty Coin
TARGET, THE	Jennings
TARGET, THE	Silver King Novelty
TARGET PRACTICE	Ad-Lee Company
TARGET PRACTICE	Banner Specialty
TARGET PRACTICE	California Sales
TARGET PRACTICE	Davis Novelty
TARGET PRACTICE	Industry Novelty
TARGET PRACTICE	Mills Novelty
TARGET PRACTICE	Pace
TARGET PRACTICE	Reliable Coin
TARGET PRACTICE	Rex Novelty
TARGET PRACTICE	Specialty Manufacturing
TARGET PRACTICE, 1920 (NO.73)	Silver King Novelty
TARGET PRACTICE, 1922 (NO.75)	Silver King Novelty
TARGET PRACTISE	Iowa Novelty
TARGETS	Pierce Tool
TAVERN	Ohio Specialty
TAVERN VENDER	Groetchen Tool
TEASER	Bally
TEN POCKET	Frederick W. Bishop
TEN TO ONE	Clawson Slot Machine
10 TO 1	Troxler Novelty
TEXAS LEAGUER	Automat Games
36 GAME	Exhibit Supply Company
36 LUCKY PLAY PEE-WEE	Monarch Sales
36 LUCKY SPOT	A. B. T.
36 LUCKY SPOT MIDGET	Fey
36 LUCKY SPOT MIDGET	Mills Sales
36 ROULETTE	A. B. T.
36 ROULETTE	Fey
36 ROULETTE	Sherman & Fey
3-A-LIKE	Arkansas Novelty
THREE ARROW	Watling
THREE BALL	Clawson Slot Machine
THREE CADETS	A. B. T.
THREE CADETS	Fey
3 DIAL FORTUNE	Murray, Spink
3 DIAL FORTUNE	Willard A. Smith
3-IN-A-ROW	A. B. T.
3-IN-1	Atlas Manufacturing
3-IN-1	Fey
3-IN-1	L. C. Graham
THREE LITTLE BONES	Northland Vending
THREE OF A KIND	Garden City Novelty
THREE PENNY	(Unknown)
3-PLAYER POKER	Crooks & Crooks
THREE SPINDLE	Fey
THREE SPINDLE	Reliance Novelty
THREE-WAY DIVIDEND PRODUCER	Great Western Products
THREE WAY SELECTIVE MINT VENDER	Superior Confection
TIA JUANA	National Coin Machine Exchange
TIC-TAC-TOE	Exhibit Supply
TICKETTE	Mills Novelty
TID-BIT	Munves Corporation
TIGER	Caille Bros.
LE TIGRE	Caille Bros.
TILT TEST	ABCO Novelty
TILT TEST	Atlas Games
TILT TEST	Auto-Bell
TINY TIM GOLF	International Mutoscope
TIP, THE	Tip Slot Machine
TIP THE BELL HOP	International Mutoscope
TIPSY TUMBLERS	A. M. Walzer
TIT-TAT-TOE	Daval
TOPSY TURVY	Kelly Novelty
TOSS A COIN - HIT HITLER	Wisconsin Deluxe
TOSS A COIN - MUSS UP MUSSOLINI	Wisconsin Deluxe
TOVILO	Tivoli Automatic Machine
TOWER WONDER WORKER	B. Madorsky
TNT	Ad-Lee Company
T.N.T.	Gist Cabinet
TOBACCO PAK	Gottlieb
TOKETTE	Paul Bennett & Company
TOKETTE	J. M. Sanders
TOP-ITS	A. B. T.
TOM MACK	Western Distributors
TOPSY	Bally
TOPSY TURVY DERBY	Swanson
TORNADO	Superior Products
TOT	Calvert Novelty
TOT	Western Products
TOT GUM VENDOR	Western Products
TOTALIZER	Scientific Machine
TOTEM	Totem Manufacturing
TOUCHDOWN	ABCO Novelty
TOUCHDOWN	Auto-Bell
TOWER, THE	Idea Novelty
TRACK REELS	Daval
TRADE STIMULATOR	Mooney & Goodwin
TRADE VENDING MACHINE	Leo Canda
TRADE VENDING MACHINE	Samuel Nafew
TRADER, THE	Industry Novelty
TRADER, THE	Mills Novelty
TRADER, THE	R. J. White
TRADER, ROYAL	Royal Novelty (San Francisco)
TREASURY BANK	Jennings
TRIANGLES	Hercules Novelty
TRI-COLOR DICE	A. and F. Engineering
TRIOGRAPH, THE	Charles C. Bishop & Company
TRI-O-PACK	Daval
TRIPLE ROLL	Fey
TRIPLE ROULETTE	Fey
TRIPLE ROULETTE	Fey
TRIPLE ROULETTE (ON THE LEVEL)	Wisconsin Novelty
TRIPLEX	Wisconsin Novelty
TROJAN	Peerless Manufacturing
TROPHY	Reliance Novelty
TRUE-DICE	Daval
TRY IT	Ad-Lee Company
TRY IT	Blue Bird Sales
TRY ME	Try Me Manufacturing
TRY-SKILL	Novix Specialties
TRY SKILL	Van-White Novelty
TRY YOUR FORTUNE	Clawson Slot Machine
TRY YOUR LUCK	Jonas D. Bell & Company
TRY YOUR LUCK	Clawson Slot Machine
TUMBLER	Cardinal
TUNNEL	Bally
TURF	Automatic Coin Machine
TURF	Garden City Novelty
TURF FLASH	Groetchen Tool
TURF TIME	Exhibit Supply Company
TURTLE SOUP	Five Boro Machine
TUXEDO	California Machine
TUXEDO	Sundwall
"21"	Daval
TWENTY ONE	Groetchen Tool
21 BLACK JACK	Jennings
21 GAME	ABCO Novelty
21 VENDER	Groetchen Tool
21 VENDER	Jennings
TWENTY SIX	Rubini Cigar
26 GAME	Atlas Novelty
26 GAME	G. F. Hochriem
20TH CENTURY PROSPECTOR	Columbia Novelty
TWICO	Twico Corporation
TWIN-ROLL	Frey
TWINS WIN	Liberty
TWO ARROW	Maley
TWO FOR ONE SKILL	Leonhardt
TWO-IN-ONE	Midwest Novelty
2-IN-1 BABY VENDER	Field
2-IN-1 BABY VENDER	Field Paper Products

TWO PENNY RACING MACHINE	Boyce Coin
2-PLAYER POKER	Crooks & Crooks

U

UMBRELLA, THE	(Unknown)
UMPIRE	Mills Novelty
UNIVERSAL	Pierce Tool
UNIVERSAL ADVERTISER	Universal Advertising
UNIVERSAL ADVERTISING MACHINE	United States Bulletin
UNIVERSAL ADVERTISING MACHINE	Universal Advertising Machine
UPRIGHT CARD	Leo Canda Manufacturing
UPRIGHT CARD MACHINE	Leo Canda
UPRIGHT CARD MACHINE	Mills Novelty
UPRIGHT FIGARO	Leo Canda
UPRIGHT PERFECTION	Hudson Moore
UPRIGHT PERFECTION	Maley
UPRIGHT PERFECTION	Mills Novelty
UPRIGHT PERFECTION	T. J. Nertney
UPRIGHT-PERFECTION	Ogden & Company
UPRIGHT PERFECTION CARD	Mills Novelty
UPRIGHT POLICY	Leo Canda

V

"V"	Blake Manufacturing
"V"	Planet Manufacturing
VAUDETTE	Davis Novelty
VEMCO CONFECTION VENDER	Vendor Manufacturing
VENDOR	Bally
VEST POCKET BASEBALL	Field Manufacturing
VEST POCKET BASKET BALL	Peo Manufacturing
VEST POCKET COUNTER TOP BASEBALL	Field manufacturing
VETERAN	Western Automatic Machine
VICTOR	W. H. Clune
VICTOR	Drobisch Brothers
VICTOR	Mills Novelty
VICTOR	Reliance Novelty
VICTOR CALIFORNIA JACK	Mills Novelty
VICTORY	Auto-Bell
VICTORY TRADE MACHINE	Jonas D. Bell & Company
VOTING MACHINE	Clarence M. Kemp

W

WAGON WHEELS	A. B. T.
WAGON WHEELS	Hawkeye Novelty
WAHOO	Osbrink Games
WALL STREET BANK	Perfection Novelty
WAMPUM	Bally
WASP	Caille Bros.
WATCH YOUR MONEY	Huntley Russell
WATCHEM	Electra Corporation
WEE GEE	Boyce Coin
WEE GEE	Peo Sales
WHEEL, THE	Overton Manufacturing
WHEEL OF FORTUNE	M. O. Griswold
WHEEL OF FORTUNE	Kellog & Company
WESTERN WHEEL	Western Company
WHIM	Factories Sales
WHIRL-O-BALL	Linbrite Specialty
WHIRL POOL	Hawkeye Novelty
WHIRL-SKILL	Garden City Novelty
WHIRL-WIND SKILL CONTEST	D. Robbins
WHIRLWIND	Field Manufacturing
WHIRLWIND	Pierce
WHIZ-BALL	Pace
WHIZ-BOWLER	Binks Industries
WHOOPEE BALL	Field Manufacturing
WHOOPEE BALL	John Goodbody
WHOOPERDOO	Western Products
WILD CHERRIES	Osbrink
WILD DEUCES	Mills Novelty
WILDCAT, THE	John Goodbody
WIN	Exhibit Supply Company
WIN A BEER	Quality Supply
WIN-A-PACK	Buckley
WIN-A-PACK	Daval
WIN-A-PACK	General Novelty
WIN-A-PACK	Jennings
WIN-A-PACK	Pierce
WIN-A-SMOKE	Daval
WIN-O	H. C. Evans
WIN OR LOSE	Bally
WIN YOUR SMOKES	Quality Supply
WINDMILL	Standard Games
WINDMILL JR.	Standard Games
WINGS	ABCO Novelty
WINGS	Groetchen Tool
WINNER	Keystone Novelty & Manufacturing
WINNER	United States Novelty
WINNER DICE	Caille Bros.
WINNER DICE	Star Novelty
WINNER DICE	Unit Sales
WINNER DICE	Watling
WINNER DICE	Winner Novelty
WINNER ROULETTE	Cowper
WINNER ROULETTE	Stirrup
WIZARD	Best Novelty
WIZARD CLOCK	Loheide
WIZARD CLOCK	Pogue, Miller
WIZARD CLOCK	Progressive Manufacturing
WIZARD CLOCK	W. M. White
WONDER BELL	Tippecanoe Manufacturing
WONDER VENDER	Chicago Slot Machine Exchange
WORLD SERIES BASE BALL	Parmly Engineering
WORLD'S FAIR	Field
WRIGLEY DICE MACHINE	Dunn Brothers
WRIGLEY PERFECTION	Dunn Brothers
WRIGLEY'S SLOT MACHINE	Jonas D. Bell & Company

X

X-RAY	Daval

Y

YALE WONDER CLOCK	Yale Wonder Clock Company
YANKEE	ABCO Novelty
YANKEE	Groetchen Tool
YOUR DEAL	Reel Profits
YOUR NEXT (YOU'RE NEXT)	Mills Novelty
YUCCA	Zeno Manufacturing

Z

ZENO	David Rosen
ZEPHYR	ABCO Novelty
ZEPHYR	Groetchen Tool
ZIG-ZAG	Ad-Lee Company
ZIG-ZAG	Groetchen Tool
ZIG-ZAG	Hutchison Engineering
ZIP	Ad-Lee Company
ZIP	Buckley
ZIP-O	Midwest Novelty
ZIPPER	Binks Industries
ZODIAC	Wain & Bryant
ZOOM	Groetchen Tool
ZULU	Hunt And Company

PRICE GUIDE

Prices indicated are based on a simple "One Shot" price structure. If you know the key "One Shot" value you can handle just about any buying or selling situation with ease with the knowledge that you are within acceptable pricing ranges. Pricing is provided by the author based on data provided by a supporting panel of dealers in the field.

Here's how the "One Shot" price structure works:

Buying or Selling

The value given is the average going price for a good to excellent condition fully working original and unrestored machine with complete graphics either purchased from or sold by a reputable antique coin machine dealer.

As-Found

Use half of the value indicated for the selling or purchase price from an original owner, picker or estate auction. It is assumed this machine is working but in need of some work and possible restoration.

Basket Case

Use one quarter of value indicated, or even less depending on condition.

Restored Machine

Restorations are generally priced on the basis of the original cost of the machine to the restorer plus the added value of the amount of work put into the machine. As a rule of thumb, a restoration that brought an "As-Found" or a "Basket Case" back to acceptable standards will follow the "One Shot" price, or slightly less. Where the machine has been quickly brought back to acceptable looks with little concern for the original in terms of plating to the point of overrestoration the value is somewhat less than that. In the case of an elegant and accurate restoration, preserving the original integrity of the machine and its graphics, the value is a step up from the "One Shot" value.

Price values are in 1997 $US.

Page	Location	Name	Value
23	CL	Cie Jost RACE TRACK	250
23	BR	H. C. Evans MINIATURE RACE COURSE	300
24	T	RACE TRACK	400
25	TL	H. C. Evans Horse Race Layout	450
25	CR	J. and E. Stevens RACE COURSE	1850
26	TL	Chicago Nickel AUTOMATIC RACE TRACK	2600
26	TR	Excelsior EXCELSIOR	2600
26	BL	Fox EXCELSIOR AUTOMATIC	1850
26	BR	Excelsior EXCELSIOR	2600
27	TL	Unknown Race Game	2400
27	TR	National Table NATIONAL TABLE	500
27	BR	Western Electric DERBY	3750
28	TL	Whitlock THE DARBY	750
28	TR	Peerless THE RACE OF HORSES AND MARBLES	600
28	BL	Western Electric SWEEP STAKES	2500
29	TL	H. C. Evans SARATOGA SWEEPSTAKES SPECIAL	650
29	CR	Great States SANDY'S HORSES	450
29	BL	Rock-Ola OFFICIAL SWEEPSTAKES BALL GUM VENDER	1100
30	TL	A.B.T. HALF MILE	375
30	CR	Maitland MISTIC DERBY	3500
30	BL	Daval DAVAL DERBY	375
31	CR	Young AUTOMATIC ROULETTE	2600
31	BL	Mansfield ROULETTE	650
32	TL	Western Automatic IMPROVED ROULETTE	750
32	BR	Merriam Collins PEERLESS ADVERTISER	3200
33	TL	National LITTLE MONTE CARLO	4200
33	BR	Mills IMPROVED LITTLE MONTE CARLO	6500
34	TL	Mills LITTLE SCARAB	10500
34	TR	Wisconsin Novelty TRIPLE ROULETTE	1200
34	BL	Mills LITTLE SCARAB VENDER	12000
34	BR	Fey TRIPLE ROLL	1400
35	TL	Williams THE ADDER	2800
35	TR	Mills Sales MIDGET ROULETTE	550
35	BL	Fey MIDGET	550
35	BR	Michigan Novelty DETROITER	750
36	TL	Monarch PEE-WEE ROULETTE	475
36	TR	Keystone ROULETTE POKER	450
36	BL	Monarch BULL DOG GUM VENDER	850
37	TL	Roll-Etto ROLL-ETTO	450
37	CR	Garden City DE-LUXE VENDER	600
37	BL	Jennings LITTLE MYSTERY	550
38	CL	Griswold WHEEL OF FORTUNE	450
38	CR	Lichty AUTOMATIC SALESMAN AND PHRENOLOGIST	3200
39	TL	Unknown CIGAR WHEEL	850
39	BR	Unknown CIGAR ONE WHEEL	1750
40	TR	Decatur FAIREST WHEEL "Small Wheel"	700
40	BL	Griswold BLACK CAT	1200
41	T	Decatur IMPROVED FAIREST WHEEL "Large Wheel"	1000
41	BR	Waddell THE BICYCLE WHEEL	1600
42	TL	Waddell THE BICYCLE DISCOUNT WHEEL	1650
42	TR	Waddell THE BICYCLE DISCOUNT WHEEL "Large Square Wheel"	3800
42	BL	Waddell THE BICYCLE	3600
42	BR	Barnes CRESCENT CIGAR WHEEL	2200
43	TL	Barnes BONUS WHEEL	1650
43	TR	Nertney COINOGRAPH SALESMAN	2100
43	CR	Brunhoff AUTOMATIC VOTE RECORDER	6250
43	BL	Sun THE BICYCLE	3800
44	TL	Bennett COURT HOUSE	750
44	TR	Bennett SHENANDOAH UP-TO-DATE	750
44	BL	Decatur FAIREST WHEEL NO. 3	600
44	BR	Waddell W. W. Works THE BICYCLE WHEEL	1200
45	TL	Waddell W. W. Works THE BICYCLE WHEEL	1600
45	CR	Caille-Schiemer BUSY BEE	7500
45	BL	Caille BUSY BEE	6500
46	T	Griswold THE BIG THREE	6750
46	BL	Griswold STAR 1902	650
46	BR	Overton THE WHEEL	625
47	TL	Wain and Bryant ZODIAC	7500
47	TR	Caille SEARCHLIGHT	7500
47	BL	Mills BULLS EYE	8500
47	BR	Brunhoff SLOTLESS CIGAR CUTTER	3800

Page	Pos	Name	Price
48	TL	Ruff THE DEWEY	2500
48	TR	Caille WASP	4200
48	BL	Poole THE BICYCLE DISCOUNT WHEEL	1800
48	BR	Griswold STAR 1905	450
49	T	Pick FAIREST WHEEL	750
49	B	Pana IMPROVED FAIREST WHEEL	950
50	T	Caille LINCOLN	6500
50	CL	Griswold SELF PAY	1350
50	BR	Smokers Supply CIGAR DICE	450
51	TL	Griswold NEW STAR 1919	600
51	TR	Bally CUB	450
51	BL	Griswold NEW STAR 1922	450
52	TR	Exhibit SWEET SALLY	250
52	BL	Stock CORK TIP	1450
53	B	Gambling Layout	350
54	TR	McLoughlin GUESSING BANK	5500
54	BL	Griswold LAYOUT	450
55	TL	McLoughlin PRETTY WAITER GIRL	15000
55	CR	Murray, Spink 3 DIAL FORTUNE	3000
56	TL	World's Fair COLUMBIAN FORTUNE TELLER	1200
56	TC	Reliance THREE SPINDLE	3800
56	TR	Drobisch VICTOR	1000
56	BL	United States Novelty JOKER	3200
56	BC	Fey KLONDIKE	4500
56	BR	Drobisch STAR ADVERTISER	1200
57	TL	Drobisch THE LEADER	1200
57	TC	Ogden DEWEY SALESMAN	9000
57	TR	Watling THE FULL DECK	3800
57	BL	Waddell THE BOOMER	1250
57	BC	Comstock PERFECTION WHEEL	1600
57	BR	Page "Drinks" SALES INCREASER	1600
58	TL	Page "Free Merchandise" SALES INCREASER	1400
58	TR	Keeney and Sons MAGIC CLOCK	250
58	BL	W. C. Steel Ball PUT AND TAKE	750
58	BR	M-R Advertising M-R	250
59	TL	Jennings SWEET MUSIC	700
59	TR	Unknown SWEET SIXTEEN	150
59	BL	Mooney and Goodwin TRADE STIMULATOR	250
59	BR	Unknown DUNSTAN'S CIGAR SELLER	450
60	TR	Amusement Machine BABY CARD MACHINE	2200
62	CL	Tibbils CARD MACHINE	1800
62	CR	Amusement Machine STANDARD	1800
63	TL	Sittman and Pitt LITTLE MODEL CARD MACHINE	1650
63	CR	Nafew LITTLE MODEL CARD MACHINE	1500
64	TL	Watson COMBINATION CARD AND DICE MACHINE	5200
64	BR	Canda GIANT	5500
65	TL	Nafew MODEL CARD MACHINE	1600
65	BL	Unknown I.X.L. JR.	1800
65	BR	Holtz CARD MACHINE	1800
66	TL	Reliance RELIANCE	2000
66	TC	Holtz BROWNIE	2600
66	TR	Reliance STANDARD	2400
66	BL	Reliance VICTOR	2300
66	BC/BR	Reliance TROPHY	2400
67	TL	Canda JUMBO	1700
67	TC	Mills THE JUMBO	1500
67	TR/CR	Columbian Automatic AUTOMATIC CARD MACHINE	7500
67	BL	Canda COUNTER JUMBO	1400
67	BC	Canda PERFECTION CARD	850
68	TL	Mills THE LITTLE DUKE	4000
68	TC	Clune VICTOR	2000
68	TR	Canda CANDA CARD MACHINE	4000
68	BL	Salm (UNKNOWN)	1200
68	BC/BR	Columbia SUCCESS	1250
69	TL	Mills JUMBO SUCCESS NO. 2	1200
69	TC	Mills YOUR NEXT	4200
69	TR	Mills JUMBO SUCCESS NO. 4	1200
69	BL	Mills JOCKEY	2400
69	BC	Mills UPRIGHT CARD MACHINE	750
69	BR	Caille QUINTETTE	5200
70	TL	Wheeland PERFECTION	0000
70	TC	Portland OREGON	7500
70	TR	Royal Novelty ROYAL TRADER	4200
70	BL	Wheeland CALIFORNIA	4500
70	BC	Oakland OAKLAND	2700
70	BR	Mills LITTLE PERFECTION "Round Top"	650
71	TL	Mills UPRIGHT PERFECTION	750
71	TC	Mills IMPROVED JOCKEY	2500
71	TR	Mills COUNTER SUCCESS NO. 6	850
71	BL	Mills BEN FRANKLIN	1650
71	BC	Mills SUCCESS NO. 6	1200
71	BR	Hamilton THE HAMILTON	1200
72	TL	Canda JUMBO GIANT	3800
72	TC	Mills KING DODO (Three-Way)	3200
72	TR	Mills KING DODO (Five-Way)	3400
72	CL	Canda AUTOMATIC CARD MACHINE	1700
72	CR	Mills RELIABLE	12500
72	BC	Mills SUCCESS NO. 8	1000
73	TL	Watling THE CLOVER	1600
73	TC	Mills COMMERCIAL	2000
73	TR	Mills SPECIAL COMMERCIAL	2300
73	BL/BC	Feinberg SUCCESS	1200
74	TL	Mills THE TRADER	3200
74	TC	Mills DRAW POKER	4000
74	TR	Caille HY-LO	4200
74	BL	Mills SUPERIOR	4500
74	BC	Caille GOOD LUCK	650
74	BR	Mills HY-LO	4000
75	TL	Foley DRAW POKER	3800
75	TC	Caille DRAW POKER	3400
75	TR	Mills VICTOR	2200
75	B	Caille GLOBE	7500
76	TL	Caille MAYFLOWER "Style A"	3200
76	TC	Unknown ROYAL CARD MACHINE	700
76	TR	Mills LITTLE PERFECTION "Flat Top"	550
76	BL	F. W. Mills JOCKEY	2900
76	BR	Mills NEW JOCKEY "Plain Jockey"	1600
77	TL	Groetchen POK-O-REEL	400
77	TC	Daval JACKPOT CHICAGO CLUB-HOUSE Model No. 3	500
77	TR	Daval JACKPOT CHICAGO CLUB-HOUSE Model No. 5	500
77	B	Happy Jack HAPPY JACK	2400
78	TL	Buckley PILGRIM VENDER	375
78	TC	Daval "Gold Medal" CHICAGO CLUB-HOUSE Model No. 7350	
78	TR	Groetchen 21 VENDER	450
78	BL	Pierce JACKPOT THE NEW DEAL	375
78	BC	Daval "Gold Medal" JACKPOT CHICAGO-CLUB HOUSE Model No. 6	500
78	BR	National Coin DRAW POKER	500
79	TL	A.B.T. ROYAL REELS	450
79	TC	A.B.T. ROYAL REELS GUM VENDER	475
79	TR	Jennings CARD MACHINE	400
79	BL	Pierce JACKPOT HIT ME	375
79	BR	Pierce HIT ME	350
80	TL	Pace THE NEW DEAL JAK-POT	500
80	TC	Pace CARDINAL JAK-POT	450
80	TR	Jennings 21 BLACKJACK	450
80	CL	Pace CARDINAL	425
80	C	National Coin DRAW POKER GUM VENDER	500
80	CR	Buchanan THE AUSTRALIA	2600
80	BC	Reel Profits YOUR DEAL	375
81	TR	Pearsall and Finkbeiner "FIRE ENGINE"	3200
81	BL	Pearsall and Finkbeiner THE (RED) STAR	3500
82	TL	Watling GOOD LUCK RACE HORSE	2000
82	CR	Sheffler COMMERCE	900
83	TR	Groetchen SOLITAIRE	600
83	BL	Pierce WHIRLWIND	750
83	BR	Groetchen GOLD RUSH	650
84	TL	Rock-Ola RADIO WIZARD	450
84	TC	Fey THREE CADETS	600
84	TR	Fey PLAY AND DRAW	850
84	BL	Fey SKILL DRAW (1935)	650
84	BC	Fey SKILL DRAW (1936)	550
84	BR	Exhibit SKILL DRAW	350
85	TL	Exhibit RED DOG	600
85	CR	Acme MATCH COLOR	275
85	BL	Star Amusement SPARKY	175
86	BL	Moore DICE BOX	1600
86	TR	Schloss DICE TOSSER AND CIGAR CUTTER	800
87	TL	Bishop THE TRIOGRAPH	4500
87	BR	Unknown GOOD LUCK	1200
88	T	Clawson AUTOMATIC DICE	4200
88	CL	Bucyrus DICE BOX	650
88	C	Griswold DICE MACHINE	3750
88	CR	DeGrain DICE GAME	250
88	BL	Keane Novelty SQUARE DEAL 1891	1400
88	BR	Keane Novelty SQUARE DEAL 1892	1500
89	TL	Colby COMBINATION LUNG TESTER	3800
89	TC	Lighton DICE SHAKER	1850

Page	Pos	Name	Price
89	TR	American Automatic AUTOMATIC DICE	1200
89	BL	Automatic Manufacturing POPE DICE MACHINE	1400
89	BR	Maley AUTOMATIC DICE MACHINE	1800
90	TL	American Automatic AUTOMATIC DICE SHAKING MACHINE	850
90	TR	Unknown "Triangle Dicer"	1200
90	BL	U. S. Novelty WINNER	950
90	BC	Unknown AUTOMATIC DICE	850
90	BR	Unknown "Spittoon Dicer"	650
91	TL	Unknown "Hex Dicer"	2400
91	TC	Nafew AUTOMATIC DICE MACHINE	1200
91	TR/CR	E.A.M. EAGLE	12000
91	BL	Drobisch NO.5 MONARCH DICE MACHINE	3400
91	BC	Erickson LOG CABIN	4400
92	TL	Mills I WILL	8500
92	TC	Ennis and Carr PERFECTION	2200
92	TR	United Automatic 6-WAY DICE	1400
92	BL	Ruff CRAP SHOOTERS DELIGHT	1500
92	BC/BR	Dunn WRIGLEY DICE	1200
93	TL	Bradford Novelty THE LARK	4800
93	TC	McClellan BOARD OF CRAPS AND GUM MACHINE	750
93	TR	Fey ON THE LEVEL	2800
93	BL/BC	Fey ON THE SQUARE	6000
93	BR	Mills ON THE LEVEL	4500
94	T	Mills CRAP SHOOTER	6500
94	C	Mills MILLS PIPPIN	12000
94	BL	Fey AUTOMATIC DICE BOX	650
94	BR	Specialty Machine PORTOLA	700
95	TL	Royal DICE	4600
95	TC	Imperial IS IT ANY OF YOUR BUSINESS	900
95	CL	Acme Sales THE ACME	900
95	C	Keystone WINNER	550
95	BC	Star Novelty WINNER DICE	600
96	TL	Exhibit DICE FORTUNE TELLER	2600
96	TR	Fey MIDGET WITH SALESBOARD	650
96	BL	Southern Novelty CHUCK-O-LUCK	550
96	BC	Mills Sales 36 LUCKY SPOT MIDGET	625
96	BR	Fey 3-IN-1	1200
97	TL	Monarch Sales 36 LUCKY PLAY PEE-WEE	625
97	TC	Keystone Novelty CHUCK-O-LUCK	550
97	TR	Superior Confection AUTO-DICE AMUSEMENT TABLE	425
97	BL	A.B.T. 36 LUCKY SPOT	650
97	BC	Exhibit JUNIOR	450
97	BR	Keystone Sales KEYSTONE	350
98	TL	Bowman DIXIE DICE	450
98	TC	Ad-Lee FOUR WAY FROLIC	575
98	TR	Chicago Coin SHOOTEM	325
98	BL	Ad-Lee CRYSTAL GAZER	675
98	BC	New Era JUMPING JACK	450
98	BR	Buckley BABY SHOES	275
99	TL	Bally BOSCO	300
99	TC	Bally DICETTE	450
99	TR	Pioneer BIG BONES	500
99	BL	Bally BALRICKEY	200
99	BC	Ad-Lee CUBA	350
99	BR	Palamedes LOTTA DOE	475
100	TL	Groetchen DICE-O-MATIC VENDER	425
100	TC	Stone KENTUCKY DERBY	250
100	TR	Stephens FLIP	450
100	BL	Garden City CHERRY JITTERS	1200
100	BC	New Era NEW ERA VENDER	450
100	BR	Exhibit SELECT-EM	275
101	TL	Fey FAIR-N-SQUARE	550
101	TC	Kalamazoo KAZOO	250
101	TR	Pacific P.D.Q.	250
101	BL	Fort Wayne LITTLE JOE	550
101	BC	Pierce PIX-IT	350
101	BR	Pacific ELECTRIC SPINNER	250
102	TL	National Coin TIA JUANA	250
102	TC	Gottlieb INDIAN DICE	650
102	TR	Withey IMPROVED SEVEN GRAND	550
102	BL	Gottlieb DAILY RACES JR.	300
102	BC	Exhibit 36 GAME	550
102	BR	Mikro-Kall-It MIKRO-KALL-IT	650
103	TL	Norris ROLL-EM	550
103	TR	Keeney JITTER BONES	240
103	BL	Baker PICK-A-PACK	550
103	BC	Victor ROLL-A-PACK	350
103	BR	Dependable DICE-MAT (10-Column)	450
104	TL	Quality Supply HORSES	150
104	TR	Jess SHAKE, RATTLE AND ROLL	125
104	BL	Quality Supply HI HAND	175
104	BR	Twico TWICO	150
105	TR	Maley NICKEL TICKLER NO. 2	1700
105	BL	Canda TRADE VENDING MACHINE	3800
106	T	Western Weighing IMPROVED NICKEL TICKLER	1900
107	TR	Unknown THE COMBINATION	1750
107	BL	Consolidated Coin 200	1850
108	CL	Canda EAGLE	2600
108	CR	Drobisch ADVERTISING REGISTER	1200
109	TL	Cook HOWARD'S FAVORITE	850
109	TC	Comstock THE PERFECTION	750
109	TR	Klondike THE KLONDIKE	350
109	BL	Hillsboro THE HILLSBORO	1250
109	BC	Bell (WRIGLEY'S) DEWEY PIN MACHINE 600	
109	BR	Unknown "Tiered Backdrop"	650
110	TL	Comstock PERFECTION WHEEL	1600
110	TC	Hamilton DAISY NO BLANK "Bread Loaf Top"	300
110	TR	Fisher (ORIGINAL) PREMIUM	500
110	C	Zeno YUCCA	250
110	BL	Miller LITTLE DREAM PLAY BASEBALL	275
110	BR	Sloan THE LEADER	350
111	TL	Hamilton DAISY "Diamond Top"	300
111	TC	Bradford LITTLE GEM	350
111	TR	Mills THE PREMIUM	475
111	BL	Hamilton DAISY "Advertising Diamond Top"	450
111	BC	Cawood PLAY BALL	350
111	BR	Knight OUR LEADER	350
112	TL	Banner Specialty LEADER	350
112	TC	Boyce WEE GEE	375
112	TR	Jackson CAN'T LOSE	150
112	BL	Unknown OUR LEADER	350
112	BC	Jackson NEVER LOSE	175
112	BR	Northwest Coin CHURCHILL DOWNS	525
113	TL	Pierce WIN-A-PACK	350
113	TC	Exhibit THE BOUNCER	400
113	TR	Atlas CIVILIAN DEFENSE	450
113	BL	Exhibit DOUBLE DICE	450
113	BC	Five Boro DUCK SOUP	225
113	BR	Baker BOMB HIT	650
114	TL	Runyan Sales KEEP EM BOMBING	750
114	TR	Unknown BOUNCERINO	220
114	BR	Frantz BULLS EYE	95
115	TR	Latimer GAME O' SKILL 1895	600
115	BL	Latimer GAME O' SKILL 1893	600
116	TL	Unknown LITTLE HELPER	575
116	BR	Fisher LEGAL	750
117	TL	Unknown GAME O' SKILL	550
117	CR	Mills TARGET PRACTICE	450
117	BR	Mills STAR TARGET PRACTICE	400
118	TL	Silver King THE TARGET	500
118	TC	National Coin NATIONAL TARGET PRACTICE	300
118	TR	Exhibit PLAY BALL	850
118	BL	Matheson ORIGINAL AUTOMATIC VENDER	450
118	BC	Unknown TARGET PRACTICE	300
118	BR	Mills NEW TARGET PRACTICE	425
119	TL	Jennings THE TARGET	475
119	TC	Jennings FAVORITE (PEANUTS)	1700
119	TR	Coin Sales MILLARD'S MINIATURE BASEBALL	450
119	BL	Pace TARGET PRACTICE	425
119	BC	Blue Bird BASE BALL TARGET	275
119	BR	Hiawatha TARGET PRACTICE	500
120	TL	Robbins AUTOMATIC BASEBALL	250
120	TC	Robbins PENNY CONFECTION VENDER	950
120	TR	Calvert SPOOK HOUSE	800
120	BL	Jennings FAVORITE (BALL-GUM)	1200
120	BC	Calvert ALL ABOARD	775
120	BR	Calvert GEM CONFECTION VENDER	850
121	TL	Atlas Indicator BASE BALL	750
121	TC	Great States BEER TARGET	850
121	TR	Great States HAPPY DAYS	2000
121	CL/BL	Great States FOOT BALL PRACTICE	550
121	BR	Skipper Sales SKIPPER	175
122	TL	Paupa and Hochriem THE ELK	3800
122	BL	Bell WRIGLEY'S SLOT MACHINE	850
123	TR	Mills SPECIAL	3000
123	BL	Paupa and Hochriem ELK	2800
124	CL	Paupa and Hochriem PILOT	3800

256

Ref	Name	Price
124 CR	Sundwall THE EAGLE	4200
125 TL	d'Abel Nau LES PETITES CHEVAUX	2000
125 BR	Mills UMPIRE	4200
126 TL	Mills L'AEROPLAN	3600
126 TC	Caille IMPROVED BASE-BALL	4500
126 TR	Silver King BASEBALL	4200
126 BL	Caille NEW SPECIAL TIGER	3800
126 BC	Caille THE COMET	2400
126 BR	F. W. Mills PREMIUM TRADER	3400
127 TL	Daycom REEL-O-BALL	2000
127 CR	Shelly SHELSPESHEL	1700
127 BL	Yendes REEL-O-BALL	2000
128 TR	Caille MAYFLOWER	2800
128 BL	Caille MERCHANT	3000
129 TR	Silver King BALL GUM VENDER	400
129 BL	Caille PENNY BALL GUM VENDER	800
130 B	Mills PURITAN BELL	550
131 TL	Silver King IMPROVED PENNY BELL	950
131 TC	Caille FORTUNE BALL GUM VENDER	850
131 CR	Engel IMP	700
131 BL	Monarch PENNY GUM VENDER	850
131 BC	Superior Confection FORTUNE BALL GUM VENDER	850
132 TL	Burnham and Mills BABY VENDER	750
132 TC	Jennings PURITAN GIRL	850
132 TR	Midwest JACK POT PURITAN BABY BELL	450
132 BL	Keystone KEYSTONE PURITAN BELL	550
132 BC	Superior Confection THREE WAY SELECTIVE MINT VENDER	1650
132 BR	Midwest THE ACE	350
133 TL	Field Paper KEYSTONE PURITAN BELL	550
133 TC	Mills THE BELL BOY	1100
133 TR	Chicago Mint PURITAN CONFECTION VENDER	600
133 BL	Giles and Simpkins HOLLYWOOD	400
133 BC	Lion LION PURITAN BABY VENDOR	400
133 BR	Pace DANDY VENDER	425
134 TL	Groetchen NEW DANDY VENDER	475
134 TC	Daval DAVAL GUM VENDER	450
134 TR	Chicago Mint JACK POT PURITAN CONFECTION VENDER	475
134 BL	Douglis DAVAL GUM VENDER	450
134 BC	Sanders BABY JACK POT VENDER	550
134 BR	Pierce WHIRLWIND	650
135 TL	Daval ULTRA MODERN DAVAL GUM VENDER JACKPOT	500
135 TR	Pace DANDY VENDER	450
135 BC	Dixie DIXIE BELL	675
135 BR	Pace JAK-POT DANDY VENDER	500
136 TL	Buckley ALWIN	400
136 TR	Bally BABY RESERVE	250
136 BL	Groetchen ZEPHYR	275
136 BC	Groetchen IMP	120
136 BR	Daval CUB	140
137 TL	Groetchen LIBERTY	200
137 TR	Western Products TOT	350
137 BL	Daval AMERICAN EAGLE	175
137 BR	Groetchen YANKEE	275
138 TR	Lion PURITAN BABY BELL	375
138 BL	Monarch PENNY GUM VENDER (Cigarette)	800
139 TR	Groetchen SILENT DANDY VENDER	500
139 BL	Superior Confection CIGARETTE BALL GUM VENDER	500
140 BL	Pace HOL-E-SMOKES	425
140 BR	Silver King LITTLE PRINCE	850
141 TL	National Coin PENNY KING	350
141 CR	Ohio Novelty THE BARTENDER (STYLE B)	500
141 BL	Jennings LITTLE MERCHANT	550
142 TL	Groetchen PENNY SMOKE	375
142 TC	Buckley CENT-A-PACK	350
142 TR	Superior Confection CIGARETTE SALESMAN MINT VENDER	625
142 BL	Pace HOL-E-SMOKES	525
142 BC	Superior Confection CIGARETTE GUM VENDER	500
142 BR	National Coin SMOKES	475
143 TL	Daval SPIN-A-PACK	450
143 TC	Bally BABY	175
143 TR	Garden City GEM	125
143 BL	Daval CENTA-SMOKE	400
143 BC	Jennings STAR PENNY PLAY	150
143 BC	Jennings STAR VENDER	150
143 BR	Groetchen SMOKE HOUSE	450
144 T	Buckley DELUXE CENT-A-PACK	450
144 CL	Groetchen ZEPHYR (Cigarette)	250
144 CC	Western Equipment MATCH-EM	150
144 CR	Garden City PRINCE	175
144 BL	Groetchen GINGER	195
145 TL	Jennings GRANDSTAND	650
145 TC	Groetchen SPARKS DELUX	275
145 TR	Daval 1940 PENNY PACK	275
145 BL	Norris LUCKY	1100
145 BC	Bally WAMPUM	300
145 BR	Daval TALLY	325
146 T	Groetchen MERCURY	200
146 CL	Sanders TOKETTE	150
146 C	Daval JIFFY	275
146 CR	Daval MERCURY VISIBILITY	225
146 BL	Daval COMET	375
147 TL	Daval MARVEL BALL GUM VENDER	225
147 TC	Groetchen SPARKS CHAMPION	250
147 C	Sanders LUCKY PACK	200
147 CR	Sanders ZIP	150
147 BL	Groetchen IMP	120
148 TL	Groetchen WINGS	300
148 TR	Groetchen ATOM	165
148 BL	Daval FREE-PLAY	350
148 BR	Comet NON-COIN COMET	175
149	Kelley THE KELLEY	1200
150 TR	Caille REGISTER	4200
150 BL	Kelley THE IMPROVED KELLEY	1450
151 TL	Cowper PURITAN	1000
151 BL	Mills IMPROVED PURITAN	700
151 BR	Caille BON-TON	4800
152 TR	Wall HUSTLER	1250
152 BL	Mills PURITAN	550
152 BR	Keeney and Sons BABY VENDER	750
153 TL	Midwest LION PURITAN BABY VENDOR	500
153 TC	A.B.C. JOCKEY CLUB	800
153 TR	Dean PENNY DRAW	225
153 BL	Midwest PURITAN BABY BELL	450
153 BC	Wonder Novelty SQUARESPIN	2200
153 BR	Rock-Ola HOLD AND DRAW	700
154 TL	Stephens MAGIC BEER BARREL	700
154 TC	Mills DIAL	1100
154 TR	Groetchen THE FORTUNE TELLER	550
154 BL	Groetchen TURF FLASH	850
154 BC	Groetchen TAVERN (STYLE A)	500
154 BR	Pace SUDS	500
155 TL	Pace NEW DEAL JAK POT	550
155 TC	Daval REEL 21	350
155 TR	Daval REEL DICE	400
155 BL	Daval CLEARING HOUSE	325
155 BC	Daval RACES	350
155 BR	Western Equipment REEL RACES	350
156 TL	Groetchen HIGH STAKES	500
156 TC	Groetchen HIGH TENSION	550
156 TR	Buckley GOLDEN HORSES	475
156 BL	Groetchen TWENTY ONE	550
156 BC	Buckley HORSES	450
156 BR	Buckley MUTUEL HORSES	475
157 TL	Groetchen DIXIE SPELLING BEE	500
157 TC	Bally THE NUGGET	150
157 TR	Garden City BABY JACK	200
157 BL	Daval REEL SPOT	350
157 BC	Bally GOLD RUSH	175
157 BR	Four Jacks CLUB JACK	200
158 TL	Daval TRACK REELS	350
158 TC	Mills WILD DEUCES	350
158 TR	Bennett DOUGH BOY	175
158 BL	Mills KOUNTER KING	400
158 BC	Bennett DEUCES WILD	175
158 BR	Bennett SKILL TEST	175
159 TL	Daval X-RAY	350
159 TC	Groetchen LIBERTY SPORTS PARADE	175
159 TR	Daval "21"	250
159 BL	Daval HEADS OR TAILS	350
159 BC	Groetchen SPARKS CHAMPION	300
159 BR	Groetchen KLIX	350
160 TL	Groetchen KLIX	300
160 BR	Comet KING	175
161 BL	Ganss GAME TABLE	500
161 TR/BR	Weston SLOT MACHINE	4500
162 TR	Clawson THE NEWARK RAINBOW	2750
162 BL	Clawson LIVELY CIGAR SELLER NO. 2	1800

Page/Loc	Maker & Model	Price
163 TR	Brunhoff SPINNING TOP	3400
163 CL/BR	Brunhoff DIAMOND EYE	3800
164 TR/B	Mills THE MANILA	10000
165 B	Yale YALE WONDER CLOCK	6500
166 TL	Kelley FLIP FLAP (LOOP-THE-LOOP)	1750
166 TC	Park RED BIRD	900
166 TR	Twentieth Century SPIRAL	1800
166 BL	Mills LITTLE KNOCKER	5500
166 BC	Loss SKILLIARD	1800
166 BR	Parker SPIRAL	1750
167 TL	Yale AUTOMATIC CASHIER AND DISCOUNT MACHINE	6500
167 TC	Dunn WRIGLEY PERFECTION	1200
167 TR	Loheide WIZARD CLOCK ("2-Column")	475
167 BL	Dunn PERFECTION	1200
167 BC	Progressive AUTOMATIC TRADE CLOCK ("2-Column")	650
167 BR	Loheide WIZARD CLOCK ("4-Column")	070
168 TL	Pogue, Miller WIZARD CLOCK ("4-Column")	1100
168 TC	Jentzsch and Meerz BAJAZZO (CLOWN)	1500
168 TR	Caille JESTER	1850
168 BL	National Institute A GOOD TURN	850
168 BC	A.B.T. PLAY BALL	1400
168 BR	Whiteside BASE BALL GAME OF SKILL	650
169 TL	Burnham and Mills PLAY BALL	1800
169 TC	Webb MIRROR OF FORTUNE	750
169 TR	R. and S. CHIP GOLF GAME	800
169 BL	National Novelty BASE BALL	750
169 BC	Erie THE EMCO (NERVE EXERCISE)	325
169 BR	Arcade Supply THE CLOWN	1350
170 TL	Calvert INDIAN SHOOTER	1100
170 TC	Unknown FLYING ACES	700
170 TR	Jentzsch und Meerz 1930 BAJAZZO	650
170 BL	Novix TRY-SKILL	225
170 BC	A.B.T. THE SIX HORSEMEN	4,500
170 BR	International Mutoscope JUGGLING CLOWN	1450
171 TL	Dean EVERREADY TRADE BOARD	650
171 TC	Superior Confection AUTOMATIC CARD TABLE	750
171 TR	Gottlieb BASEBALL ROLL-IT	375
171 BL	Unknown CHAMPION SPEED TESTER	1600
171 BC	Peo LITTLE WHIRLWIND	275
171 BR	Gottlieb ROLL-IT	275
172 TL	Unknown LARK	325
172 TC	Lilliput LILLIPUT GOLF	2500
172 TR	Gottlieb MINIATURE BASEBALL	475
172 BL	B. Madorsky FOOTBALL	1400
172 BC	Pace WHIZ-BALL	250
172 BR	Parmly THE LUNATIC	500
173 TL	Goodbody THE WILDCAT	300
173 TC	Pace DEUCES WILD	350
173 TR	E. E. Junior PLAY BASKETBALL "3 Shots"	1350
173 BL	Peo PLAY POKER	275
173 BC	E. E. Junior PLAY FOOTBALL	1450
173 BR	International Mutoscope TIP THE BELL HOP	750
174 TL	Field VEST POCKET COUNTER TOP BASEBALL	300
174 TC	Sloan THE POKERALL GAME	375
174 TR	Mills CATCH THE BALL	650
174 BL	Marcus SCOTCH GOLF	375
174 BC	Bally BALLY	475
174 BR	Field DING THE DINGER	600
175 TL	Dean VAUDETTE	400
175 TC	Pierce BUGG HOUSE	550
175 TR	Peerless ODD PENNY MAGNET	325
175 BL	Buddick PAR-KET	750
175 BC	Stephens ODD PENNY MAGNET	275
175 BR	Hercules BUY AMERICAN	400
176 TL	Garden City PICK A PACK	550
176 TC	Pierce CIG-O-ROLL	275
176 TR	Pierce GYPSY	350
176 BL/BC	Royal BOUNCING BALL	575
176 BR	Stephens LUCKY COIN TOSSER	375
177 TL	Lu-Kat Novelty LUCAT	6500
177 TC	Berger CATCH-N-MATCH	450
177 TR	Unknown PLACE THE COPPERS	725
177 BL	Groetchen PUNCHETTE	350
177 BC	Sands SKILL KATCH	375
177 BR	Unknown QUEEN TOP	350
178 TL	Mueller Specialties CHANGE MASTER	275
178 TC	Star FLIP FLOP FLOOZIE	325
178 TR	Fortune Sales LUCKY ROLL	800
178 BL/BC	Kelly PENNY JITTER BALL	275
178 BR	Munves TID-BIT	500
179 TL	Mills FLIP SKILL	650
179 TC	Erie Machine BOMBER	450
179 TR	Scientific TOTALIZER	300
179 BL	Genco PUNCH A BALL	125
179 BC	Western Products OOMPH	800
179 BR	Howard Sales BRODI LIBERTY	25
180 TL	Groetchen PIKES PEAK	400
180 TC	Standard Coin SKILL-A-RETTE	450
180 TR	Pioneer Coin SMILEY	250
180 BL	Groetchen ZOOM	450
180 BC	Groetchen KILL THE JAP	3500
180 BR	Standard Games WINDMILL	375
181 TL	Standard Games WINDMILL JR.	375
181 TC/TR	Central HI FLY	475
181 CL/BC	Amusement Enterprises PITCHEM	400
181 BR	Central HIT A HOMER	475
182 TL	Associated Amusements BASKETBALL STARS	400
182 TC	Daval BEST HAND	350
182 TR	Daval OOMPH	600
182 B	Bonus Advertising ?JACK-POT BONUS?	350
183 TL	Marvel SLUGGER	400
183 TC	Binks WHIZ-BOWLER	350
183 TR	Marvel LUCKY HOROSCOPE	120
183 BL	Auto-Bell VICTORY	300
183 BC	Frantz POT OF GOLD	275
183 BR	UPI GOLD MINE	350
190 TL	Gorski/Clawson AUTOMATIC DICE	2500
190 TR	Amusement Machine STANDARD	1200
190 BL	Canda PERFECTION	650
192 TL	H. C. Evans EVANS BABY BELL	350
192 BL	Silver King PENNY BASE BALL	2200
193 TL	German "Allwin"	450
193 BL	British WIN AND PLACE	450
193 BR	Buffalo Toy JACK POT GAME NO. 265/266	55
194 TL	Waddell ROOLO	300
195 BL	Automatic Manufacturing LUCKY DICE	800
198 TR	Almy AUTOMATIC CASHIER AND DISCOUNT MACHINE	8500
199 BR	Daval SAFE HIT	400

INDEX

A.B.C. Coin Machine Company
    JOCKEY CLUB, 150, 153
Abel, Bernard, and Co.
    SQUARE DEAL, 86
A.B.T. Manufacturing Co., 33
    HALF MILE, 30
    MIRROR OF FORTUNE, 169
    PLAY BALL, 168
    ROYAL REELS, 79
    ROYAL REELS GUM VENDER, 79
    SIX HORSEMEN, THE, 170
    SKILL DRAW, 82
    36 LUCKY SPOT, 33, 97, 191
    36 ROULETTE, 33
    THREE CADETS, 82
Acme Game Co.
    DIZZY DISKS, 82
    MATCH COLOR, 82, 85
Acme Novelty Works, The
    ROULETTE, 32
Acme Sales Novelty Co.
    ACME, THE, 95
Ad-Lee Company
    CRYSTAL GAZER, 8, 98
    CUBA, 99
    FOUR WAY FROLIC, 98
    TARGET PRACTICE, 116, 117
Almy Manufacturing Co.
    AUTOMATIC CASHIER AND DISCOUNT MACHINE, 198
Ambercrombie and Fitch
    RACE TRACK, 24
American Automatic Machine Co.
    AUTOMATIC DICE, 89
    AUTOMATIC DICE SHAKING MACHINE, 90
    AUTOMATIC ROULETTE, 32
    IMPROVED AUTOMATIC ROULETTE, 32
American manufacturing Co.
    AUTOMATIC DICE, 89
    POPE DICE MACHINE, 89
American Mechanical Toy Co., 23
Amusement Enterprises
    PITCHEM, 181
Amusement machine Co., 8, 12, 13
    ARROW, 54
    BABY CARD MACHINE, 13, 60, 62, 65
    CARD MACHINE, 60, 61
    POLICY, 60
    "Rotating Toy", 12
    STANDARD, 61, 62, 190
Anthony Cigar Co.
    ECLIPSE, 54
Anthony and Smith
    ECLIPSE, 55
Arcade Supply Co.
    CLOWN, THE, 169
Associated Amusement
    BASKETBALL STARS, 182
Atlas Games
    CIVILIAN DEFENSE, 113
    TILT-TEST, 165
Atlas Indicator Works
    BASE BALL, 121
Atlas Manufacturing Co.
    IMPROVED ROULETTE, 32
August Grocery Co.
    HOO DOO, THE, 46
Australia, 80

Auto-Bell Novelty Co.
    HIT-A-HOMER, 183
    TILT-TEST, 165
    TOUCHDOWN, 183
    VICTORY, 183
Automatic Cash Discount Register Co.
    PROFIT SHARER, THE, 55
Automatic Machine Co.
    AUTOMATIC DICE, 204
    EMPIRE, 201
Automatic Manufacturing Co.
    LUCKY DICE, 195
Baker Novelty Co.
    BOMB HIT, 108, 113
    KICKER-CATCHER, 165
    PICK-A-PACK, 103
    SKILL-A-RETTE, 180
Bally Manufacturing Co., 8, 9
    BABY, 143
    BABY RESERVE, 143
    BALLY, 136
    BALRICKEY, 99
    BOSCO, 99
    CUB, 51
    DICETTE, 99
    GOLD RUSH, 157
    LITE-A-PAX, 162, 163
    NUGGET, THE, 157
    PONIES, 163
    SPINNERINO, 55
    WAMPUM, 145
Banner Specialty Co.
    BANNER TARGET PRACTICE, 116
    LEADER, 108, 112
Barnes, Monroe, Manufacturer, 39
    BONUS CIGAR WHEEL, 39, 42, 43
    CIGAR WHEEL, 39
    CRESCENT (CIGAR WHEEL), 39, 42
    FAIREST WHEEL, 39
Barr Novelty Co.
    DOUBLE TARGET, 117
Bat-A-Peny Corp.
    BAT-A-PENY, 202
Behn, Henry A., 23, 27
Bell, Jonas D.
    DEWEY, 107, 109
    HOW IS YOUR LUCK, 122
    NICKELSCOPE, 122
    PENNYSCOPE, 122
    WRIGLEY TRY YOUR LUCK, 122
    WRIGLEY'S SLOT MACHINE, 109, 122
Bennett and Co.
    COURT HOUSE, 44
    SHENANDOAH UP-TO-DATE, 44
Bennett, Paul, and Co.
    DEUCES WILD, 158
    DOUGH BOY, 158
    SKILL TEST, 158
Berger Manufacturing Co.
    CATCH-N-MATCH, 165, 177
Billiard Parlors, 18
Binks Industries
    WHIZ-BOWLER, 183
Bishop, Charles C, , and Co.
    TRIOGRAPH, THE, 87
Bishop, Frederick W.
    NINE POCKET, 107
    TEN POCKET, 107

Blue Bird Products Co., 117
Blue Bird Sales Corp., 117
    BASE BALL TARGET, 119
Boduwil
    PROMOTION SALES REGISTER, 163
Bonus Advertising System
    ?JACK-POT BONUS?, 182
Bowman Specialty Co.
    DIXIE DICE, 98
Boyce Coin machine Amusement Corp.
    WEE-GEE, 112
Bradford Novelty Co.
    LITTLE GEM, 111
    LITTLE GEM FORTUNE TELLER, 111
Bradford Novelty Machine Co.
    LARK, THE, 93
Britain, William H. (British), 23
    FOUR-HORSE RACE, 23
    MECHANICAL WALKING RACE, THE, 23
Brunhoff Manufacturing Co.
    AUTOMATIC VOTE RECORDER AND CIGAR SELLER, 43
    DIAMOND EYE, 161, 163
    SLOTLESS (CIGAR CUTTER), 47
    SPINNING TOP, 161, 163
Bryant Pattern and Novelty Co.
    ZODIAC, 41
Buchanan, A.O. (Australia)
    AUSTRALIA, THE, 80
Buckley Manufacturing Co., 8, 9, 64, 150
    ALWIN, 136
    BABY SHOES, 98
    CENT-A-PACK, 142
    DELUXE CENT-A-PACK, 144
    GOLDEN HORSES, 150, 151, 156
    HORSES, 150, 156
    IMPROVED PURITAN VENDER, 151, 152
    MUTUEL HORSES, 151, 156
    PILGRIM VENDOR, 64, 78
    PINBOARD TARGET PRACTICE, 117,
    PURITAN BELL (Beer), 151
Bucyrus
    DICE BOX, 88
Buddick Engineering Co.
    PAR-KET, 175
Buffalo Toy and Tool Works
    JACK POT GAME, 193
Burnham and Mills
    BABY VENDER, 132
    PLAY BALL, 169
Caille Bros. Co., The, 8, 15, 128
    BASE-BALL, 124
    BON-TON, 151, 202
    BUSY BEE, 41, 45
    CHECK PAY PURITAN, 149
    CLOWN, 165
    COMET, THE, 126
    DRAW POKER, 75
    ELK, 123
    FORTUNE BALL GUM VENDER, 131, 150
    GLOBE, 60, 75
    GOOD LUCK, 19, 74
    HY-LO, 60, 74
    IMPROVED AUTOMATIC CHECK PAYING CARD MACHINE, 123
    IMPROVED BASE BALL, 126
    INDIAN PIN POOL, 107
    JESTER, 165, 168

LINCOLN, 41, 50
LITTLE DREAM, 107
MAYFLOWER, 60, 76, 128
MERCHANT, 128
NEW SPECIAL TIGER, 126
PENNY BALL GUM VENDER, 129, 131
PURITAN, 149
PURITAN BELL, 149
QUINTETTE, 18, 69
REGISTER, 150, 197
SEARCHLIGHT, 41, 47
SPECIAL (AUTOMATIC CHECK PAYING CARD MACHINE), 124
SPECIAL ELK, 124
TIGER, THE, 124
WASP, 41, 48

Caille, Cie (French)
   LA COMETE, 124
   LE TIGRE, 124

Caille-Schiemer Co., 11
   BUSY BEE, 45
   LOG CABIN, 11

Calvert Manufacturing Co., 117
   ALL ABOARD, 120
   GEM CONFECTION VENDER, 120
   INDIAN SHOOTER, 170
   SPOOK HOUSE, 120

Canda, Leo, (Manufacturing) Co., 8, 32, 62, 63
   AUTOMATIC CARD MACHINE, 60, 72
   CANDA CARD MACHINE, 68
   COUNTER GIANT DICE, 62
   COUNTER GIANT POLICY, 62
   COUNTER JUMBO, 67
   COUNTER PERFECTION, 62
   DICE MACHINE, 62
   EAGLE, THE, 107, 108
   FIGARO, 62, 150
   FIGARO CHECK, 62
   GIANT, 14, 64
   GIANT CARD, 62
   GIANT COUNTER CARD, 62
   GIANT DICE, 62
   GIANT POLICY, 62
   JUMBO, 67
   JUMBO GIANT, 72
   MODEL CARD MACHINE, 62
   NEW CARD MACHINE, 62
   PERFECTION CARD, 13, 62, 63, 67, 190
   POLICY MACHINE, 62, 150
   TRADE VENDING MACHINE, 105, 107
   UPRIGHT FIGARO, 62

Cawood Novelty Co.
   PLAY BALL, 111
   PANAMA CANAL, 111

Central Manufacturing Co.
   HI FLY, 181
   HIT A HOMER, 181

Clawson, Clement C., 9, 32
Clawson Slot Machine Co.
   AUTOMATIC DICE (SHAKER), 7, 11, 87, 88, 162
   AUTOMATIC FORTUNE TELLER, 10, 87
   DICE TOSSER NO.1, 87
   FAIR SELLING MACHINE, 106
   HEADS AND TAILS, 106
   LIVELY CIGAR SELLER NO.1, 106
   LIVELY CIGAR SELLER NO.2, 106, 162
   NEWARK RAINBOW, THE, 162

Clune, William H., Manufacturer
   VICTOR, 68

Coin Sales Corp.
   MILLARD'S MINIATURE BASEBALL, 116, 119

Colby Specialty Supply Co.
   COMBINATION LUNG TESTER, 89

Collins, Merriam, and Co.
   PEERLESS ADVERTISER, 32

Columbia Manufacturing Co.
   SUCCESS, 68

Columbian Automatic Card Machine Co.
   AUTOMATIC CARD MACHINE, 67, 199

Comet Industries
   COMET, 148
   KING, 160

Comstock Novelty Works
   PERFECTION, THE, 54, 107, 109
   PERFECTION WHEEL, 54, 57, 110

Condon and Co.
   GAME O'SKILL, 116

Consolidated Coin Control Co.
   200, 107

Cook, S. A., and Co.
   HOWARD'S FAVORITE, 109

Cowper Manufacturing Co., 8, 11
   ELK, THE, 123
   PURITAN, 151

Coyle and Rogers
   AUTOMATIC RACE COURSE, 24
   ELECTRICAL DICE, 87

Daval Manufacturing Co., 8, 9, 16
   AMERICAN EAGLE, 137
   BEST HAND, 164, 182
   CENTA-SMOKE, 143
   CHICAGO CLUB-HOUSE, 78
   CLEARING HOUSE, 152, 155
   COMET, 146
   CUB, 136
   DAVAL GUM VENDER, 134
   DOUBLE DECK, 21
   FREE PLAY, 148
   HEADS OR TAILS, 159
   JACKPOT CHICAGO CLUB HOUSE, 77
   JIFFY, 146
   MARVEL BALL GUM VENDER, 147
   MARVEL VISIBILITY, 146
   MEXICAN BASEBALL, 164
   OOMPH, 164, 182
   (1938) PENNY PACK, 21
   1940 PENNY PACK, 145
   RACES, 155, 198
   REEL DICE, 151, 155
   REEL SPOT, 151, 157
   REEL 21, 155
   SAFE HIT, 199
   SPIN-A-PACK, 143
   TALLY, 145
   TIT-TAT-TOE, 151
   TRACK REELS, 158
   "21", 159
   ULTRA MODERN DAVAL GUM VENDER JACKPOT, 135
   X-RAY, 159

Daycom, Inc.
   REEL-O-BALL, 125, 127

Dean Novelty Co.
   EVERREADY TRADE BOARD, 171
   PENNY DRAW, 153
   VAUDETTE, 175

Decatur Fairest Wheel Co., 8, 39, 40
Decatur Fairest Wheel Works, 13, 14, 39
   DISCOUNT WHEEL, 39
   FAIREST WHEEL, THE "Large Wheel", 13, 14, 39, 40, 54, 192
   FAIREST WHEEL, THE "Small Wheel", 40, 41, 192
   FAIREST WHEEL NO.2, 39, 44
   FAIREST WHEEL NO.3, 17, 39, 44, 49
   IMPROVED FAIREST WHEEL "Large Wheel", 41

DeGrain
   DICE GAME, 88

Dennings, William
   GAME WHEEL NO.1, 122
   GAME WHEEL NO.2, 122

Dependable Enterprises
   DICE-MAT (10-Column), 103

Dixie Manufacturing Co.
   DIXIE BELL, 135

Douglis, A.S. and (Machine) Co., 16
   DAVAL GUM VENDER, 134

Drobisch Brothers and Co., 8, 39
   ADVERTISING REGISTER, 107, 108
   LEADER, THE, 54, 57
   NO.5 MONARCH DICE MACHINE, 91
   STAR ADVERTISER, 54, 56
   VICTOR, 14, 54, 56

Dunn Brothers
   PERFECTION, 167
   WRIGLEY DICE, 92
   WRIGLEY PERFECTION, 167

E.A.M.
   EAGLE, 91

Electronic poker, 16

Engel Manufacturing Co.
   IMP, 131

Ennis and Carr
   PERFECTION, 92

Erickson
   LOG CABIN, 91

Erie Machine Co.
   BOMBER, 179

Erie Manufacturing Co.
   EMCO (NERVE EXERCISE), THE, 169

Evans, H.C., and Co., 8, 23, 24, 25
   EVANS BABY BELL, 192
   MINIATURE RACE COURSE, 23
   SARATOGA SWEEPSTAKES, 24, 29
   SARATOGA SWEEPSTAKES SPECIAL, 29

Excelsior Race Track Co., 26, 27
   EXCELSIOR, 26, 200

Exhibit Supply Co., 8
   BOUNCER, THE, 113
   DICE FORTUNE TELLER, 96
   DOUBLE DICE, 113
   JUNIOR, 97, 196
   OPERATORS DICE MACHINE, .96
   PLAY BALL, 117, 118
   RED DOG, 82, 85
   SELECT-EM, 100
   SKILL DRAW, 82, 84
   SWEET SALLY, 52
   36 GAME, 102
   TURF TIME, 82

Feinberg, A., Co.
   SUCCESS, 73

Fey, Charles A., 8, 33, 128
Fey, Charles, (Manufacturing) and Co., 8
   AUTOMATIC DICE BOX, 94
   FAIR-N-SQUARE, 101
   KLONDIKE, 56
   LIBERTY BELL, 128
   MIDGET, 33, 35
   MIDGET WITH SALESBOARD, 96
   MIDGET 36, 33
   NEW 36 GAME, 33
   ON THE LEVEL, 17, 33, 93
   ON THE SQUARE, 93
   PEE-WEE, 33
   PLAY AND DRAW, 82, 84
   SKILL-DRAW, 82, 84
   SKILL ROLL, 33
   THREE CADETS, 82, 84
   3-IN-1, 96
   36 LUCKY SPOT MIDGET, 33
   36 ROULETTE, 33
   TRIPLE ROLL, 34

Field Manufacturing Corp., 174
   VEST POCKET COUNTER TOP BASEBALL, 174

Field Paper Products Co., 8
   KEYSTONE PURITAN, 133

Fisher, A.J., and Co.
   LEGAL, 116
   (ORIGINAL) PREMIUM, 110

Five Boro Machine Manufacturing Co.
   DUCK SOUP, 113

Flour City Manufactory, 23
   AUTOMATIC RACE COURSE, 23, 27

Foley, J.L., Machinist
   DRAW POKER, 75

Fort Wayne Novelty Manufacturing Co.
    LITTLE JOE, 101
Fortune Sales Co.
    LUCKY ROLL, 178
Four Jacks Corp.
    CLUB JACK, 157
Fox, Richard K., 23
    EXCELSIOR AUTOMATIC, 22, 26
    FRENCH RACE GAME, 22
France, 22
Frantz, J.F., Manufacturing Co., 165
    BIG TOP, 188
    BULLS EYE, 114
    KICKER & CATCHER, 188
    LITTLE LEAGUE, 188
    LONG SHOT, 188
    POT OF GOLD, 183, 188
"French Race Game", 22, 23
Frey, James P.
    PUNCH BOARD MACHINE, 189
Friedman and Co.
    SONG DICK, 107
Fun Industries
    POT OF GOLD, 183
Galton, Sir Thomas (Galton Board), 105, 106, 115, 162
Gambling, 53, 61
Ganss, Clarke and Dengler
    GAME TABLE, 161
Garden City Novelty manufacturing Co.
    BABY JACK, 157
    CHERRY JITTERS, 100
    DE-LUXE VENDER, 37
    GEM, 143
    PICK A PACK, 176
    PRINCE, 144
    SPIN-O, 164
Genco, Inc., 164
    PUNCH A BALL, 179
Giles and Simpkins
    HOLLYWOOD, 133
Goodbody, John
    WILDCAT, THE, 173
Gorski, Mike
    AUTOMATIC DICE, 190
Gottlieb, D., and Co., 164
    BASEBALL ROLL-IT, 171
    DAILY RACES JR., 102
    INDIAN DICE, 102
    MINIATURE BASEBALL, 172
    ROLL-IT, 171
Grand Rapids Slot Machine Co.
    ECLIPSE, 55
Great States Manufacturing Co., 8
    BEER TARGET, 117, 121
    FOOTBALL PRACTICE, 117, 121
    HAPPY DAYS, 117, 121
    SANDY'S HORSES, 29
Griswold, Milton O., and Co., 8, 39
    BIG THREE, THE, 46, 82
    BLACK CAT, 81
    DICE MACHINE, 87, 88
    LAYOUT, 54
    NEW STAR, 51
    SELF PAY, 50
    STAR (1902), 46
    STAR (1905), 48, 82
    WHEEL OF FORTUNE, 38, 81
Groetchen Tool (and Manufacturing) Co., 8, 9, 15, 16
    ATOM, 148
    DANDY VENDER, 16
    DICE-O-MATIC VENDER, 100
    DIXIE SPELLING BEE, 157
    FORTUNE TELLER, THE, 154
    GINGER, 21, 144
    GOLD RUSH, 82, 83
    HIGH STAKES, 21, 156
    HIGH TENSION, 150, 156

    IMP, 20, 136, 192
    IMP (CIGARETTE), 147
    KILL THE JAP, 165, 180
    KLIX, 159
    KLIX (1949), 160
    LIBERTY, 137, 146
    LIBERTY SPORTS PARADE, 152, 159
    MERCURY, 146
    NEW DANDY VENDER, 134
    PENNY SMOKE, 142
    PIKES PEAK, 165, 180
    POK-O-REEL, 64, 77
    PUNCHETTE, 163, 177
    ROYAL FLUSH, 21
    SILENT DANDY VENDER, 139
    SMOKE HOUSE, 143
    SOLITAIRE, 82, 83
    SPARKS, 21
    SPARKS CHAMPION, 147, 159
    SPARKS DELUX, 145
    TAVERN STYLE A, 141, 152, 154
    TAVERN STYLE B, 141, 152
    TURF FLASH, 154
    TWENTY ONE, 156
    21 VENDER, 64, 78
    WINGS, 148
    YANKEE, 137
    ZEPHYR, 21, 136
    ZEPHYR (Cigarette), 144
    ZOOM, 180
Hamilton Manufacturing Co., 19
    DAISY "Diamond Top", 19, 108, 111
    DAISY NO BLANK "Bread Top", 108, 110
    HAMILTON, THE, 71
    SUCCESS, 71
Hammond and Jones
    HORSE RACE, 24
Happy Jack Co.
HAPPY JACK, 77
Hawes, Butman and Co., 23
Hercules Novelty Co.
    BUY AMERICAN, 175
Hiawatha Amusement Co., The
    TARGET PRACTICE, 119
Hillsboro Wooden Ware Co.
    HILLSBORO, THE, 109
Hochriem, G.F., Manufacturer
    BOOSTER, 162
Holtz, T.F. and Co.
    BROWNIE, 66
Holtz, T.F., Novelty Machine Works
    CARD MACHINE, 65
Howard Sales Co.
    BRODI LIBERTY, 179
Hudson Moore Co., 23
Huffman, James G., 13, 14
Imperial Manufacturing Co.
    IMPERIAL, 95
    IS IT ANY OF YOUR BUSINESS, 95
Industry Novelty Co., 8
    1918 BASEBALL, 124
    TARGET PRACTICE, 116
"Iron Card" machines, 63
Jackson Co.
    CAN'T LOSE, 108, 112
    HAVE SOME FUN, 108
    NEVER LOSE, 108, 112
Jennings, O.D. and Co., 8
    CARD MACHINE, 79
    CLUB VENDER, 143
    FAVORITE, 116, 119
    FAVORITE (BALL GUM), 120
    FAVORITE (CANDY), 119
    FAVORITE (PEANUTS), 119
    GRANDSTAND, 145
    LITTLE MERCHANT, 141
    LITTLE MYSTERY, 37
    PURITAN GIRL, 132, 149
    STAR PENNY PLAY, 143

    STAR VENDER, 143
    SWEET MUSIC, 59
    TARGET, THE, 116, 119
    21 BLACK JACK, 80
Jentzsch und Meerz (German)
    BAJAZZO (CLOWN), 165, 168
    1930 BAJAZZO, 170
Jess Manufacturing Co.
    SHAKE, RATTLE AND ROLL, 104
Johnson Act, 16
Jones, William C.
    WIZARD CLOCK, 163
Jost, Cie (French), 23
    RACE TRACK, 23
Junior, E.E., Manufacturing Co.
    PLAY BASKETBALL, 173
    PLAY FOOTBALL, 173
Kalamazoo Automatic Music Co.
    KALAMAZOO, 87, 101
    KAZOO, 101
    KAZOO-ZOO, 87, 101
Keane Novelty Co.
    SQUARE DEAL, 86, 88
Keeney, J.H. and Co., 164
    JITTER BONES, 103
Keeney and Sons
    BABY VENDER, 152
    MAGIC CLOCK, 58
Kelley Manufacturing Co.
    BICYCLE, THE, 40
    FLIP-FLOP (LOOP-THE-LOOP), 162, 166
    IMPROVED KELLEY, THE, 150, 196
    KELLEY, THE, 14, 18, 149, 150
Kellog and Co.
    WHEEL OF FORTUNE, 54
Kelly, W.H. and Co.
    PENNY JITTER BALL, 178
Kemp, Clarence M.
    VOTING MACHINE, 38
Kennedy and Diss Machinists, 11, 12
    AUTOMATIC RACE TRACK, 12, 24
Keystone Novelty and Manufacturing Co. (Philadelphia)
    CHUCK-O-LUCK, 97
    WINNER, 95
Keystone Novelty Manufacturing Co. (Chicago)
    ROULETTE POKER, 36
Keystone (Novelty and) Sales Co.
    KEYSTONE, 97
    KEYSTONE PURITAN BELL, 132
Klondike Slot machine Co.
    KLONDIKE GUM MACHINE, THE, 109
Knight Novelty Co.
    OUR LEADER, 111
Latimer and Co.
    GAME O'SKILL, 115
Lewis Manufacturing Co., 23
Lichty, Norman, Manufacturing Co.
    AUTOMATIC SALESMAN AND PHRENOLOGIST, 38
Lighton, John, machine Co.
    DICE SHAKER, 87, 89
Lilliput manufacturing Co.
    LILLIPUT GOLF, 172
Lion Manufacturing Co.
    LION PURITAN BABY VENDER, 133, 150
    PURITAN BABY BELL, 138
Little Giant manufacturing Co.
    LITTLE GIANT, 106
Loheide Manufacturing Co.
    WIZARD CLOCK ("4-Column"), 167
    WIZARD CLOCK ("2-Column"), 14, 167
Loss (Novelty Co.)
    SKILLIARD, 166
Lu-Kat Novelty Co.
    LUCAT, 177
Lumley, Edwin J., 24
Madorsky, B.
    FOOTBALL, 172

Maitland Manufacturing Co.
   MISTIC DERBY, 30, 163
Maley, Charles T., Novelty Co., 8, 9, 13, 15
   AUTOMATIC DICE (SHAKING SLOT) MACHINE, 9, 87, 89
   CASHIER, 107
   ECLIPSE, 55
   NICKEL TICKLER, 13
   NICKEL TICKLER NO.2, 9, 105, 107
   PENNY TICKLER, 107
Mansfield Brass Foundry
   ROULETTE, 31, 32
Marcus, M. M, , and Co.
   SCOTCH GOLF, 174
Marean, Josiah T., 23
   RACE COURSE, 23
Marvel Manufacturing Co.
   LUCKY HOROSCOPE, 183
   SLUGGER, 183
Matheson Novelty and Manufacturing Co.
   KITZMILLER'S AUTOMATIC SALESMAN, 116
   ORIGINAL AUTOMATIC VENDER, 118
McClellan, William
   BOARD OF CRAPS AND GUM MACHINE, 93
McLoughlin, Edward S.
   BANKER WHO PAYS, 54
   DRINKS, 54
   GUESSING BANK, 54
   PRETTY WAITER GIRL, 54, 55
McManus, Nichols
   RACE TRACK, 23
Menu Wheel Co.
   MENU WHEEL, 54
Michigan Novelty Co.
   DETROITER, 35
Midwest Novelty Co., 6
   ACE, THE, 132
   JACK POT PURITAN BABY BELL, 132
   LION PURITAN BABY VENDER, 6, 153
   PURITAN BABY BELL, 153
Mikro-Kall-It
   MIKRO-KALL-IT, 102
Miller Novelty Co.
   LITTLE DREAM PLAY BASEBALL, 110
Mills, F.W., Manufacturing Co.
   JOCKEY, 76
   PREMIUM TRADER, 125, 126
Mills Novelty Co., 8, 13, 14, 24, 33, 63, 64, 128
   ARROW (CIGAR SALESMAN), 55
   BELL BOY, 133
   BEN FRANKLIN, 71
   BULL'S EYE, 41, 47
   CATCH THE BALL, 174
   CHECK BOY, 124, 125
   CHECK CARD, 63
   CHECK FIGARO, 63
   CHECK JUMBO, THE, 63
   CHECK POLICY, 63
   CHECK UPRIGHT CARD MACHINE, 63
   COMMERCIAL, 19, 60, 73
   COUNTER SUCCESS NO.6, 71
   CRAP SHOOTER, 94
   DIAL, 154
   DRAW POKER, 74
   EAGLE, 124
   ELK, 123, 191
   FLIP SKILL, 179
   GAME O'SKILL, 115, 116
   GEM, 60
   GIANT, THE, 63
   HY-LO, 60, 74
   I WILL, 92
   IMPROVED JOCKEY, 71
   IMPROVED PURITAN, 149, 151
   JOCKEY, 18, 19, 69
   JUMBO, THE, 63, 67
   JUMBO GIANT, 63, 72
   L'AEROPLAN, 124, 126
   JUMBO SUCCESS NO.2, 69
   JUMBO SUCCESS NO.4, 69
   KING DODO (5-Way), 72, 191, 203
   KING DODO (3-Way), 72, 203
   KOUNTER KING, 158
   LITTLE DREAM, 107
   LITTLE DUKE, THE, 60, 68
   LITTLE KNOCKER, 166
   LITTLE MONTE CARLO, 33
   LITTLE MONTE CARLO, IMPROVED, 33
   LITTLE PERFECTION "Round Top", 13, 19, 63, 70
   LITTLE PERFECTION "Flat Top", 13, 76, 191
   LITTLE SCARAB, 34
   LITTLE SCARAB VENDER, 345
   MANILA, THE, 164, 165
   MILLS PIPPIN, 94
   NATIVE SON, 60
   NEW IDEA CIGAR MACHINE, 54
   NEW JOCKEY, 76
   NEW TARGET PRACTICE, 118
   ON THE LEVEL, 93
   PERFECTION CARD, 63
   PILOT, 124
   PREMIUM, THE, 111
   PROFIT SHARING REGISTER STYLE A, 55
   PROFIT SHARING REGISTER STYLE B, 55
   PROFIT SHARING REGISTER STYLE C, 55
   PURITAN, 60, 130, 152
   PURITAN BELL, 130, 149
   QUARTOSCOPE, 11
   RACE HORSE PIANO, 24, 27
   RELIABLE, 72, 197
   SILENT SALESMAN, 107, 108
   SPECIAL, 123
   SPECIAL COMMERCIAL, 73
   STAR TARGET PRACTICE, 117
   SUPERIOR, 60, 74
   SUCCESS NO.6 "Little Success", 71
   SUCCESS NO.8 "Little Success", 72
   TARGET PRACTICE (Iron), 15, 116, 117
   TARGET PRACTICE (Aluminum), 116
   TICKETTE, 163
   TRADER, THE, 60, 74
   UMPIRE, 124, 125
   UPRIGHT CARD MACHINE, 69
   UPRIGHT PERFECTION "Square Top", 63, 71
   UPRIGHT PERFECTION CARD, 63
   VICTOR, 75
   WILD DEUCES, 158
   YOUR NEXT, 69
Mills Sales Co.
   MIDGET ROULETTE, 35
   36 LUCKY SPOT MIDGET, 96
Monarch Card machine Co.
   MONARCH CARD MACHINE, 65
Monarch Manufacturing and Sales Co.
   BULL DOG GUM VENDER, 36
   PENNY GUM VENDER, 131, 138
Monarch Sales Co.
   PEE-WEE ROULETTE, 36, 97
   36 LUCKY PLAY PEE-WEE, 97
Moon and Avers
   FLICKER, 162
Mooney and Goodwin
   TRADE STIMULATOR, 59, 204
Moore, M.E.
   DICE BOX, 86
M-R Advertising System
   M-R, 58
Mueller Specialties
   CHANGE MASTER, 178
Munves Corp., The
   TID-BIT, 117, 178
Murphy, William H.
Murray, Spink and Co.
   3 DIAL FORTUNE, 55
Mutoscope, International, Reel Co.
   JUGGLING CLOWN, 170
   TIP THE BELL HOP, 173

Nafew, Samuel, Co.
   AUTOMATIC DICE MACHINE, 91
   LITTLE MODEL CARD MACHINE, 63
   MODEL CARD MACHINE, 65
National Automatic Device Co., 23
   AUTOMATIC RACE COURSE, 23
National Coin machine Co.
   NATIONAL TARGET PRACTICE, 117, 118
National Coin Machine Exchange
   DRAW POKER, 64, 78
   DRAW POKER GUM VENDER, 80
   PENNY KING, 141
   SMOKES, 142
   TIA JUANA, 102
National Institute (English)
   A GOOD TURN, 168, 193
National Manufacturing Co.
   LITTLE MONTE CARLO, 33
National Novelty Manufacturing Co.
   BASE BALL, 169
National Table Co.
   NATIONAL TABLE, 27
Nau, Pierre-Abel (French)
   L'AIGLE, 124
   L'ELAN, 124
   LES PETITES CHEVAUX, 124, 125
   SPECIAL, 124
   3 COULEURS, 124
Nertney, Thomas J., Manufacturing Co., 43
   COINOGRAPH SALESMAN, 43
New Era Manufacturing Co.
   JUMPING JACK, 98
   NEW ERA VENDER, 100
Norris Manufacturing Co.
   LUCKY, 145
   ROLL-EM, 103
Northwest Coin Machine Co.
   CHURCHILL DOWNS, 112
Novix Specialties
   INDOOR AVIATOR, 116
   INDOOR BASEBALL, 116
   KOIN-KICK BASEBALL, 116
   TRY-SKILL, 170
Oakland Novelty Co.
   OAKLAND, 70
O'Donoghue, James D., 23
Ogden and Co.
   DEWEY SALESMAN, 57
Ohio Specialty Co.
   BARTENDER STYLE A, THE, 141
   BARTENDER STYLE B, THE, 141
Oliver, William S., 23
Overton Manufacturing Co.
   WHEEL, THE, 46
Pace Manufacturing Co., 8, 64, 128, 164
   CARDINAL, 80
   CARDINAL JAK-POT, 80
   DANDY VENDER, 133, 135
   DEUCES WILD, 173
   HIT ME, 64
   HOL-E-SMOKES, 140
   HOL-E-SMOKES (1935), 142
   JAK-POT DANDY VENDER, 135
   NEW DEAL, THE, 64, 80
   NEW DEAL JAK-POT, THE, 80, 155
   SUDS, 152, 154
   TARGET PRACTICE, 117, 119
   WHIZ-BALL, 172
Pacific Amusement Manufacturing Co.
   ELECTRIC SPINNER, 101
   P.D.Q., 101
Page Manufacturing Co.
   SALES INCREASER "Drinks", 55, 57
   SALES INCREASER "Free Merchandise", 55, 58
   SALES INCREASER "Profits Shared", 55
Palamedes Sales Co.
   LOTTA DOE, 99
Pana Enterprise Co., 39

IMPROVED FAIREST WHEEL, 39, 49
Park Novelty Co.
    RED BIRD, 161, 162, 166
Parker, E. D., Co.
    SPIRAL, 166
Parmly Engineering Co.
    LUNATIC, THE, 172
Patents, 12
Paupa and Hochriem Co., 8, 123
    BASE BALL, 165
    ELK, 122, 123, 149
    IMPROVED ELK, 124, 149
    MANILA, THE, 164, 165
    PILOT, 124, 125
Payout odds, 61
Pearsall and Finkbeiner
    "FIRE ENGINE", 81
    STAR, THE (RED), 81
Peerless products Co.
    ODD PENNY MAGNET, 175
    RACE OF HORSES AND MARBLES, THE, 28
Peo Manufacturing Co.
    BARD YARD GOLF, 164
    LITTLE WHIRLWIND, 164, 171
    PLAY POKER, 173
Percival, Orin L., Manufacturer
    CIGAR WHEEL, 39
Pick, Albert, and Co.
    FAIREST WHEEL, 49, 191
Pierce Tool and Manufacturing Co.
    BUGG HOUSE, 162, 175
    CIG-O-ROLL, 176
    FOUR LEAF CLOVER, 82
    GYPSY, 176
    HIT ME, 64, 79
    JACKPOT HIT ME, 79
    JACKPOT THE NEW DEAL, 78
    NEW DEAL, THE, 64
    PIX-IT, 101
    WHIRLWIND, 83, 134
    WIN-A-PACK, 113
Pioneer Coin Machine Co.
    SMILEY, 108, 180
Pioneer Games Co.
    BIG BONES, 99
Pogue, Miller and Co.
    WIZARD CLOCK, 168
Poole Brothers Manufacturers
    BICYCLE, THE, 40, 191
    BICYCLE DISCOUNT WHEEL, THE, 48
Portland Novelty Co.
    OREGON, 60, 70
Progressive Manufacturing Co.
    AUTOMATIC TRADE CLOCK, 167
    WIZARD CLOCK, 163
Progressive Novelty Co., 39
    FAIREST WHEEL, THE, 39
Puritan Machine Co. Ltd.
    CHECK PAY PURITAN, 149
    MAYFLOWER, 60
    PILGRIM, 60
    PURITAN, 60, 149
Quality Supply Co.
    ADD'EM, 104
    BEAT THE HOUSE, 104
    FOUR OF A KIND, 104
    HAZARD, 104
    HI-LOW SEVEN, 104
    HORSES, 104
    WIN A BEER, 104
    WIN YOUR SMOKES, 104
Reel Profits, Inc.
    YOUR DEAL, 80
Reel-O-Ball Co.
    REEL-O-BALL, 125
Reliable Coin machine Exchange
    TARGET PRACTICE, 117
Reliance Novelty Co.
    IMPROVED RELIANCE, 17

    THREE SPINDLE, 56
    RELIANCE, 66, 201
    STANDARD, 66
    TROPHY, 66
    VICTOR, 66
Repeal, 8, 20, 64
Robbins, D., and Co., 117
    AUTOMATIC BASEBALL, 120
    EVERLASTING AUTOMATIC SALESBOARD, 163
    PENNY CONFECTION VENDER, 120
Robinson, Earl A., Novelties
    NEW PIANO GAME, THE, 116
Rock-Ola Manufacturing Co. (Corp.)
    HOLD-AND-DRAW, 153
    OFFICIAL SWEEPSTAKES, 29
    OFFICIAL SWEEPSTAKES BALL GUM VENDER, 29
    RADIO WIZARD, 84
Roever Manufacturing Co.
    MULTIPHONE, 32
Rosenfield, William E., 12
Rothschild's and Sons, 22, 25
    AUTOMATIC RACE TRACK, 25
Royal Manufacturing Co.
    BOUNCING BALL, 176
Royal Novelty Co. (Indianapolis)
    TARGET, 116
Royal Novelty Co. (San Francisco)
    DICE, 95
    ROYAL TRADER, 70
R. and S. Co.
    CHIP GOLF GAME, 169
Ruff, F. A.
    CRAP SHOOTERS DELIGHT, 92
    DEWEY, THE, 48
Runyan Sales Co.
    KEEP 'EM BOMBING, 108, 114
Salm, J., Manufacturing, 68
    "Iron Card", 68
Saloons, 8, 10, 11, 17, 22, 55, 64
Sanders, J. M., Manufacturing Co., 8, 15
    BABY JACK POT VENDER, 134
    BABY VENDER, 15
    LUCKY PACK, 147
    TOKETTE, 146
    ZIP, 147
Sands Manufacturing Co.
    SKILL KATCH, 177
Schall, Daniel N., and Co., 39
    CIGAR ONE WHEEL, 39
Schiemer-Yates Co.
    HY-LO, 60
Schloss and Co.
    DICE TOSSER AND CIGAR CUTTER, 86
Scientific Machine Corp.
    TOTALIZER, 179
Seeburg, J. P., Piano Co., 24
    GRAYHOUND RACE PIANO, 24, 27
Sheffler Brothers, Inc.
    COMMERCE, 82
Shelley, Charles, Pty. Ltd. (Australia)
    SHELSPESCHEL, 125, 127
Shipman manufacturing Co.
    SPIN-IT, 162, 164
Seirsdorfer, M., and Co.
    COIN TARGET BANK, 165
Silver King Novelty Co., 8, 128
    BAll GUM VENDER, 15, 129
    BASE BALL (AMUSEMENT), 124, 126, 150
    IMPROVED BABY BELL, 131
    LITTLE PRINCE, 140
    PENNY BASE BALL, 192
    TARGET, THE, 116, 118
Sittman and Pitt, 15
    LITTLE MODEL CARD MACHINE, 15, 63
Skipper Sales Co.
    SKIPPER, 121
Sloan Novelty Co.

    LEADER, THE, 110
Sloan, T. H., Manufacturing and Sales
    POKERALL GAME, THE, 174
Smith, Willard A.
    LITTLE JOKER STYLE A, 54
    LITTLE JOKER STYLE B, 54
    3 DIAL FORTUNE, 54
Smokers Supply
    CIGAR DICE, 50
Specialty Coin Machine Builders
    TARGET, 117
Specialty Machine Works
    PORTOLA, 94
Specialty manufacturing Co.
    TARGET PRACTICE BALL GUM VENDER, 117
Standard
    STANDARD, 54
Standard Coin Machine Co.
    SKILL-A-RETTE, 180
Standard Games Co.
    WINDMILL, 180
    WINDMILL JR., 180
Star Amusement Co.
    SPARKY, 85
Star Manufacturing and Sales Co.
    FLIP FLOP FLOOZIE, 178
Star Novelty Co.
    WINNER DICE, 95
Star Trade Register Co.
    STAR TRADE REGISTER, 162
Stephens, A. J., and Co.
    FLIP, 100
    LUCKY COIN TOSSER, 176
    ODD PENNY MAGNET, 175
    MAGIC BEER BARREL, 154
    PENNY DRAW, 153
Stevens, J. and E., Co., 23, 25
    BIG RACE COURSE, 23, 25
    RACE AGAINST TIME, 23, 25
    RACE COURSE, 23, 25
Stirrup, J. W., Manufacturing Co.
    AUTOMATIC CIGAR SELLER, 32
    FAIREST ROULETTE, 32
    WINNER ROULETTE, 32
Superior Confection Co., 20
    AUTO-DICE AMUSEMENT TABLE, 97
    AUTOMATIC CARD TABLE, 171
    CIGARETTE BALL GUM VENDER, 139
    CIGARETTE GUM VENDER, 142
    CIGARETTE SALESMAN MINT VENDER, 142
    FORTUNE BALL GUM VENDER, 20, 131
    THREE WAY SELECTIVE MINT VENDER, 132
Supermarket, 16, 21
Thompson, Edmund A., 23
    RACE COURSE, 23
Tibbils Manufacturing Co., 14
    CARD MACHINE, 14, 62
    SUCCESS, 68, 73
Twentieth Century Novelty Co.
    SPIRAL, 162, 166
Twico Corp.
    TWICO, 87, 104
United Automatic Machine Co.
    6-WAY DICE, 92
United States Novelty Co., 16
    JOKER, 56
    WINNER, 16, 90
Universal
    AUTOMATIC RACE TRACK, 23
Unknown manufacturers
    APPAREIL ELECTRIQUE MUSICAL (French), 194
    ALLWIN (German), 193
    AUTOMATIC DICE, 90
    BOUNCERINO, 108, 114
    CHAMPION SPEED TESTER, 171
    COMBINATION, THE, 107
    FLYING ACES, 170
    GAME O'SKILL, 117

GOOD LUCK, 87
"Hex Dicer", 91
I.X.L. JR, 60, 65
LARK, 172
LITTLE HELPER, 115, 116
OUR LEADER, 112
PLACE THE COPPERS, 177
QUEEN TOP (French), 177
ROYAL CARD MACHINE, 76
"Spittoon Dicer", 90
SWEET SIXTEEN, 59
"Tiered Backdrop", 109
"Triangle Dicer", 90
WIN AND PLACE (British), 193
UPI GOLD MINE, 183
Victor Vending Corp.
    ROLL-A-PACK, 103
Waddell, John M., Manufacturing Co., 13, 16, 40
    BICYCLE, THE, 42
    BICYCLE DISCOUNT WHEEL, THE, 16, 40, 42
    BICYCLE WHEEL, THE, 40, 41, 195
    BOOMER, 54, 57
    DISCOUNT BICYCLE WHEEL "Large Square Wheel", 40, 42
    ROOLO, 194
Waddell Wooden Ware Works, 8, 13, 40, 189
    BICYCLE, THE, 13, 17, 40, 162
    BICYCLE WHEEL, THE, 13, 44, 45
Wain and Bryant Co.
    ZODIAC, 41, 47
Wall Novelty Co.
    HUSTLER, 152
Watling Manufacturing Co., 8, 128
    BASEBALL, 124
    BUFFALO JR., 41
    CLOVER, THE, 73
    FULL DECK, THE, 57
    GOOD LUCK GYPSY, 82
    GOOD LUCK RACE HORSE, 82
    IMPROVED ELK, 123
    JACK POT "Blue Seal", 20
    MOOSE, THE, 122
    PILOT, 124
    SPECIAL ELK, 124
Watson, John J., 64
    COMBINATION CARD AND DICE MACHINE, 64, 201
W. C. Steel Ball Table Co.
    PUT AND TAKE, 58
Webb Novelty Co.
    MIRROR OF FORTUNE, 169
Western Automatic Machine Co. (Cincinnati)
    ECLIPSE, 55
    IMPROVED ROULETTE, 32
    NICKEL TICKLER, 107
Western Automatic Machine Co. (San Francisco)
    FAIR-N-SQUARE, 82
Western Electric Piano Co., 24
    DERBY, 24, 27, 28
    SWEEP STAKES, 28
Western Equipment and Supply Co., 137
    MATCH-EM, 144
    REEL RACES, 155
Western Products Co.
    OOMPH, 179
    TOT, 137
Western Weighing Machine Co.
    ECLIPSE, 55
    IMPROVED NICKEL TICKLER, 106
Weston Slot Machine Co.
    SLOT MACHINE, 161, 162
Wheeland Novelty Co.
    CALIFORNIA, 70
    PERFECTION, 70
White Manufacturing Co.
    GAME O'SKILL, 116
    TRADER NO.1, THE, 193
Whiteside Specialty Co.
    BASE BALL GAME OF SKILL, 168
Whitlock, J. H., Co.
    DARBY, THE, 28
Williams Electronic Games, 65
    TOUCHMASTER, 65
Williams Manufacturing Co.
    ADDER, THE, 35
Wisconsin Novelty Manufacturing Co.
    TRIPLE ROULETTE, 34
Withey, B. A., Co.
    AUTOMATIC SALESBOARD, 163
    IMPROVED SEVEN GRAND, 102
Wonder Novelty
    SQUARESPIN, 153
World's Fair Slot Machine Co., The
    COLUMBIAN FORTUNE TELLER, 56
Yaggy, Levi W.
    FUNNY FACES, 83
Yale Wonder Clock Co.
    AUTOMATIC CASHIER AND DISCOUNT MACHINE, 162, 166
    YALE WONDER CLOCK, 161, 162, 165, 166
Yendes Manufacturing and Sales Co., 125
Yendes Service, Inc., 125
    REEL-O-BALL, 125, 127
Young, Charles P.
    AUTOMATIC ROULETTE, 31, 32
Zeno Manufacturing Co.
    YUCCA, 110